New Symmetries and Integrable Models

Proceedings of XIV Max Born Symposium

New Symmetries and Integrable Models

Karpacz, Poland 21–24 September 1999

Edited by

Andrzej Frydryszak

Jerzy Lukierski

Ziemowit Popowicz

University of Wrocław, Poland

World Scientific

Singapore • New Jersey • London • Hong Kong

Published by

World Scientific Publishing Co. Pte. Ltd.
P O Box 128, Farrer Road, Singapore 912805
USA office: Suite 1B, 1060 Main Street, River Edge, NJ 07661
UK office: 57 Shelton Street, Covent Garden, London WC2H 9HE

British Library Cataloguing-in-Publication Data
A catalogue record for this book is available from the British Library.

NEW SYMMETRIES AND INTEGRABLE MODELS

ISBN 981-02-4270-0

Printed in Singapore by Uto-Print

PREFACE

Max Born Symposia started in 1991 as the continuation of series of seminars in theoretical physics Leipzig–Wrocław, which were held in seventies and eighties. Since 1991 we had already thirteen these scientific meetings in different fields of theoretical physics. The present XIV-th Max Born Symposium is devoted to new developments in the theory of fundamental interactions and integrable models. The programme deals with new results in supersymmetry, in particular supersymmetric field–theoretic models, the formalism of quantum groups and quantum deformations, as well as with some other new ideas. From the list of speakers I see that the scientific programme of the meeting should be a considerable success.

This meeting is organized by Institute for Theoretical Physics of Wrocław University, and supported financially by several sources: Wrocław University, Ministry of Education of Sciences as well as Bogolubov–Infeld Foundation. The last source provided grants for the lecturers from former Soviet Union. I would like also to add that in preparation of this meeting I had invaluable help from the members of our Organizing Committee: dr. A. Frydryszak, dr. M. Mozrzymas and prof. Z. Popowicz.

I hope all of you - lecturers and participants - will enjoy your stay in Karpacz. Lot of interesting lectures and the prospects of pleasant walks around the site of our Symposium makes me sure that you shall be well satisfied.

Let me announce the XIV-th Max Born Symposium as opened.

Jerzy Lukierski
The Head of the
Organizing Committee
of XIV-th Max Born Symposium

Contents

Noncommutative Geometry and Quantum Groups

EXTERIOR CALCULUS WITH $\mathbf{d^3 = 0}$ ON A FREE ASSOCIATIVE ALGEBRA AND REDUCED QUANTUM PLANE[a]

V. ABRAMOV AND N. BAZUNOVA

Institute of Pure Mathematics, University of Tartu, Vanemuise 46, 51014 Tartu, Estonia

In this paper we study an exterior calculus with differential d satisfying $d^3 = 0$ on a free associative algebra $\mathcal{A}$ generated by $x^i, i = 1, \ldots, n$ and on the reduced quantum plane. Given a homomorphism ξ from an algebra $\mathcal{A}$ to the algebra of n by n matrices over $\mathcal{A}$ and assuming $d^2 \neq 0$, we introduce the second order differentials of x^i and study their commutation relations with elements of $\mathcal{A}$. We find a homomorphism ξ such that mentioned above commutation relations are similar to those between the first order differentials of x^i and elements of $\mathcal{A}$. This allows us to construct an exterior algebra with $d^3 = 0$ over $\mathcal{A}$. We also show that two kinds of cohomology induced by the differential d on the reduced quantum plane are trivial.

1 Introduction

In a recent series of papers [3,5,7] a general apparatus for the $d^N = 0, N \geq 2$ generalization of classical exterior calculus has been worked out and studied. A distinguishing feature of this apparatus is that it uses the higher order differentials of variables. This feature raises an intriguing question about the higher order differential structure of a theory induced by the higher order differentials, which are missing in classical exterior calculus.

In this paper we construct an exterior algebra with a differential d satisfying $d^3 = 0$ on a free associative algebra $\mathcal{A}$ generated by $x^i, 1 \leq i \leq n$. An essential component of our construction is a coordinate calculi.[1] This is a first order differential calculi which includes a homomorphism ξ from an algebra $\mathcal{A}$ to the algebra of n by n matrices over $\mathcal{A}$. Supposing $d^2 \neq 0$, we introduce the second order differentials d^2x^i. The relation between right and left structures of a bimodule over $\mathcal{A}$ generated by the second order differentials is studied and the homomorphism ξ such that the commutation relations between elements of $\mathcal{A}$ and second order differentials are homogeneous, i.e. they do not contain terms dx^idx^j, is found. Following the paper,[3] we consider a graded q-differential algebra with $d^3 = 0$ on the reduced quantum plane and show that two kinds of cohomology induced by the differential d are trivial.

---

[a]Talk given by N. Bazunova

2 Exterior Calculus with $\mathbf{d^3 = 0}$ on a Free Associative Algebra

Let $\mathcal{A}$ be an unital free associative algebra generated by the variables $x^1, x^2, \ldots, x^n$. If $\xi : \mathcal{A} \to \mathcal{A}_{n\times n}$ is a homomorphism from the algebra $\mathcal{A}$ to the algebra of $(n \times n)$-matrices over $\mathcal{A}$ and $\partial_i : \mathcal{A} \to \mathcal{A},\ i = 1, \ldots, n$ are linear maps such that $\partial_i(fh) = \partial_k(f)\xi_i^k(h) + f\partial_i(h),\ \forall f, h \in \mathcal{A}$, then according to a coordinate calculi[1] the map

$$d : f \to df = \partial_i(f)dx^i\,, \qquad i = 1, \ldots, n\,,$$

where dx^i, $i = 1, \ldots, n$ are the first order differentials, is a coordinate differential. This means that d is a linear map $d : \mathcal{A} \to D_\xi(\mathcal{A})$, where $D_\xi(\mathcal{A})$ is a free left module over $\mathcal{A}$ generated by $dx^1, dx^2, \ldots, dx^n$, satisfying the Leibniz rule

$$d(fh) = d(f)\,h + f\,d(h)\,, \qquad \forall f, h \in \mathcal{A}\,. \tag{1}$$

The right module structure of $D_\xi(\mathcal{A})$ is defined by the commutation rule

$$dx^i f = \xi_k^i(f)\,dx^k\,. \tag{2}$$

In order to construct a generalization of exterior calculus with exterior differential d satisfying $d^2 \neq 0$, let us introduce the second order differentials $d^2x^1, d^2x^2, \ldots, d^2x^n$. Now we define $\Omega_\xi(\mathcal{A})$ to be a free left module over $\mathcal{A}$ generated by all monomials which can be composed from first and second order differentials by means of the following relations:

$$dx^i dx^j dx^k = 0\,, \qquad \forall\, i, j, k = 1, \ldots, n\,, \tag{3}$$

$$d^2x^i dx^j = q^2 dx^i d^2x^j\,, \qquad \forall\, i, j = 1, \ldots, n\,. \tag{4}$$

Assuming as usual $(dx^i)^0 = 1$, we add the unit element of $\mathcal{A}$ to the set of possible monomials composed from $dx^1, dx^2, \ldots, dx^n$. Thus $\mathcal{A} \subset \Omega_\xi(\mathcal{A})$.

The module $\Omega_\xi(\mathcal{A})$ becomes a N-graded module over $\mathcal{A}$ if we assign grading zero to elements of $\mathcal{A}$, and gradings 1, 2 respectively to the differentials dx^i, d^2x^i, $i = 1, \ldots, n$. Let us denote the submodule of homogeneous elements of grading m by $\Omega_\xi^m(\mathcal{A})$. Then $\Omega_\xi^1(\mathcal{A}) = D_\xi(\mathcal{A})$, $\Omega_\xi^0(\mathcal{A}) = \mathcal{A}$. It follows from the second part of commutation relations (4) that one can change the positions of the first and second order differentials in any monomial composed from them. This means that any element of the module $\Omega_\xi(\mathcal{A})$ can be written in the form $\sum_{k,l} f_{kl}R_kS_l$, where $f_{kl} \in \mathcal{A}$, and R_k, S_l are monomials composed respectively from first and second order differentials. Since there are no relations between the second order differentials any monomial of grading $2k$ on these differentials can be represented in the form

$$F_{2k}^\mu = (d^2x^{i_1})^{m_1}(d^2x^{i_2})^{m_2} \ldots (d^2x^{i_r})^{m_r}\,,$$

where $k \geq 1$, $m_1 + \ldots + m_r = k$, $r \leq k$ and μ is a multi-index entirely determined by $(i_1, \ldots, i_r, m_1, \ldots, m_r)$. The first part of the commutation relations (3) shows that any monomial $dx^i dx^j dx^k$ vanishes. Thus any element $\omega \in \Omega_\xi^{2k}(\mathcal{A})$, $k \geq 1$, can by written in the form

$$\omega = f_\mu F_{2k}^\mu + h_{ij\nu} dx^i dx^j F_{2(k-1)}^\nu \,, \qquad f_\mu,\, h_{ij\nu} \in \mathcal{A}\,. \tag{5}$$

Analogously any element θ of odd grading can be expressed in terms of basic monomials as follows

$$\theta = f_{i\mu} dx^i F_{2k}^\mu \,, \qquad f_{i\mu} \in \mathcal{A}\,. \tag{6}$$

Now we extend the differential d to the whole module $\Omega_\xi(\mathcal{A})$ by defining it on the elements of even (5) and odd (6) gradings by the following formulae:

$$d\omega = \partial_i(f_\mu) dx^i F_{2k}^\mu - h_{ij\mu} dx^i F_{2k}^{j\mu} \,, \qquad d\theta = \partial_i(f_{j\mu}) dx^i dx^j F_{2k}^\mu \,, \tag{7}$$

where $F_{2k}^{j\mu} = d^2x^j F_{2(k-1)}^\mu$.

Proposition 1 *The differential d defined by the formulae (7) is an endomorphism of grading 1 of $\Omega_\xi(\mathcal{A})$, i.e. $d : \Omega_\xi^m(\mathcal{A}) \to \Omega_\xi^{m+1}(\mathcal{A})$, satisfying $d^3\omega = 0$ for any $\omega \in \Omega_\xi(\mathcal{A})$.*

The peculiar property of the above construction is an appearance of the second order differentials d^2x^i, which are missing in the classical theory of differential forms. The module $\Omega_\xi(\mathcal{A})$ becomes an associative algebra if we add a commutation relation between an element of $\mathcal{A}$ and a second order differential to defining relations (2,3,4). This relation can be derived by differentiating (2) with regard to q-Leibniz rule. This gives the relation

$$d^2x^i f = \xi_k^i(f) d^2x^k + [\partial_m(\xi_k^i(f)) - q\xi_m^i(f)(\partial_k(f))]\, dx^m dx^k\,. \tag{8}$$

The right-hand side of the above formula is not homogeneous because of the term containing $dx^m dx^k$ and it is natural to impose on a homomorphism ξ the following condition:

$$\partial_m(\xi_k^i(f)) - q\xi_m^i(f)(\partial_k(f)) = 0, \quad \forall f \in \mathcal{A}. \tag{9}$$

The solution to this condition is the homomorphism ξ with entries

$$\xi_k^i(x^m) = qx^i \delta_k^m\,. \tag{10}$$

Now the commutation relation between an element of $\mathcal{A}$ and a second order differential takes on the form $d^2x^i f = \xi_k^i(f)\, d^2x^k$, which is similar to the form of the relation (2) between an element of $\mathcal{A}$ and a first order differential. Differentiating this relation once more time one finds that the entries of the homomorphism (10) must q-commute with derivatives ∂_m, i.e.

$$\partial_m(\xi^i_k(f)) - q\,\xi^i_k\,(\partial_m(f)) = 0\,. \tag{11}$$

The above condition leads to the only choice for the homomorphism ξ, which is

$$\xi^i_i(x^i) = qx^i\,, \quad i = 1, \ldots, n\,,$$

and $\xi^i_k(x^m) = 0$ in all other cases. Now $\Omega_\xi(\mathcal{A})$ is an unital associative N-graded algebra generated by $x^1, \ldots, x^n, dx^1, \ldots, dx^n, d^2x^1, \ldots, d^2x^n$ satisfying the following commutation relations:

$$\begin{aligned} dx^i x^j &= q\delta^{ij} x^j dx^i\,, & d^2x^i x^j &= q\delta^{ij} x^j d^2x^i\,, \\ dx^i dx^j dx^k &= 0, & d^2x^i dx^j &= q^2 dx^i d^2x^j\,, \end{aligned}$$

where there is no summation with regard to superscripts i, j.

3 Graded q-Differential Algebra on Reduced Quantum Plane

Let $\mathcal{C}$ be an associative unital algebra generated by two elements x and y, which satisfy the commutation relation $xy = qyx$. This algebra is known as "the algebra of polynomials over quantum plane". If one imposes the additional condition $x^3 = y^3 = 1$ then the vector space of this algebra becomes finite-dimensional and it is spanned by the monomials $\{1, x, y, x^2, y^2, xy, x^2y, xy^2, x^2y^2\}$. In this case $\mathcal{C}$ is called the algebra of polynomials over reduced quantum plane. [2] It should be mentioned that the algebra of polynomials over reduced quantum plane is a particular case of a generalized Clifford algebra. [4]

In order to construct a graded q-differential algebra [5] on reduced quantum plane we shall use the construction described in the paper. [3] Let us assign grading 0 to the unit element of $\mathcal{C}$ and grading 1 to the generators x, y. The grading of any monomial composed from the generators is the sum of gradings of its constituents modulo 3. The endomorphism $d : \mathcal{C} \to \mathcal{C}$ is defined as follows

$$d = d' + d'', \quad d'f = xf - q^{gr(f)} fx, \quad d''f = yf - q^{gr(f)} fy, \quad f \in \mathcal{C}\,.$$

It can be shown by a straightforward calculation that the endomorphism d satisfies the q-Leibniz rule and $d^3 = 0$.

According to the paper [6] there are two kinds of cohomology in the case of $d^3 = 0$. Therefore let us define the following subspaces of the algebra $\mathcal{C}$: $\mathbf{Z}_p = Ker\ d^p, \quad \mathbf{B}_p = Im\ d^{3-p}, \quad p = 1, 2.$

Any element $f \in \mathbf{Z}_1$ can be written in the form $f = s_1 - s_2(-x^2+y^2+qxy)$. Then the element

$$h = t_1+t_2x+t_3y+t_4x^2+t_5y^2-qt_4xy+t_6x^2y-\frac{1}{3}s_2xy^2+\Big(q(t_2+t_3)-\frac{1}{3}s_1\Big)x^2y^2\,,$$

satisfies $d^2h = f$. This proves that the sequence $\mathbf{0} \longrightarrow \mathcal{C} \xrightarrow{d^2} \mathbf{Z}_1 \xrightarrow{d} \mathbf{0}$ is exact and the cohomology of first kind $\mathbf{Z}_1/\mathbf{B}_1$ is trivial.

Similarly any element $f \in \mathbf{Z}_2$ can be expressed in terms of basic monomials as follows $f = s_1 + s_2x + s_3y + s_4y^2 - qs_5xy + s_6x^2y + q(s_2+s_3)x^2y^2$. Then the equation $dh = f$ can be solved and the solution h can be written in the form

$$\begin{aligned} h \;&=\; t_1 + \frac{1}{1-q}s_5x + \frac{s_4 - t_4(1-q^2)}{1-q}y + t_3x^2 + \Big(\frac{s_1}{1-q^2} - t_4\Big)y^2 \\ &+\Big(\frac{s_6}{1-q} - qt_4\Big)xy + \frac{s_3}{1-q^2}x^2y - \frac{s_2}{1-q^2}xy^2 + t_5x^2y^2\,. \end{aligned}$$

This proves that the sequence $\mathbf{0} \longrightarrow \mathcal{C} \xrightarrow{d} \mathbf{Z}_2 \xrightarrow{d^2} \mathbf{0}$ is exact and the cohomology $\mathbf{Z}_2/\mathbf{B}_2$ of second kind is also trivial.

Acknowledgments

The second author is very grateful to the organizers of XIVth Max Born Symposium for financial support and hospitality. The both authors would like to acknowledge the financial support of the Estonian Science Foundation under the grants No.3308 and No.1134.

References

1. A. Borowiec, V.K. Kharchenko, Z. Oziewicz in *Non Associative Algebra and Its Applications*", Ed. S. Gonzalez Jimeneza, Ser. Mathematics and its Applications, **303**, 46 (1994), (Kluwer Academic Publishers, Dordrecht 1994).
2. R. Coquereaux, A.O. Garcia, R. Trinchero, math-ph/9807012.
3. R. Kerner, B. Niermeyer, *Lett. Math. Phys.* **2**, 161 (1998).
4. N. Fleury, M. Raush de Traubenberg, *Adv. Appl. Cliff. Alg.* **4**, 123 (1994).
5. M. Dubois-Violette, *AMS Contemp. Math.* **219**, 69 (1998).
6. M.M. Kapranov, q-alg/9611005.
7. V.Abramov, R. Kerner, *Rep. Math. Phys.*, **43**, 179 (1999).

NONCOMMUTATIVE GAUGE FIELDS ON POISSON MANIFOLDS[a]

I.YA. AREF'EVA AND I.V. VOLOVICH
Steklov Mathematical Institute, Gubkin St.8, 117966, Moscow, Russia
e-mail:arefeva@mi.ras.ru, volovich@mi.ras.ru

It is shown by Connes, Douglas and Schwarz that gauge theory on noncommutative torus describes compactifications of M-theory to tori with constant background three-form field. This indicates that noncommutative gauge theories on more general manifolds also can be useful in string theory. We discuss a framework for noncommutative quantum gauge theory on Poisson manifolds by using the deformation quantization. The Kontsevich formula of the star product was given originally in terms of the perturbation expansion and it leads to a non-renormalizable quantum field theory. We propose to use the Cattaneo and Felder nonperturbative formulation of the star product formula for the construction of noncommutative quantum field theory on Poisson manifolds. Some other aspects of classical and quantum noncommutative field theory are also discussed.

1 Introduction

In a remarkable paper Connes, Douglas and Schwarz [1] have shown that supersymmetric gauge theory on noncommutative torus is naturally related to compactification of Matrix theory. [2] This shows that the framework of noncommutative geometry [3] can be useful in string theory. For some reviews and further developments see Aref'eva et al. and Seiberg et al. [4,5] One of the natural questions is whether noncommutative gauge theory can be used to describe more general compactifications in string theory. In order to have progress in this direction one has to develop the theory of noncommutative quantum gauge fields not only on the torus but also on more general manifolds. In this note we discuss a framework for noncommutative gauge theory on Poisson manifolds by using the recently developed deformation quantization. In particular the question of renormalizability of noncommutative quantum theory is discussed.

Gauge theory on *noncommutative* torus is equivalent to *noncommutative* gauge theory on the *commutative* (i.e. ordinary) torus. Motivated by investigations of quantum group symmetries in two dimensional integrable models [6] a proposal of consideration of noncommutative gauge symmetry on ordinary manifolds (quantum group gauge theory) has been suggested in Arf'eva and Volovich, [7] see also Isaev et al. and Castellani [8,9] for further discussions. Such a theory could be used as an alternative to the conventional Higgs mechanism

[a]Talk given by I.Ya. Aref'eva

of symmetry breaking. It was also shown [12] that a noncommutative gauge theory (quantum Boltzmann theory) describes the large N limit in QCD. Other approaches to field theory on noncommutative spaces have been discussed in Lukierski and Borowiec. [10,11]

A connection of noncommutative geometry with non-Archimedian geometry at the Planck scale and p-adic mathematical physics [13] has been considered in Aref'eva and Volovich. [14] A proposal to replace the picture of spacetime as a manifold by the theory of motives with motivic Galois group as gauge group has been done in Volovich [15] as a natural extension of p-adic string theory and p-adic gravity. [16,17] In particular L-function of Deligne's motive has been interpreted as the partition function of string theory. Applications of motives to quantum field theory also have been discussed recently in Kontsevich. [18] A general discussion of relations between number theory and physics is given in Manin. [19] The approach of noncommutative geometry is in fact, closely related with ideas of noncommutative (quantum) probability, see Accardi et al. [20]

Noncommutative gauge theory uses the deformed product (the "star product") instead of the ordinary product. The star product [21] is a generalization of the Moyal product well known in quantum mechanics. Kontsevich [22] has found an explicit formula for the star product on Poisson manifolds (nondegenerate Poisson structure is reduced to the symplectic structure). Quantum deformations of the Poisson-Lie structures have been considered in Faddeev et al. and Aref'eva et al. [23,24]

The formula of the star product was given in Kontsevich [22] in terms of a perturbation series and if one attempts to use it in noncommutative quantum field theory then one gets a non-renormalizable quantum theory. However it was pointed out [22] that the formula can be viewed as a perturbation series for a topological quantum field theory coupled with gravity. Recently Cattaneo and Felder [27] have shown that Kontsevich's star product formula is equivâlent to the perturbative expansion of the path integral of a two-dimensional topological quantum field theory.

In this paper we discuss the application of the Cattaneo and Felder formulation of the star product formula to the construction of noncommutative field theory on Poisson manifolds.

The paper is organized as follows. In the next section we will recall some notions of noncommutative geometry, then discuss the Moyal product and its applications in quantum field theory. In Section 5 the deformation quantization is described and finally in Section 6 field theory on Poisson manifolds is discussed.

2 Noncommutative Geometry and Gauge Fields

Noncommutative geometry appears in physics in works of the founders of quantum mechanics. Heisenberg and Dirac have proposed that the phase space of quantum mechanics must be noncommutative and it should be described by quantum algebra. After works of von Neumann and more recently by Connes mathematical and physical investigations in noncommutative geometry became very intensive.

Noncommutative geometry uses a generalization of the known duality between a space and its algebra of functions. [3,1] If one knows the associative commutative algebra $\mathcal{A}(M)$ of complex-valued functions on topological space M then one can restore the space M. Therefore all topological notions can be expressed in terms of algebraic properties of $\mathcal{A}(M)$.For example, the space of continuous sections of vector bundle over M can be regarded as a projective $\mathcal{A}(M)$-module. So vector bundle over compact space M can be identified with projective modules over $\mathcal{A}(M)$.

Now let $\mathcal{A}$ be an abstract noncommutative algebra which we will interpret as an "algebra of functions" on (nonexisting) noncommutative space. One can introduce many geometrical notions in this setting. For example, a vector bundle is by definition a projective module over $\mathcal{A}(M)$ and one can develop a theory of such bundles generalized the standard theory. In particular one can define a connection in the following way. Let $\mathcal{G}$ be a Lie algebra of derivations of $\mathcal{A}$ and $\alpha_1, ...\alpha_d$ be generators of $\mathcal{G}$. If V is a projective module over $\mathcal{A}$ one defines a connection in V as the set of linear operators $\nabla_1, ...\nabla_d$ on V satisfying $\nabla_i(a\phi) = a\nabla_i(\phi) + \alpha_i(a)\phi$, where $a \in \mathcal{A}$, $\phi \in V$, $i = 1, ...d$.

In the case when $\mathcal{A}$ is an algebra of smooth functions on R^d or on the torus T^d we get the standard notion of connection in a vector bundle. In this case the abelian algebra $\mathcal{G} = R^d$ acts on R^d or T^d and correspondingly on $\mathcal{A}$ by means of translations. The curvature of connection ∇_i, $F_{ij} = \nabla_i\nabla_j - \nabla_j\nabla_i - f_{ij}^k\nabla_k$ belongs to the algebra of endomorphisms of the $\mathcal{A}$-module V.

The d-dimensional noncommutative torus is defined by its algebra $\mathcal{A}_\theta$ with generators $U_1, ...U_d$ satisfying the relations $U_iU_j = e^{2i\pi\theta_{ij}}U_jU_i$, where $i, j = 1, ...d$ and $\theta = (\theta_{ij})$ ia a real antisymmetric matrix. The algebra $\mathcal{A}_\theta$ is equipped with an antilinear involution $*$ obeying $U_i^* = U_i^{-1}$ (i.e. $\mathcal{A}_\theta$ is a *-algebra). An element of $\mathcal{A}_\theta$ is a power series $f = \sum f(p_1, ...p_d)U_1^{p_1}...U_d^{p_d}$, where $p = (p_1, ...p_d) \in Z^d$ and the sequence of complex coefficients $f(p_1, ...p_d)$ decreases faster than any power of $|p| = |p_1| + ... + |p_d|$ when $|p| \to \infty$. The function $f(p)$ is called the symbol of element f. We denote U^p the product $U_1^{p_1}...U_d^{p_d}$. Then one has $U^pU^k = e^{2i\pi\varphi(p,k)}U^{p+k}$, where $\varphi(p, k) = \sum \varphi_{ij}p_ip_j$ and φ_{ij} is a matrix obtained from θ after deleting all its elements below the diagonal.

To simplify the product rule we replace U^p by $e^{i\pi\varphi(p,q)}U^p$ so that we have $U^pU^k = e^{i\pi\theta(p,k)}U^{p+k}$. If f and g are two elements of $\mathcal{A}_\theta$, $f = \sum_p f(p)U^p$, $g = \sum_k g(k)U^k$ then the product

$$fg = \sum_{p,k} f(p)g(k)U^pU^k = \sum_{p,k} f(p)g(k)e^{2i\pi\theta(p,k)}U^{p+k} = \sum_q (f \star g)(q)U^q \,, \quad (1)$$

where the star-product $(f \star g)(q)$ of symbols $f(p)$ and $g(k)$ is $(f \star g)(q) = \sum_p f(p)g(q-p)e^{i\pi\theta(p,q-p)}$. The differential calculus on the noncommutative torus is introduced by means of the derivations ∂_j defined as $\partial_j U^p = ip_jU^p$, $j = 1,...d$. They satisfy the Leibniz rule $\partial_j(fg) = \partial_j f \cdot g + f \cdot \partial_j g$ for any $f, g \in \mathcal{A}_\theta$.

The integral of $f = \sum f(p)U^p$ is defined as $\int f = f(0)$, which is in correspondence with the commutative case. The integral has the property of being the trace on the algebra $\mathcal{A}_\theta$, i.e. $\int fg = \int gf$ for any $f, g \in \mathcal{A}_\theta$. Moreover one has $\int \partial_j f \cdot g = -\int \partial_j g \cdot f$.

The gauge field A_i on the noncommutative torus is defined as $A_i = \sum_{p\in Z^d} A_i(p)U^p$, $i = 1,...d$. Here $A_i(p)$ is a sequence of $N \times N$ complex matrices indexed by a spacetime index. It corresponds to the Fourier representation of the ordinary gauge theory on commutative torus. A_i is an element of a matrix algebra with coefficients in $\mathcal{A}_\theta$ and its curvature is defined as $F_{ij} = \partial_i A_j - \partial_j A_i + [A_i, A_j]$, where ∂_i is a derivative on $\mathcal{A}_\theta$ defined above. The Yang-Mills action is $S = -\frac{1}{4}\int \mathrm{tr}(F_{ij}F^{ij})$. The action is invariant under gauge transformations $A_i \to \Omega A_i \Omega^{-1} + \Omega\partial_i\Omega^{-1}$, $F_{ij} \to \Omega F_{ij}\Omega^{-1}$, where Ω is a unitary element of the algebra of matrices over $\mathcal{A}_\theta$.

3 Moyal Product

We discuss here the well known deformation quantization for the simplest case of one-dimensional quantum mechanics in the Hilbert space of functions on the real line $L^2(\mathbf{R})$ with ordinary canonical operators of position $\hat{q}$ and momenta $\hat{p}$, $[\hat{p}, \hat{q}] = -i\hbar$, acting as $\hat{q}\psi(x) = x\psi(x)$, $\hat{p}\psi(x) = -i\hbar d\psi(x)/dx$. If a function of two real variables $f(q,p)$ is given in terms of its Fourier transform $f(q,p) = \int e^{i(rq+sp)}\tilde{f}(r,s)drds$, then one can associate with it an operator $\hat{f}$ in $L^2(\mathbf{R})$ by the following formula $\hat{f} = \int e^{i(r\hat{q}+s\hat{p})}\tilde{f}(r,s)drds$. This procedure is called the Weyl quantization and the function $f(q,p)$ is called the *symbol* of the operator $\hat{f}$. One has the correspondence $\hat{f} \longleftrightarrow f = f(q,p)$. If two operators $\hat{f}_1$ and $\hat{f}_2$ are given with symbols $f_1(q,p)$ and $f_2(q,p)$ then the symbol of product $\hat{f}_1\hat{f}_2$ is given by the Moyal product $f_1 \star f_2 = (f_1 \star f_2)(p,q)$ as

$$(f_1 \star f_2)(p,q) = e^{i\hbar L}(f_1(q_1,p_1)f_2(q_2,p_2))|_{q_1=q_2=q,\; p_1=p_2=p} \,, \quad (2)$$

where $L = \frac{1}{2}(\frac{\partial^2}{\partial q_1 \partial p_2} - \frac{\partial^2}{\partial q_2 \partial p_1})$. Introducing the constant Poisson structure on the plain $\omega^{\mu\nu} = -\omega^{\nu\mu}$, $\omega^{12} = 1$ and notations $x = (q,p)$, $x_i = (q_i, p_i)$, $i = 1, 2$, $x_1 \omega x_2 = q_1 p_2 - q_2 p_1$ one writes

$$(f_1 \star f_2)(x) = \frac{1}{(2\pi)^2} \int f_1(x_1) f_2(x_2) e^{2i(x\omega x_1 + x_1 \omega x_2 + x_2 \omega x)} dx_1 dx_2 . \quad (3)$$

One also has $2i(x\omega x_1 + x_1 \omega x_2 + x_2 \omega x) = 2i \int_\triangle pdq$, where $\triangle$ is the triangle on the plane with vertices $x = (p,q)$, $x_1 = (p_1, q_1)$ and $x_2 = (p_2, q_2)$ and there is the path integral representation $e^{2i\int_\triangle pdq} = \int e^{2i\int_\triangle pdq} Dx$. One integrates over trajectories $x = x(\tau)$, $0 \leq \tau \leq 1$ in the phase plane with the boundary conditions $x(0) = x(1) = x$, $x(1/3) = x_1$, $x(2/3) = x_2$.

Therefore we obtain that the Moyal bracket is represented in the form

$$(f \star g)(x) = \int K(x, x_1, x_2) f(x_1) g(x_2) dx_1 dx_2 , \quad (4)$$

where the kernel $K(x, x_1, x_2)$ has a path integral representation $K(x, x_1, x_2) = \int e^{iS} Dx$ where $S = 2 \int pdq$ and the path integral is taken over the trajectories $x = x(\tau)$, $0 \leq \tau \leq 1$ subject to the conditions $x(0) = x(1) = x$, $x(1/3) = x_1$, $x(2/3) = x_2$.

4 Scalar Field

To study divergencies it is instructive to study an example of quartic interaction of a scalar complex field. We consider the action

$$S = \int d^d x [\frac{1}{2} |\partial_\mu \phi|^2 + \frac{m^2}{2} |\phi|^2 + \frac{g^2}{4} (\{\bar{\phi}, \phi\}_{M.B.})^2] , \quad (5)$$

$d = 2n$, $\{\bar{\phi}, \phi\}_{M.B.} = \bar{\phi} \star \phi - \phi \star \bar{\phi}$. Note that the interaction in (16) can be also represented by using an auxiliary field χ, $L_{int} = g\{\bar{\phi}, \phi\}\chi$ and a vertex including auxiliary field χ can be written in Fourier representation as

$$\int \chi(x)(\bar{\phi} \star \phi)(x) dx = \int \chi(p) \bar{\phi}(q) \phi(r) \delta(p + q + r) e^{ip\omega r} dp dq dr . \quad (6)$$

Let us consider the one-loop diagram describing "mass"-renormalization of the auxiliary field χ

$$\Sigma_\epsilon(p, \theta) \;=\; \Sigma_\epsilon(p, 0) + \Sigma_\epsilon^+(p, \theta) + \Sigma_\epsilon^-(p, \theta)$$

$$= \frac{1}{2}\int \frac{1-e^{2i\theta k\omega p}-e^{-2i\theta k\omega p}}{(k^2+m^2)((p-k)^2)+m^2}dk\,, \tag{7}$$

here $d=2n-2\epsilon$. Let us prove that the contributions of the last two terms in (18) are finite due to oscillation factors. To show this it is convenient to use the standard α-representation. For example, for the second term in (18) we have

$$\Sigma^{+}_{\epsilon}(p,\theta) = -\frac{(\pi)^{d/2}}{2}\int_0^1 dx \int_0^\infty \frac{da}{a^{d/2-1}} e^{i[(p^2x(1-x)+m^2)a-\theta^2p^2/4a]}\,. \tag{8}$$

For $\theta=0$ we get the standard divergences for $d\geq 4$, that can be regularized assuming $d=2n-2\epsilon$, $\epsilon>0$. For $\theta\neq 0$, $d=4$ $a=0$ is not a dangerous point. In fact , for $d=4$

$$\int_0^1 \frac{da}{a}e^{iB/a} = \int_1^\infty \frac{db}{b^2}e^{iBb} < \infty, \quad for \quad B\neq 0\,. \tag{9}$$

Let us note that it is evident that the index of absolute divergence of a diagram with extra exponential factor is the same as the index of the corresponding diagram with local interaction. However this estimation is not enough to guarantee the renormalizability since to have the renormalizability we have to prove that the counterterms has the form of the oroginal Lagrangian.

5 Deformation Quantization

Here we describe the Cattaneo and Felder path integral approach [27] to the Kontsevich [22] deformation quantization [21] formula. Let M be a manifold with a Poisson structure on the algebra $C^\infty(M)$ of functions on M, $\{f,g\}(x)=p^{ij}(x)\partial_i f(x)\partial_j g(x)$ given by a skew-symmetric tensor p^{ij}, obeying the Jacobi identity.

The formal deformation quantization is given by means of the star-product, i.e. an associative product on $C^\infty(M)[[\hbar]]$, such that for $f,g\in C^\infty(M)$

$$f\star g\,(x) = f(x)g(x) + \frac{i\hbar}{2}\{f,g\}(x) + O(\hbar^2)\,. \tag{10}$$

Kontsevich [22] gave a formula for the star-product in terms of diagrams. The coefficient of $(i\hbar/2)^n$ in $f\star g$ is given by a sum of terms labeled by diagram of order n. To each diagram Γ of order n there corresponds a bidifferential operator D_Γ whose coefficients are differential polynomials, homogeneous of degree n in the components p^{ij} of the Poisson structure. Kontsevich 's formula is

$$f \star g = fg + \sum_{n=1}^{\infty} \left(\frac{i\hbar}{2} \right)^n \sum_{\Gamma \,\text{of order}\, n} w_\Gamma D_\Gamma(f \otimes g)\,. \tag{11}$$

The weight w_Γ is the integral of a differential form over the configuration space of n ordered points on the upper half plane.

Cattaneo and Felder [27] have shown that this formula can be interpreted as the perturbative expansion of the path integral of a topological sigma model. The model has two real bosonic fields X and a. X is a map from the disc $D = \{u \in R^2, |u| \leq 1\}$ to M and a is a differential 1-form on D taking values in the pull-back by X of the cotangent bundle of M. In local coordinates, X is given by d functions $X^i(u)$ and a by d differential 1-forms $a_i(u) = a_{i,\mu}(u)du^\mu$, $i = 1, ..., d$, $\mu = 1, 2$. The action reads

$$S[X, a] = \int_D a_i(u) \wedge dX^i(u) + \frac{1}{2} p^{ij}(X(u)) a_i(u) \wedge a_j(u)\,. \tag{12}$$

The boundary condition for a is that for $u \in \partial D$, $a_i(u)$ vanishes on vectors tangent to ∂D.

This model was considered in Ikeda. [25] It is a generalization of a model of two-dimensional gravity with dynamical torsion. [26,25]

The star product is given by the semiclassical expansion of the path integral

$$f \star g\,(x) = \int f(X(1)) g(X(0)) e^{\frac{i}{\hbar} S[X,a]} \prod_{X(\infty)=x} dX\, da\,. \tag{13}$$

Here 0, 1, ∞ are any three cyclically ordered points on the unit circle. The path integral is over all X and a subject to the boundary conditions $X(\infty) = x$, $a(u)(n) = 0$ if $u \in \partial D$ and n is tangent to ∂D.

The model is quantized [27] by using the Batalin-Vilkovisky method. To the fields X^i, a_j one adds ghost β_i, antighost γ^i and the scalar Lagrange multiplier λ^i together with their antifields $X_i^+, a^{+j}, \beta^{+i}, \gamma_i^+$ and λ_i^+ with complementary ghost number and degree as differential forms on D. The Batalin-Vilkovisky action in a fixed gauge can be written in terms of superfields as

$$S = \int_D d\theta^2 \tilde{a}_i D\tilde{X}^i + \frac{1}{2} p^{ij}(\tilde{X}) \tilde{a}_i \tilde{a}_j - \int_D \lambda^i \gamma_i^+\,. \tag{14}$$

Here $D = \theta^\mu \partial / \partial u^\mu$,

$$\tilde{X}^i = X^i + \theta^\mu a_\mu^{+i} - \frac{1}{2} \theta^\mu \theta^\nu \beta_{\mu\nu}^{+i},\ \tilde{a}_i = \beta_i + \theta^\mu a_{i,\mu} + \frac{1}{2} \theta^\mu \theta^\nu X_{i,\mu\nu}^+ \tag{15}$$

and $\Psi = -\int_D \lambda^i \gamma_i^+$ is the gauge fixing function corresponding to the Lorenz-type gauge $d{*}a_i = 0$ ($*$ is the Hodge operator). In the path integral one has to integrate over the Lagrangian submanifold defined by equations $\phi^+_\alpha = \vec{\partial}_{\phi^\alpha}\Psi$. Then one has $X^+ = \beta^+ = \lambda^+ = 0$, $\gamma_i^+ = d{*}a_i$ and $a^{+i} = {*}d\gamma^i$. The action in component fields is then

$$\begin{aligned} S &= \int_D a_i \wedge dX^i + \frac{1}{2} p^{ij}(X) a_i \wedge a_j - {*}d\gamma^i \wedge (d\beta_i + \partial_i p^{kl}(X) a_k \beta_l) \\ &\quad - \frac{1}{4} {*}d\gamma^i \wedge {*}d\gamma^j \partial_i \partial_j p^{kl}(X) \beta_k \beta_l - \lambda^i d{*}a_i \,. \end{aligned} \tag{16}$$

The perturbation expansion was obtained in Cattaneo and Felde [27] by taking $X(u) = x + \xi(u)$ with a fluctuation field $\xi(u)$ with $\xi(\infty) = 0$. The Feynman propagators were deduced from the kinetic part

$$S_0 = \int_D a_i \wedge d\xi^i - {*}d\gamma^i \wedge d\beta_i - \lambda^i d{*}a_i = \int_D a_i \wedge (d\xi^i + {*}d\lambda^i) + \beta_i d{*}d\gamma^i \,, \tag{17}$$

of the gauge fixed action. The other terms of S are considered as perturbations. In terms of superfields $S = S_0 + S_{int}$ where the kinetic part

$$S_0 = \int_D \int d^2\theta\, \tilde{a}_j D\tilde{X}^j - \int_D \lambda^i \gamma_i^+ \,, \tag{18}$$

and

$$S_{int} = \frac{1}{2} \int_D \int d^2\theta\, p^{ij}(\tilde{X}(u,\theta)) \tilde{a}_i(u,\theta) \tilde{a}_j(u,\theta) \,. \tag{19}$$

Here $\tilde{a}_j(u,\theta) = \beta_j(u) + \theta^\mu a_{j,\mu}(u)$, $\tilde{\xi}^k(w,\theta) = \xi^k(u) + \theta^\mu a^{+j}_\mu(u)$, with $a^{+j} = {*}d\gamma^j$.

The formula for the star product (22) is obtained then as the expansion of the path integral in powers of S_{int}.

6 Noncommutative Field Theory on Poisson Manifolds

In this section we discuss noncommutative field theory on Poisson manifolds. We would like to use the star product as a generalization of the Moyal product to define noncommutative field theory on general Poisson manifolds by analogy with Sects. 4,5. However if one uses the perturbation formula (22) we immediately run into difficulties because this quantum field theory will be nonrenormalizable. One can see this already on the torus if one expands the exponent in Moyal product into the series. Then one gets vertices including momenta in higher powers which spoil the renormalizability.

Equations of motion for the action (23)

$$dX^i + p^{ij}(X)da_j = 0, \quad da_i - \frac{1}{2}\partial_i p^{lm}(X)a_l \wedge a_m = 0\,, \tag{20}$$

have the classical solution $X(u) = x$, $a_i(u) = 0$. The perturbation expansion from the previous section reproduces the Kontsevich formula (22) however this expansion is not an expansion around the classical solution $X(u) = x$. In particular if we apply this expansion to the constant Poisson structure then we get a series representing the expansion of the Moyal product.

We can obtain the semiclassical expansion if we use the Taylor expansion of the Poisson structure

$$p^{ij}(x+\xi(u)) = p^{ij}(x) + \partial_k P^{ij}(x)\xi^k(u) + ... = p^{ij}(x) + b^{ij}(x,\xi(u)) \tag{21}$$

and then set $S = S_0 + S_{int}$ where

$$S_0 = \int_D \int a_i \wedge (d\xi^i + *d\lambda^i) + \beta_i d * d\gamma^i + \frac{1}{2}p^{ij}(x)a_i \wedge a_j \tag{22}$$

and

$$S_{int} = \frac{1}{2}\int\int d^2\theta b^{ij}(x,\tilde{\xi}(u))\tilde{a}_i\tilde{a}_j\,. \tag{23}$$

Now expanding the path integral in powers of S_{int} we get the semiclassical expansion around the classical solution $X(u) = x$. Investigation of this diagram technique and its applications to the field theory on Poisson manifolds will be the subject of a further work.

Acknowledgments

This work is supported in part by INTAS grant 96-0698, I.Ya.A. is supported also by RFFI-99-01-00166 and I.V.V. by RFFI-99-01-00105 and by grant for the leading scientific schools.

References

1. A. Connes, M.R. Douglas, A. Schwarz, *JHEP* **02**, 003 (1998).
2. T. Banks, W. Fischler, S.H. Shenker and L. Susskind, *Phys. Rev.* D **55**, 5112 (1997).
3. A. Connes, Noncommutative Geometry, Academic Press, 1994.
4. I.Ya. Aref'eva and I.V. Volovich, hep-th/9907114.
5. N. Seiberg and E. Witten, *JHEP* **9909**, 032 (1999).

6. L.D. Faddeev, N. Reshetishin and L. Taxhtajain, *Algebra and Analysis,* **1**, 1 (1987).
7. I.Ya. Aref'eva and I.V. Volovich, *Mod. Phys. Lett.* **A6**, 893 (1991).
8. A.P. Isaev and Z. Popowicz, *Phys. Lett.* **B307**, 353 (1993).
9. L. Castellani, *Mod. Phys. Lett.* **9**, 2835 (1994).
10. J. Lukierski, hep-th/9812063.
11. A. Borowiec, q-alg/9710006.
12. I.Ya. Aref'eva and I.V. Volovich. *Nucl. Phys.* B **462**, 600 (1996).
13. V.S. Vladimirov, I.V. Volovich and E.I. Zelenov, *P-Adic Analysis and Mathematical Physics*, (World Scientific, Singapore, 1994).
14. I.Ya. Aref'eva and I.V. Volovich. *Phys. Lett.* **B268**, 179 (1991).
15. I.V. Volovich, Trudi Steklov Mathemat. Inst., 203(1994)41; hep-th/9608137.
16. I.V. Volovich, *Class. Quant. Grav.* **4**, L83 (1987).
17. I.Ya. Aref'eva, P. Frampton, B. Dragovich and I.V. Volovich, *Intern. J. Mod. Phys* **A6**, 4341 (1991).
18. M. Kontsevich, math.QA/9904055.
19. Yu.I. Manin, in *Conformal Invariace and String Theory*, eds. P. Dita and V.Gergescu, (Academic Press, Boston, 1989) pp.293-303.
20. L. Accardi, Y.G. Lu and I.V. Volovich, *Quantum Theory and Its Stochastic Limit*, (Springer-Verlag, 2000, in press).
21. F. Bayen, M. Flato, C. Fronsdal, A. Lichnerowicz, D. Sternheimer, *Ann. Phys.* **111**, 1 (1978).
22. M. Kontsevich, q-alg/9709040.
23. L.D. Faddeev and A.Pyatov, hep-th/9402070.
24. I.Ya. Aref'eva, G.E. Arutyunov and P.M. Medvedev, *Jour. Math. Phys.* **35**, 6658 (1994); I.Ya. Aref'eva, G.E. Arutyunov and P.M. Medvedev, in *Quantum groups, Formalism and Applications* (Polish Scientific Publishers PWN, Poland).
25. N. Ikeda, *Ann. Phys.* **235**, 453 (1994);
P.Schaller and T.Strobl, *Mod. Phys. Lett.* **A9**, 329 (1994).
26. M.A.Katanaev and I.V.Volovich, *Phys. Lett.* **B175**, 413 (1986).
27. A. S. Cattaneo and G. Felder, math.QA/9902090.

NEW q-BOSON REALIZATIONS OF THE QUANTUM GROUPS $U_q(C_3)$[a]

Č. BURDÍK
Department of Mathematics, Faculty of Nuclear Science,
Czech Technical University,
Trojanova 13, 120 00 Prague, Czech Republic

O. NAVRATIL
Department of Mathematics, Faculty of Transportation Sciences,
Czech Technical University,
Konviktská 20, 110 00 Prague, Czech Republic

We give explicit realizations for the quantum enveloping algebras $U_q(C_3)$. In these formulae the generators of the algebra are expressed by means of 6 canonical q-boson pairs and one auxiliary representation of $U_q(gl(3))$

1 Introduction

In 1989, in works of Biedenharn [1] and Macfarlane, [2] the generalization of Jordan-Schwinger [3] map for the quantum group $SU_q(2)$ is given. The commutation relation in $SU_q(2)$ are $[J_z, J_\pm] = \pm J_\pm$ and $[J_+, J_-] = [2J_z]$, where $[x] = (q^x - q^{-x})/(q - q^{-1})$. In those works the original boson operators were changed to q-deformed boson operators a^+, a, N for which we have the commutation relations $[N, a^+] = a^+$, $[N, a] = -a$ and $aa^+ - q^{-1}a^+a = q^N$. The Jordan-Schwinger map for quantum group $SU_q(2)$ is given by $J_+ = a_1a_2^+$, $J_- = a_2a_1^+$ and $J_z = 2(N_1 - N_2)$.

In physical applications, we can meet not only these boson realizations but Holstein-Primakoff[4] and Dyson [5] boson realizations too. In work of Okubo [6] such realizations for the algebra $U(n)$ were constructed. We can find the application in solid state physics [7,8] and in nuclear physics. [3,9] The general method for their construction is described in work of Burdík. [10]

The starting point in this construction is Verma modules. These realizations were generalized in the case of quantum groups too. The simplest cases of quantum algebras were studied by using different methods. [11,12] The boson realizations of Dyson type were obtained for the first time in our work [13] for $U_q(sl(2))$ by using the standard boson and in the general case of $U_q(sl(n))$, in work of Burdík et al. [14] It is possible but the expression uses infinite sums. In conclusion to our work, [14] the connection with a q-deformed boson is men-

[a]Talk given by Č. Burdík

tioned. In work of Fu Hang-Chen and Ge Mo-Lin [15] the authors used the method of Verma modules for the explicit construction of q-deformed boson realizations of Dyson type for $U_q(sl(2))$ a $U_q(sl(3))$ and in our work [16] we treated the general case $U_q(sl(n))$. Practically, the same Dyson realizations were published in the recent work [17] and the general case $U_q(B_n)$, $U_q(C_n)$ and $U_q(D_n)$ are are studied in our works. [18–20]

In some physical applications, we can change algebras to q-deformed algebras. Thus in the theory we obtain the new parameter q and some results depend on that parameter. [21]

From a mathematical point of view, the q-boson realizations have applications in the representation theory. We can represent q-deformed pairs in a different ways and obtain different representations of q-deformed algebra. In this place we would like to remember that in the case $q^n = 1$ (root of unity) the q-deformed boson operators have finite dimensional representations and by using them we obtained finite dimensional representations at the root of unity.

2 Preliminaries

The q-deformed Weyl algebra W_2^q, is defined as the associative algebra over C generated by a^+, $a = a^-$, $q^{\pm x}$ and obeys the commutation relations, [1,2]

$$\begin{aligned} aa^+ - q^{-1}a^+a &= q^x\,, \qquad aa^+ - qa^+a = q^{-x}\,, \\ q^x a^+ q^{-x} &= qa^+\,, \qquad q^x a q^{-x} = q^{-1}a\,, \\ q^x q^{-x} &= q^{-x}q^x = 1\,. \end{aligned} \tag{1}$$

We will denote $W_{2n}^q = W_2^q \otimes W_2^q \otimes \ldots$ n-times.

Let $U_q(\mathcal{L})$ be the quantum algebra and $U_q(\mathcal{L}_0)$ its subalgebra. A realization of the quantum algebra $U_q(\mathcal{L})$ is homomorphism

$$\rho : U_q(\mathcal{L}) \to W_{2n}^q \otimes U_q(\mathcal{L}_0)\,.$$

In this article, we will use the definition of a quantum group [22] given by relations among its Chevalley generators. The Cartan matrix of the Lie algebra C_3 is

$$\mathbf{A} = (a_{ij}) = \begin{pmatrix} 2, & -1, & 0 \\ -1, & 2, & -2 \\ 0, & -1, & 2 \end{pmatrix}.$$

Starting from this Cartan matrix we obtain for $U_q(C_3)$ the commutation relations

$$
\begin{aligned}
&K_iE_jK_i^{-1} = q^{a_{ij}}E_j\,, \quad K_iF_jK_i^{-1} = q^{-a_{ij}}F_j\,, \qquad i=1,2\,;\ j=1,2,3\,,\\
&K_3E_jK_3^{-1} = q^{2a_{3j}}E_j\,, \quad K_3F_jK_3^{-1} = q^{-2a_{3j}}F_j\,, \qquad j=1,2,3\,,\\
&E_1^2E_2-(q+q^{-1})E_1E_2E_1+E_2E_1^2=0\,,\\
&E_1E_3=E_3E_1\,,\\
&E_2^2E_1-(q+q^{-1})E_2E_1E_2+E_1E_2^2=0\,,\\
&E_2^3E_3-(q^2+1+q^{-2})E_2^2E_3E_2+(q^2+1+q^{-2})E_2E_3E_2^2-E_3E_2^3=0\,,\\
&E_3^2E_2-(q^2+q^{-2})E_3E_2E_3+E_2E_3^2=0\,,
\end{aligned} \tag{2}
$$

and similar ones for F_1, F_2 and F_3.

3 Construction of Realizations

For a construction similar to Reference here, [10,13,15,16,18–20] we will take the quantum subalgebra $U_q(gl(3))$ generated by $\{E_1, E_2, F_1, F_2, K_i, K_i^{-1}, i = 1,2,3.\}$ We adjoin to this subalgebra the generator F_3 and obtain again a subalgebra of $U_q(C_3)$ which we will denote by A. If we put $\varphi(F_3)v = 0$ for all $v \in U_q(gl_3)$, we extend the left regular representation $U_q(gl_3)$ to the representation φ of A.

For our construction we will use a generalized Verma representation ρ which is induced representation of $U_q(C_3)$ on representation space $U_q(C_3)/A \otimes U_q(gl(3))$ by means of representation φ.

To obtain the explicit form of this representation we define more Cartan-Weyl generators of $U_q(C_3)$. Let

$$
\begin{aligned}
X_{33} &= E_3\\
X_{23} &= E_2X_{33} - q^{-2}X_{33}E_2\\
X_{22} &= E_2X_{23} - X_{23}E_2\\
X_{13} &= E_1X_{23} - q^{-1}X_{23}E_1\\
X_{12} &= E_2X_{13} - q^{-1}X_{13}E_2\\
X_{11} &= E_1X_{12} - X_{12}E_1\\
Y_{23} &= E_2\\
Y_{12} &= E_1\\
Y_{13} &= E_1E_2 - q^{-1}E_2E_1
\end{aligned} \tag{3}
$$

$$
|x; v_i\rangle = |x_{33}, x_{23}, x_{22}, x_{13}, x_{12}, x_{11}; v_i\rangle = X_{33}^{x_{33}}X_{23}^{x_{23}}X_{22}^{x_{22}}X_{13}^{x_{13}}X_{12}^{x_{12}}X_{11}^{x_{11}} \otimes v_i,
$$

where v_i are basis elements of $U_q(gl(3))$. On this space we can define the representation Γ of W_{12}^q by

$$\begin{aligned}\Gamma(a_{ij}^+)\,|x;v_i\rangle &= |x+1_{ij};v_i\rangle\\ \Gamma(a_{ij})\,|x;v_i\rangle &= [x_{ij}]\,|x-1_{ij};v_i\rangle\\ \Gamma([x_{ij}+\alpha])\,|x;v_i\rangle &= [x_{ij}+\alpha]\,|x;v_i\rangle\end{aligned}\tag{4}$$

Using the commutation relations from Appendix A we now obtain the explicit form of ρ for the Chevalley generators
Theorem: The form of the representation ρ is given by

$$\begin{aligned}
\tau(E_3)\,|x\rangle\otimes v &= |x+1_{33}\rangle\otimes v\\
\tau(E_2)\,|x\rangle\otimes v &= [x_{33}]_2\,|x-1_{33}+1_{23}\rangle\otimes v\\
&+q^{x_{23}-2x_{33}-1}\,[x_{23}]_1\,|x-1_{23}+1_{22}\rangle\otimes v\\
&+q^{2x_{22}-2x_{33}}\,[x_{13}]_1\,|x-1_{13}+1_{12}\rangle\otimes v\\
&+q^{2x_{22}-2x_{33}+x_{12}-x_{13}}\,|x\rangle\otimes\varphi(E_2)v\\
\tau(E_1)\,|x\rangle\otimes v &= [x_{23}]_1\,|x-1_{23}+1_{13}\rangle\otimes v\\
&+(q+q^{-1})q^{x_{13}-x_{23}}\,[x_{22}]_2\,|x-1_{22}+1_{12}\rangle\otimes v\\
&+q^{-2x_{22}+x_{13}-x_{23}+x_{12}-1}\,[x_{12}]_1\,|x-1_{12}+1_{11}\rangle\otimes v\\
&+q^{2x_{11}-2x_{22}+x_{13}-x_{23}}\,|x\rangle\otimes\varphi(E_1)v\\
\tau(K_1)\,|x\rangle\otimes v &= q^{2x_{11}+x_{13}-2x_{22}-x_{23}}\,|x\rangle\otimes\varphi(K_1)v\\
\tau(K_2)\,|x\rangle\otimes v &= q^{2x_{22}+x_{12}-2x_{33}-x_{13}}\,|x\rangle\otimes\varphi(K_2)v\\
\tau(K_3)\,|x\rangle\otimes v &= q^{4x_{33}+2x_{23}+2x_{13}}\,|x\rangle\otimes\varphi(K_3)v\\
\tau(F_1)\,|x\rangle\otimes v &= q^{2x_{22}-2x_{11}}\,[x_{13}]_1\,|x+1_{23}-1_{13}\rangle\otimes\varphi(K_1^{-1})v\\
&+q^{x_{12}-2x_{11}-1}\,[x_{12}]_1\,|x+1_{22}-1_{12}\rangle\otimes\varphi(K_1^{-1})v\\
&+(q+q^{-1})\,[x_{11}]_2\,|x+1_{12}-1_{11}\rangle\otimes\varphi(K_1^{-1})v+|x\rangle\otimes\varphi(F_1)v\\
\tau(F_2)\,|x\rangle\otimes v &= (q+q^{-1})q^{x_{23}-2x_{22}+x_{13}-x_{12}-1}\,[x_{23}]_1\,|x+1_{33}-1_{23}\rangle\\
&\otimes\varphi(K_2^{-1})v+(q+q^{-1})q^{x_{13}-x_{12}}\,[x_{22}]_2\,|x+1_{23}-1_{22}\rangle\\
&\otimes\varphi(K_2^{-1})v+[x_{12}]_1\,|x+1_{13}-1_{12}\rangle\\
&\otimes\varphi(K_2^{-1})v+|x\rangle\otimes\varphi(F_2)v\\
\tau(F_3)\,|x\rangle\otimes v &= -\frac{[x_{33}]_2}{q^2-q^{-2}}\,|x-1_{33}\rangle\\
&\otimes\left(q^{2(x_{13}+x_{23}+x_{33}-1)}\varphi(K_3)-q^{-2(x_{13}+x_{23}+x_{33}-1)}\varphi(K_3^{-1})\right)v\\
&-q^{x_{12}+x_{13}+2x_{22}+x_{23}-1}\,[x_{23}]_1\,|x-1_{23}\rangle\otimes\varphi(K_3E_2)v\\
&-q^{2(x_{23}+x_{13}-2)}\begin{bmatrix}x_{23}\\2\end{bmatrix}_1|x-2_{23}+1_{22}\rangle\otimes\varphi(K_3)v
\end{aligned}$$

$$
\begin{aligned}
&-q^{x_{23}+2x_{13}+2x_{22}-3}[x_{23}]_1[x_{13}]_1|x-1_{23}-1_{13}+1_{12}\rangle\otimes\varphi(K_3)v\\
&-(q-q^{-1})q^{2x_{22}+2x_{12}-1}[x_{22}]_2|x-1_{22}\rangle\otimes\varphi(K_3E_2^2)v\\
&-(q^2-q^{-2})q^{x_{12}+x_{13}+2x_{22}-2}[x_{22}]_2[x_{13}]_1|x-1_{22}-1_{13}+1_{12}\rangle\\
&\otimes\varphi(K_3E_2)v-(q-q^{-1})q^{2x_{22}+2x_{13}-5}[x_{22}]_2[x_{13}]_1[x_{13}-1]_1\\
&\cdot|x-1_{22}-2_{13}+2_{12}\rangle\otimes\varphi(K_3)v\\
&-q^{2x_{11}+x_{12}+x_{13}-1}[x_{13}]_1|x-1_{13}\rangle\\
&\otimes\varphi(K_3Y_{13})v-(q-q^{-1})q^{2x_{12}+x_{13}-3}[x_{12}]_1[x_{13}]_1\\
&\cdot|x-1_{13}-1_{12}+1_{11}\rangle\otimes\varphi(K_3E_2)v-q^{2x_{12}+2x_{13}-4}\begin{bmatrix}x_{13}\\2\end{bmatrix}_1\\
&\cdot|x-2_{13}+1_{11}\rangle\otimes\varphi(K_3)v\\
&-(q-q^{-1})q^{2x_{11}+x_{12}}[x_{12}]_1|x-1_{12}\rangle\otimes\varphi(K_3E_2Y_{13})v\\
&-(q-q^{-1})^2q^{2x_{12}-3}\begin{bmatrix}x_{12}\\2\end{bmatrix}_1|x-2_{12}+1_{11}\rangle\otimes\varphi(K_3E_2^2)v\\
&-(q-q^{-1})q^{2x_{11}-1}[x_{11}]_2|x-1_{11}\rangle\otimes\varphi(K_3Y_{13}^2)v
\end{aligned}
$$

A straightforward calculation shows that this representations ρ can be rewritten by means of the representations of W_{12}^q. Because this representation of W_{12}^q is faithful, we obtain finally the next theorem:

Theorem: The mapping τ defined by formulae

$$
\begin{aligned}
\rho(E_3)&=a_{33}^+\\
\rho(E_2)&=a_{33}a_{23}^+\{x_{33}\}+q^{x_{23}-2x_{33}}a_{23}a_{22}^+\\
&\quad+q^{2x_{22}-2x_{33}}a_{13}a_{12}^++q^{x_{12}-x_{13}+2x_{22}-2x_{33}}\varphi(E_2)\\
\rho(E_1)&=a_{23}a_{13}^++(q+q^{-1})q^{x_{13}-x_{23}}a_{22}a_{12}^+\{x_{22}\}\\
&\quad+q^{x_{12}+x_{13}-2x_{22}-x_{23}}a_{12}a_{11}^++q^{2x_{11}+x_{13}-2x_{22}-x_{23}}\varphi(E_1)\\
\rho(K_1)&=q^{2x_{11}+x_{13}-2x_{22}-x_{23}}\varphi(K_1)\\
\rho(K_2)&=q^{x_{12}-x_{13}+2x_{22}-2x_{33}}\varphi(K_2)\\
\rho(K_3)&=q^{2x_{13}+2x_{23}+4x_{33}}\varphi(K_3)\\
\rho(F_1)&=q^{2x_{22}-2x_{11}}a_{13}a_{23}^+\varphi(K_1^{-1})+q^{x_{12}-2x_{11}}a_{12}a_{22}^+\varphi(K_1^{-1})\\
&\quad+(q+q^{-1})a_{11}a_{12}^+\{x_{11}\}\varphi(K_1^{-1})+\varphi(F_1)\\
\rho(F_2)&=(q+q^{-1})q^{-x_{12}+x_{13}-2x_{22}+x_{23}}a_{23}a_{33}^+\varphi(K_2^{-1})\\
&\quad+(q+q^{-1})q^{-x_{12}+x_{13}}a_{22}a_{23}^+\{x_{22}\}\varphi(K_2^{-1})+a_{12}a_{13}^+\varphi(K_2^{-1})+\varphi(F_2)\\
\rho(F_3)&=-\frac{1}{q^2-q^{-2}}\left(q^{2(x_{13}+x_{23}+x_{33})}\varphi(K_3)-q^{-2(x_{13}+x_{23}+x_{33})}\varphi(K_3^{-1})\right)\\
&\quad a_{33}\{x_{33}\}-q^{x_{12}+x_{13}+2x_{22}+x_{23}}a_{23}\varphi(K_3E_2)
\end{aligned}
$$

$$
\begin{aligned}
&-\frac{1}{q+q^{-1}}q^{2x_{23}+2x_{13}}a_{23}^2a_{22}^+\varphi(K_3)\\
&-q^{2x_{13}+2x_{22}+x_{23}}a_{13}a_{23}a_{12}^+\varphi(K_3)\\
&-(q-q^{-1})q^{2x_{12}+2x_{22}+1}a_{22}\{x_{22}\}\varphi(K_3E_2^2)\\
&-(q^2-q^{-2})q^{x_{12}+x_{13}+2x_{22}}a_{22}a_{13}a_{12}^+\{x_{22}\}\varphi(K_3E_2)\\
&-(q-q^{-1})q^{2x_{22}+2x_{13}+1}a_{22}a_{13}^2\left(a_{12}^+\right)^2\{x_{22}\}\varphi(K_3)\\
&-q^{2x_{11}+x_{12}+x_{13}}a_{13}\varphi(K_3Y_{13})\\
&-(q-q^{-1})q^{2x_{12}+x_{13}}a_{12}a_{13}a_{11}^+\varphi(K_3E_2)\\
&-\frac{1}{q+q^{-1}}q^{2x_{12}+2x_{13}}a_{13}^2a_{11}^+\varphi(K_3)\\
&-(q-q^{-1})q^{2x_{11}+x_{12}+1}a_{12}\varphi(K_3E_2Y_{13})\\
&-\frac{(q-q^{-1})^2}{q+q^{-1}}q^{2x_{12}+1}a_{12}^2a_{11}^+\varphi(K_3E_2^2)\\
&-(q-q^{-1})q^{2x_{11}+1}a_{11}\{x_{11}\}\varphi(K_3Y_{13}^2)
\end{aligned}
$$

$$\textit{where}\ \{n\}=\frac{q^n+q^{-n}}{q+q^{-1}}$$

is a homomorphism $U_q(C_3)$ to $W_{12}^q\otimes U_q(gl(3))$.

Appendix

Using the relations (2) and (3), it is easy to check that the following relations hold in the case of $U_q(C_3)$

$$
\begin{aligned}
E_2X_{33}&=X_{23}+q^{-2}X_{33}E_2\\
E_2X_{23}&=X_{22}+X_{23}E_2\\
E_2X_{22}&=q^2X_{22}E_2\\
E_2X_{13}&=X_{12}+q^{-1}X_{13}E_2\\
E_2X_{12}&=qX_{12}E_2\\
E_2X_{11}&=X_{11}E_2
\end{aligned}
$$

$$
\begin{aligned}
E_1X_{33}&=X_{33}E_1\\
E_1X_{23}&=X_{13}+q^{-1}X_{23}E_1\\
E_1X_{22}&=(q+q^{-1})X_{12}+q^{-2}X_{22}E_1\\
E_1X_{13}&=qX_{13}E_1
\end{aligned}
$$

$$E_1 X_{12} = X_{11} + X_{12} E_1$$
$$E_1 X_{11} = q^2 X_{11} E_1$$
$$X_{23} E_3 = q^2 E_3 X_{23}$$
$$X_{22} E_3 = E_3 X_{22} + q(q - q^{-1}) X_{23}^2$$
$$X_{13} E_3 = q^2 E_3 X_{13}$$
$$X_{12} E_3 = E_3 X_{12} + q(q - q^{-1}) X_{23} X_{13}$$
$$X_{11} E_3 = E_3 X_{11} + q(q - q^{-1}) X_{13}^2$$

$$X_{22} X_{23} = q^2 X_{23} X_{22}$$
$$X_{13} X_{23} = q X_{23} X_{13}$$
$$X_{13} X_{22} = X_{22} X_{13}$$
$$X_{12} X_{23} = q X_{23} X_{12} + (q - q^{-1}) X_{22} X_{13}$$
$$X_{12} X_{22} = q^2 X_{22} X_{12}$$
$$X_{12} X_{13} = q X_{13} X_{12}$$
$$X_{11} X_{23} = X_{23} X_{11} + q(q^2 - q^{-2}) X_{13} X_{12}$$
$$X_{11} X_{22} = X_{22} X_{11} + q(q^2 - q^{-2}) X_{12}^2$$
$$X_{11} X_{13} = q^2 X_{13} X_{11}$$
$$X_{11} X_{12} = q^2 X_{12} X_{11}$$

$$E_1 E_2 = Y_{13} + q^{-1} E_2 E_1$$
$$E_1 Y_{13} = q Y_{13} E_2$$
$$Y_{13} E_2 = q E_2 Y_{13}$$
$$Y_{13} E_3 = X_{13} + q^{-2} E_3 Y_{13}$$
$$Y_{13} X_{13} = X_{11} + X_{13} Y_{13}$$
$$Y_{13} X_{12} = q X_{12} Y_{13} + (q - q^{-1}) X_{11} E_2$$
$$Y_{13} X_{11} = q^2 X_{11} Y_{13}$$

$$F_1 X_{13} = X_{13} F_1 + X_{23} K_1^{-1}$$
$$F_1 X_{12} = X_{12} F_1 + X_{22} K_1^{-1}$$
$$F_1 X_{11} = X_{11} F_1 + (q + q^{-1}) X_{12} K_1^{-1}$$
$$F_2 X_{23} = X_{23} F_2 + (q + q^{-1}) X_{33} K_2^{-1}$$
$$F_2 X_{22} = X_{22} F_2 + (q + q^{-1}) X_{23} K_2^{-1}$$

$$F_2X_{13} = X_{13}F_2$$
$$F_2X_{12} = X_{12}F_2 + X_{13}K_2^{-1}$$
$$F_2X_{11} = X_{11}F_2$$
$$F_3X_{33} = X_{33}F_3 - \frac{K_3 - K_3^{-1}}{q^2 - q^{-2}}$$

$$F_3X_{23} = X_{23}F_3 - K_3E_2$$
$$F_3X_{22} = X_{22}F_3 - q(q - q^{-1})K_3E_2^2$$
$$F_3X_{13} = X_{13}F_3 - K_3Y_{13}$$
$$F_3X_{12} = X_{12}F_3 - q(q - q^{-1})K_3E_2Y_{13}$$
$$F_3X_{11} = X_{11}F_3 - q(q - q^{-1})K_3Y_{13}^2$$

From these relations it is possible to derive by induction the following ones

$$E_2X_{33}^n = q^{-2n}X_{33}^nE_2 + [n]_2X_{33}^{n-1}X_{23}$$
$$E_2X_{23}^n = X_{23}^nE_2 + q^{n-1}[n]_1X_{23}^{n-1}X_{22}$$
$$E_2X_{13}^n = q^{-n}X_{13}^nE_2 + [n]_1X_{13}^{n-1}X_{12}$$
$$E_1X_{23}^n = q^{-n}X_{23}^nE_1 + [n]_1X_{23}^{n-1}X_{13}$$
$$E_1X_{22}^n = q^{-2n}X_{22}^nE_1 + (q + q^{-1})[n]_2X_{22}^{n-1}X_{12}$$
$$E_1X_{12}^n = X_{12}^nE_1 + q^{n-1}[n]_1X_{12}^{n-1}X_{11}$$
$$F_1X_{13}^n = X_{13}^nF_1 + [n]_1X_{23}X_{13}^{n-1}K_1^{-1}$$
$$F_1X_{12}^n = X_{12}^nF_1 + q^{n-1}[n]_1X_{22}X_{12}^{n-1}K_1^{-1}$$
$$F_1X_{11}^n = X_{11}^nF_1 + (q + q^{-1})[n]_2X_{12}X_{11}^{n-1}K_1^{-1}$$
$$F_2X_{23}^n = X_{23}^nF_2 + (q + q^{-1})q^{n-1}[n]_1X_{33}X_{23}^{n-1}K_2^{-1}$$
$$F_2X_{22}^n = X_{22}^nF_2 + (q + q^{-1})[n]_2X_{23}X_{22}^{n-1}K_2^{-1}$$
$$F_2X_{12}^n = X_{12}^nF_2 + [n]_1X_{13}X_{12}^{n-1}K_2^{-1}$$
$$Y_{13}X_{12}^n = q^nX_{12}^nY_{13} + (q - q^{-1})q^{2(n-1)}[n]_1X_{12}^{n-1}X_{11}E_2$$
$$F_3X_{33}^n = X_{33}^nF_3 - \frac{[n]_2}{q^2 - q^{-2}}X_{33}^{n-1}\left[q^{2(n-1)}K_3 - q^{-2(n-1)}K_3^{-1}\right]$$
$$F_3X_{23}^n = X_{23}^nF_3 - q^{n-1}[n]_1X_{23}^{n-1}K_3E_2 - q^{2(n-2)}\begin{bmatrix} n \\ 2 \end{bmatrix}_1 X_{23}^{n-2}X_{22}K_3$$
$$F_3X_{22} = X_{22}^nF_3 - (q - q^{-1})q^{2n-1}[n]_2X_{22}^{n-1}K_3E_2^2$$
$$F_3X_{13}^n = X_{13}^nF_3 - q^{n-1}[n]_1X_{13}^{n-1}K_3Y_{13} - q^{2(n-2)}\begin{bmatrix} n \\ 2 \end{bmatrix}_1 X_{13}^{n-2}X_{11}K_3$$

$$F_3X_{12}^n = X_{12}^nF_3 - (q-q^{-1})q^n[n]_1X_{12}^{n-1}K_3E_2Y_{13}$$
$$-q^{2n-3}(q-q^{-1})^2\begin{bmatrix} n \\ 2 \end{bmatrix}_1 X_{12}^{n-2}X_{11}K_3E_2^2$$
$$F_3X_{11}^n = X_{11}^nF_3 - (q-q^{-1})q^{2n-1}[n]_2X_{11}^{n-1}K_3Y_{13}^2$$

References

1. L.C. Biedenharn, *J. Phys. A.: Math. Gen.* **22**, L873 (1989).
2. A.J. Macfarlane, *J. Phys. A.: Math. Gen.* **22** 4581 (1989).
3. L.C. Biedenharn and J.D. Louk, Angular momentum in quantum physics (Encyclopedia of Mathematics and Its Application 8)(reading MA Addison-Wesley) (1981).
4. T. Holstein and H. Primakoff, *Phys. Rev.* **58**, 1098 (1940).
5. F.J. Dyson, *Phys. Rev.* **102**, 1217 (1956).
6. S. Okubo, *J. Math. Phys.* **33** 528 (1975).
7. C. Kittel, Quantum Theory of Solids.
8. W.J. Gaspers, *Spin Systems*, Singapore, World Scientific, 1989.
9. P. Rink and P. Shuck, *The Nuclear Many-body Problem*, New York, Wiley, 1963.
10. Č. Burdík, *J. Phys. A: Math. Gen.* **18**, 3101 (1985).
11. J. Katriel and A.I. Solomon, *J. Phys. A: Math. Gen.* **24**, 2093-2105 (1991).
12. R. Chakrabarti and R. Jagannathan, *J. Phys. A: Math. Gen.* **24**, L711-L718 (1991).
13. Č. Burdík and O. Navrátil, *J. Phys. A: Math. Gen.* **23**, L1205 (1990).
14. Č. Burdík, M. Havlíček and A. Vančura, *Commun. Math. Phys.* **148**, 417-423 (1992).
15. Fu Hang-Chen and Ge Mo-Lin, *J. Math. Phys.* **33**, 427 (1992).
16. Č. Burdík, L. Černý, and O. Navrátil, *J. Phys. A: Math. Gen.* **26**, L83 (1993).
17. T.D. Palev, *J. Phys. A: Math. Gen.* **31**, 5145 (1998).
18. Č. Burdík and O. Navrátil, *Czech. J. Phys.* **48**, 1301-1306 (1998).
19. Č. Burdík and O. Navrátil, *J. Phys. A: Math. Gen*, **32**, 6141 (1999).
20. Č. Burdík and O. Navrátil, *Int. Mod. Phys. Lett.*, **A14**, No. 28, 4491 (1999).
21. A. Del Sol Mesa, G. Loylola, M. Moshinsky and V. Valesquez, *J. Phys. A: Math. Gen.* **26**, 1147-1160 (1993).
22. M. Jimbo, *Lett. Math. Phys.* **11**, 247 (1986).

QUANTUM FIELD THEORY ON NONCOMMUTATIVE SPACE-TIMES: THE RELATION BETWEEN ULTRAVIOLET BEHAVIOUR AND TOPOLOGY[a]

M. CHAICHIAN[1,2], A. DEMICHEV[1,3] AND P. PREŠNAJDER[1,4]
[1] *High Energy Physics Division, Department of Physics, University of Helsinki, P.O. Box 9, FIN-00014 Helsinki, Finland*

[2] *Helsinki Institute of Physics, P.O. Box 9, FIN-00014 Helsinki, Finland*

[3] *Nuclear Physics Institute, Moscow State University, 119899, Moscow, Russia*

[4] *Department of Theoretical Physics, Comenius University, Mlynská dolina, SK-84215 Bratislava, Slovakia*

Abstract

We study properties of a scalar quantum field theory on two-dimensional noncommutative space-times. Contrary to the common belief that noncommutativity of space-time would be a key to remove the ultraviolet divergences, we show that field theories on a noncommutative plane with the most natural Heisenberg-like commutation relations among coordinates or even on a noncommutative quantum plane with $E_q(2)$-symmetry have ultraviolet divergences, while the theory on a noncommutative cylinder is ultraviolet finite. Thus, ultraviolet behaviour of a field theory on noncommutative spaces is sensitive to the topology of the space-time, namely to its compactness.

1 Introduction

The standard concept of a geometric space is based on the notion of a manifold $\mathcal{M}$ whose points $x \in \mathcal{M}$ are locally labelled by a finite number of real coordinates $x^\mu \in \mathbb{R}^4$. However, it is generally believed that the picture of space-time as a manifold $\mathcal{M}$ should break down at very short distances of the order of the Planck length $\lambda_P \approx 1.6 \times 10^{-33}\, cm$. This implies that the mathematical concepts for high energy (small distance) physics have to be changed, or more precisely, our classical geometrical concepts may not be well suited for the description of physical phenomena at small distances. No convincing

[a] Talk given by A. Demichev

alternative description of physics at very short distances is known, though different routes to progress have been proposed. One such direction is to try to formulate physics on some noncommutative space-time. There appear to be too many possibilities to do this, and it is difficult to see what the right choice is. There have been investigations in the context of Connes' approach [1] to gravity and the Standard Model of electroweak and strong interactions, [2,3] relation between measurements at very small distances and black hole formations [4] and string theory. [5] It has been shown that the noncommutative geometry naturally appears in string theory with a nonzero antisymmetric B-field. This result provides with a solid background for the study of field theories on noncommutative space-times which supposedly correspond to the low energy limit of such strings theories. One more possibility is based on quantum group theory. [6] The simplest noncommutative space is the fuzzy sphere, [7,8,9] *i.e.* the noncommutative analog of a two-dimensional sphere. As is well known from the standard quantum mechanics, a quantization of any *compact* space, in particular a sphere, leads to finite-dimensional representations of the corresponding operators, so that in this case any calculation is reduced to manipulations with finite-dimensional matrices and thus there is simply no place for UV-divergences. [9] Things are not so easy in the case of non-compact manifolds. The quantization leads to infinite-dimensional representations and we have no guarantee that noncommutativity of the space-time coordinates removes the UV-divergences. A relatively simple type of noncommutative geometry of non-compact Minkowski space is defined by Heisenberg-like commutation relations among coordinates. [4] Later it was shown [11] that this model has the same UV-behaviour as an ordinary field theory on the commutative space-time. In the next section we shall use this example, reduced to two-dimensional space, for a general discussion of construction of a quantum field theory on a noncommutative space (NC-QFT). In Section 3 we consider more complicated example of NC-QFT, defined on a noncommutative cylinder. As is well known, a scalar two-dimensional QFT on a commutative cylinder has the same UV-properties as QFT on a plane (in particular, divergent tadpole Feynman diagram): UV-behaviour is related to small distances and, hence, insensitive to the topology of a coordinate manifold. On the contrary, quantization procedure is very sensitive to the topology and this leads to essentially different properties of NC-QFT on the noncommutative cylinder and on the noncommutative plane. While the latter retains the divergent tadpoles (as an ordinary QFT), the former proves to be UV-finite. It is known that a cylinder is homeomorphic to a plane with one punctured point. In the noncommutative case, this map corresponds to transition to a quantum plane. In Section 4 we proceed to the study of a quantum plane with $E_q(2)$-symmetry. Now the coordinate and momentum operators form a subalgebra of ordinary quantum mechanical algebra, so that some operators have discrete spectra. This fact could inspire the belief that the corresponding NC-QFT has an improved UV-behaviour. However, the explicit calculation shows that this is not the case and the QFT on the quantum plane has the usual UV-divergences. Section 5 is devoted to the summary of the results. Further details of the calculations can be found in our papers.[10]

2 Two–Dimensional Quantum Field Theory on Noncommutative Space–Time with Heisenberg–Like Commutation Relations

As an introductory example, we consider the simplest case of a two-dimensional Euclidean field theory with the noncommutativity of the Heisenberg type. There exists a possibility to introduce in the two-dimensional Euclidean *coordinate* plane $P^{(2)}$ an additional Poisson structure. It is defined by the elementary bracket

$$\{x_i, x_j\} = \varepsilon_{ij} \ , \ i,j = 1,2 \ , \tag{1}$$

and extended by Leibniz rule to all smooth functions on $P^{(2)}$ (here ε_{ij} is the antisymmetric tensor, $\varepsilon_{12} = 1$). The noncommutative version $P^{(2)}_\lambda$ of the plane is obtained by the deformation (quantization) of this Poisson structure. In the noncommutative approach we replace the commuting parameters by the hermitian operators $\hat{x}_i$, $(i,j = 1,2)$ satisfying commutation relations

$$[\hat{x}_i, \hat{x}_j] = i\lambda^2 \varepsilon_{ij} \ , \qquad i,j = 1,2 \ , \tag{2}$$

where λ is a positive constant of the dimension of *length*. For all operators of the form

$$\hat{f} = \frac{\lambda^2}{(2\pi)^2} \int d^2k \ \tilde{f}(k) e^{ik\hat{x}} \tag{3}$$

(with a suitable smooth decreasing $\tilde{f}(k)$) we introduce the integral (linear functional) $I_\lambda[\hat{f}]$ as follows:

$$I_\lambda[\hat{f}] = \mathrm{Tr}\hat{f} = \tilde{f}(0) \ . \tag{4}$$

Here Tr denotes the trace in the Fock space and $k\hat{x} = k_1\hat{x}_1 + k_2\hat{x}_2$. The last equality in (4) indicates that $I_\lambda[\cdot]$ is the noncommutative analog of the usual integral on $P^{(2)}$. The noncommutative analogs of field derivatives $\partial_i\hat{\varphi}$, $i = 1,2$ are defined as

$$\partial_i\hat{\varphi} = \varepsilon_{ij} \frac{i}{\lambda^2} [\hat{x}_j, \hat{\varphi}] \ , \qquad i = 1,2 \ . \tag{5}$$

They satisfy the Leibniz rule and reduce to the usual derivatives in the commutative limit. In the noncommutative case we define the Euclidean action of self-interacting scalar noncommutative quantum field theory (NC-QFT):

$$S^{(\lambda)}[\hat{\varphi},\hat{\varphi}^{\dagger}] = S_0^{(\lambda)}[\hat{\varphi},\hat{\varphi}^{\dagger}] + S_{int}^{(\lambda)}[\hat{\varphi},\hat{\varphi}^{\dagger}] \; , \tag{6}$$

with the free part of the action

$$S_0^{(\lambda)}[\hat{\varphi},\hat{\varphi}^{\dagger}] = I_{\lambda}[(\partial_i\hat{\varphi})^{\dagger}(\partial_i\hat{\varphi}) \; + \; m^2\hat{\varphi}^{\dagger}\hat{\varphi}] \; . \tag{7}$$

The calculation of Green functions and other quantities in NC-QFT can be carried out in simple and natural way by the use of operator symbols. An operator symbol, [13] as it is defined in the ordinary Quantum Mechanics, is a function on the phase space which is constructed by a definite rule from a given operator. Different rules for the construction produce different symbols (*e.g.*, Weyl, normal symbols *etc.*). Sets of such functions endowed with a so-called star-product form algebras which are isomorphic to the initial operator algebra. Knowledge of explicit forms of star-products ($\star$-products) allows to make calculations in an easy and short way. In fact, transition to the momentum representation $\varphi(\widehat{x}) \rightarrow \widetilde{\varphi}(k)$ in eq. (3) is the first step in the construction of the corresponding *Weyl symbol.* If, in addition, we make now inverse *ordinary* Fourier transformation

$$\varphi_W(x) = \frac{\lambda^2}{(2\pi)^2}\int d^2k\,\widetilde{\varphi}(k)\mathrm{e}^{ikx} \; , \tag{8}$$

we obtain just what is called the Weyl symbol $\varphi_W(x)$ of the operator $\varphi(\widehat{x})$. Notice that the symbol is defined on the *classical (commutative) analog* of the noncommutative space. Operator product corresponds to the $\star$-product in the set of the symbols: $\varphi(\widehat{x})\psi(\widehat{x}) \longrightarrow \varphi_W \star \psi_W$. Thus, the Weyl symbols or their Fourier transforms (which plays the role of the momentum representation for a field on the noncommutative plane $P_{\lambda}^{(2)}$) are in one-to-one correspondence with the set of fields (operators) $\varphi(\widehat{x})$ on the noncommutative space. This correspondence is based on the relation $\mathrm{Tr}\,\exp\{ik_i\widehat{x}_i\} = 2\pi\lambda^{-2}\delta^{(2)}(k)$. The trace of an operator $f(\widehat{x})$ is needed, in particular, for construction of the action. It is expressed via its Weyl symbol as follows:

$$\mathrm{Tr}\; f(\widehat{x}) = \frac{1}{2\pi\lambda^2}\int d^2x\, f_W(x) = I_{\lambda}[f(\widehat{x})] \; .$$

Now any action for NC-QFT can be obtained from the corresponding classical action by the substitution of the ordinary point-wise function multiplication by the $\star$-product. For example,

$$\mathrm{Tr}\sum_i [\widehat{x}_i, \varphi(\widehat{x})]^2 = \int d^2x\, (\partial_i\varphi_W \star \partial_i\varphi_W)(x)\ .$$

Equivalently, one may use the Fourier transform $\widetilde{\varphi}(p)$ of the Weyl symbol (momentum representation). The "second quantization" in the Euclidean case amounts to calculation of the path integral over set of operator symbols which gives the generating functional $Z[J]$ for Green functions

$$Z[J] = N^{-1}\int \mathcal{D}\varphi_W(x) \exp\left\{-S[\varphi_W, J, \star]\right\}\ , \tag{9}$$

where $S[\varphi_W, J, \star]$ is the operator action (6) on the noncommutative space expressed in terms of symbols, or, in other words, the usual classical action in which the ordinary point-wise multiplication of the fields is substituted by the star-product (N is the normalization constant). When considering the Green functions $G_W(x,y) \equiv \langle\varphi_W(x)\varphi_W(y)\rangle$, one should take into account that a value of an operator symbol at a point on the classical counterpart of a noncommutative space has not direct physical and even mathematical meaning. Thus we must consider a quantum field in NC-QFT as a map from set of states on the corresponding noncommutative space into the algebra of secondary quantized operators, so that the physically meaningful object in NC-QFT is the mean values of the field operators: $\langle\Psi|\,\hat{\varphi}\,|\Psi\rangle\ ,\ |\Psi\rangle \in \mathcal{H}$. Of course, we can choose any complete set of states, but for clear physical interpretation and comparing with the commutative limit, the set should satisfy the following requirements: *i*) the states must be localized in space-time; *ii*) as the parameter of noncommutativity λ goes to zero, the states must shrink to a point. This consideration shows that in order to convert the Green function $G_W(x,y)$ into the physically meaningful object, we must average it over some localized state. States which correspond to optimal rotationally invariant localization around the point $x = (x_1, x_2)$ of the plane are uniquely (up to a phase factor) given as the coherent states $|\xi\rangle$ for the creation and annihilation operators made of the operators $\widehat{x}_1$, $\widehat{x}_2$: $|\xi\rangle = \exp\left\{\xi\hat{\alpha}^\dagger\right\}|0\rangle$ ($|0\rangle$ is the vacuum state in the Fock space $\mathcal{F}$; $\xi = (x_1 + \mathrm{i}x_2)/(\sqrt{2}\lambda)$).

3 Quantum Field Theory on a Noncommutative Cylinder

Below we shall consider UV-properties of the field theory defined on a noncommutative cylinder which can be considered as a quantization of a co-adjoint orbit of the Euclidean group $E(2)$ in the vector space $e^*(2)$ (dual to the Lie algebra $e(2)$). A standard cylinder with radius ρ can be identified with the set of points

$$C_\rho = \{(x,t),\ t \in \mathbb{R},\ x = \rho e^{i\phi},\ \rho = \text{const}\}\ . \tag{10}$$

We shall interpret C_ρ as a space-time manifold. Any function (field) on C_ρ can be expanded as

$$\varphi(x,t) = a_0(t) + \sum_{k=1}^{\infty} \left[a_k^{(+)}(t)x_+^k + a_k^{(-)}(t)x_-^k\right]\ , \tag{11}$$

where $x_+ = x$, $x_- = x^*$, $a_k^{(\pm)}(t) = a_{\pm k}(t)\rho^{-k}$, $k = 1,2,\ldots$. The action for the field $\varphi(x,t)$ we choose as follows

$$S[\varphi,\varphi^*] = \frac{1}{2\pi\rho}\int_{C_\rho} dx\, dt\, [\varphi^*(-\partial_t^2 + \partial_\phi^2 - m^2)\varphi - V(\varphi^*\varphi)]\ . \tag{12}$$

The elementary Poisson brackets are given as

$$\{x_0, x_\pm\} = \pm i x_\pm\ , \qquad \{x_+, x_-\} = 0\ . \tag{13}$$

These are exactly $e(2)$ Lie algebra relations. The point is that the function x_+x_- is central: $\{x_0, x_+x_-\} = \{x_\pm, x_+x_-\} = 0$. Therefore, the constraint $x_+x_- = \rho^2$ is consistent with the $e(2)$ Lie algebra structure. It can be straightforwardly shown that

$$\partial_t^2\varphi = \rho^2\{x_+,\{x_-,\varphi\}\}\ , \qquad \partial_\phi^2\varphi = -\{x_0,\{x_0,\varphi\}\}\ , \tag{14}$$

so that

$$\Box\varphi = \{x_0,\{x_0,\varphi\}\} + \rho^2\{x_+,\{x_-,\varphi\}\}\ . \tag{15}$$

The starting point for the noncommutative space generalization is the replacement of commutative variables $x_0, x_\pm$ by the noncommutative ones $\hat{x}_0, \hat{x}_\pm$, satisfying the $e(2)$ Lie algebra commutation relations

$$[\hat{x}_0, \hat{x}_\pm] = \pm\lambda\hat{x}_\pm\ , \qquad [\hat{x}_+, \hat{x}_-] = 0\ . \tag{16}$$

The $e(2)$ possesses one series of infinite dimensional unitary representations π_ρ in which

$$\hat{x}_+\hat{x}_- = \rho^2, \qquad \rho > 0\ . \tag{17}$$

This representation can be realized in the Hilbert space $\mathcal{L}^2(S^1, d\phi/2\pi)$ as follows

$$\hat{x}_0 = -i\lambda\partial_\phi\ , \qquad \hat{x}_\pm = \rho e^{\pm i\phi}\ . \tag{18}$$

In the basis $|n\rangle = \mathrm{e}^{in\phi}$, $n \in \mathbb{Z}$, the operator $\widehat{x}_0$ is diagonal

$$\widehat{x}_0 \, |n\rangle = \lambda n \, |n\rangle \ . \tag{19}$$

Notice, that there exist a set of inequivalent representations $|n\rangle_\alpha = \mathrm{e}^{\mathrm{i}(n+\alpha)\phi}$, $0 \leq \alpha < 1$. For simplicity, we shall use only the representation (19) which corresponds to $\alpha = 0$. The field $\hat{\varphi} = \varphi(\widehat{x}_0, \widehat{x}_\pm)$ is an operator in $\mathcal{L}^2(S^1, d\phi/2\pi)$ and we suppose that it possesses the expansion

$$\hat{\varphi} = a_0(\widehat{t}) + \sum_{k=1}^{\infty} \left[a_k^{(+)}(\widehat{t})\widehat{x}_+^k + a_k^{(-)}(\widehat{t})\widehat{x}_-^k\right] \ , \tag{20}$$

the noncommutative integral $I_\lambda[\varphi]$ being defined by

$$I_\lambda[\hat{\varphi}] \stackrel{\text{def}}{\equiv} \lambda \mathrm{Tr}\, \hat{\varphi} = \lambda \sum_{n\in\mathbb{Z}} a_0(\lambda n) \ . \tag{21}$$

The free field action has the form similar to that in the commutative case:

$$S_0[\hat{\varphi}, \hat{\varphi}^*] = I_\lambda\big[-\varphi^*(\Box_\lambda + m^2)\varphi\big] \ , \tag{22}$$

where the noncommutative analog $\Box_\lambda$ of the operator $\Box$ is defined as (*cf.* (15))

$$\Box_\lambda \hat{\varphi} = -\frac{1}{\lambda^2}[x_0, [x_0, \varphi]] + \frac{1}{\lambda^2 \rho^2}[x_+, [x_-, \varphi]] \ . \tag{23}$$

In the noncommutative case the interaction term we take in the form $V = \frac{g}{4} : (\varphi^*\varphi)^2 :$, where the ordering $: \cdots :$ means that all expansion coefficients of fields $a_k(\widehat{t}), a_k^\dagger(\widehat{t})$ are collected as a right (or equivalently left) factor. This guarantees that there are no artificial time derivatives in the potential: all operators $a_k(\widehat{t}), a_k^\dagger(\widehat{t})$ in the potential term appear at the same time $t = \lambda n$. The tadpole contribution is *finite*

$$G_0(x,t;x,t) = \int_0^{\pi/\lambda} d\omega \, \frac{\cot\left(\sqrt{\Omega_\lambda^2(\omega) - m^2 + \mathrm{i}\varepsilon}\right)}{\sqrt{\Omega_\lambda^2(\omega) - m^2 + \mathrm{i}\varepsilon}} < \infty \ , \tag{24}$$

where $\Omega_\lambda(\omega) = \frac{2}{\lambda}\sin\frac{\omega\lambda}{2}$. In the two-dimensional scalar field theory on commutative space the tadpole is the only divergent contribution to Feynman diagrams. We have shown that transition to the noncommutative cylinder leads to *UV-finite field theory.*

4 Quantum Mechanics and Quantum Field Theory on a Noncommutative Quantum Plane with $E_q(2)$–Symmetry

In this section we consider Quantum Mechanics induced by a quantum group structure. Recall that in the case of ordinary Lie group G, the group structure defines a unique symplectic structure on the cotangent bundle T^*_G to the group manifold G and, hence, the corresponding canonical quantization (via substitution of Poisson brackets by the corresponding commutators). A similar construction with necessary generalizations, can be carried out for Lie-Poisson groups, which after the quantization procedure become quantum groups (see, *e.g.*, Chaichian and Demichev [6] and refs. therein). Thus instead of starting from ordinary $E(2)$, we shall use below its quantum version [14] $E_q(2)$. In order to simplify notation, we shall omit, in what follows, the sign "hat" over operators. We start from the quantum group $E_q(2)$ generated [14] by elements $\bar{v}$, v, $\bar{t}$, t. The elements $\bar{z}$, z and $\bar{\partial}_q$, ∂_q defines the q-deformed algebra of functions on $P_q^{(2)}$ together with the q-deformed left-invariant vector fields (derivatives). Its defining relations read as

$$\begin{aligned} &z\bar{z} = q^2\bar{z}z\ , \qquad \partial_q\bar{\partial}_q = q^2\bar{\partial}_q\partial_q\ , \\ &\partial_q z = 1 + q^{-2}z\partial_q\ , \qquad \bar{\partial}_q\bar{z} = 1 + q^2\bar{z}\bar{\partial}_q\ , \\ &\bar{\partial}_q z = q^2 z\bar{\partial}_q\ , \qquad \partial_q\bar{z} = q^{-2}\bar{z}\partial_q\ . \end{aligned} \tag{25}$$

If we put $q = 1$ and define $p = -i\hbar\partial$, $\bar{p} = -i\hbar\bar{\partial}$, the relations (25) become the usual canonical commutation relations for a particle in two-dimensional space. The requirement of consistency with antipode dictates the following conjugation rule for the q-derivatives

$$\partial_q^\dagger = -q^2\bar{\partial}_q\ , \qquad \bar{\partial}_q^\dagger = -q^{-2}\partial_q\ . \tag{26}$$

We consider the relations (25) as a q-deformation of the canonical commutation relations. It is well-known that the Casimir operator generating the centrum of the enveloping algebra $\mathcal{U}_q e(2)$ is given by the formula $C = \bar{T}e^{2J}T$. On a quantum plane it is the Laplace operator $\Delta = \lambda(C)$, or more explicitly,

$$\Delta = \bar{\partial}_q q^{2\bar{N}-2N}\partial_q = -q^2\bar{P}P\ . \tag{27}$$

Our aim is to find eigenfunctions of Δ which further will be used for construction of a field theory on the q-plane. We search for the functions $\Phi_{kk'}(z,\bar{z})$ on the q-plane which are common eigenfunctions of two commuting operators Δ and ∂_q:

$$\Delta\Phi_{kk'}(z,\bar{z}) = kk'\Phi_{kk'}(z,\bar{z})\ , \qquad \partial_q\Phi_{kk'}(z,\bar{z}) = k\Phi_{kk'}(z,\bar{z})\ . \tag{28}$$

The solutions of this eigenvalue problem have the form

$$\Phi_{kk'} = \mathrm{e}_{q^{-2}}^{kz} \tilde{e}_{q^2}^{k'\bar{z}} , \ k, k' \in \mathbb{C} . \tag{29}$$

where

$$\mathrm{e}_{q^{-2}}^{kz} = \sum_{n=0}^{\infty} \frac{(kz)^n}{[n;q^{-2}]!} , \qquad \tilde{e}_{q^2}^{k'\bar{z}} = \sum_{n=0}^{\infty} \frac{(k'\bar{z})^n q^{-n(n+1)}}{[n;q^2]!} . \tag{30}$$

These eigenfunctions form a basis of the Hilbert space $\mathcal{H}_q$ of square integrable "functions" on the q-plane $P_q^{(2)}$. We define on the quantum plane the field action for a complex field $\varphi(z, \bar{z})$ in a standard way

$$S[\varphi, \varphi^\dagger] = \int^q d\bar{z}dz \ [\varphi^\dagger(z, \bar{z})(-\Delta + m^2)\varphi(z, \bar{z}) \ - \ V(\varphi, \varphi^\dagger)] , \tag{31}$$

where $\int^q d\bar{z}dz$ is the invariant q-deformed integral, [15] and expand the field $\varphi(z, \bar{z})$ into the plane waves

$$\varphi(z, \bar{z}) = \int_q d\bar{k}dk \, a(k, \bar{k})\Phi_k(z, \bar{z}) . \tag{32}$$

Inserting this into (31) we can diagonalize the free field action:

$$S_0[\varphi, \varphi^\dagger] = \int_q d\bar{k}dk \ \bar{a}(k, \bar{k})(k\bar{k} + m^2)a(k, \bar{k}) . \tag{33}$$

The last formula is valid due to the completeness and orthogonality properties of the distorted plane waves. From (33) we see that the free field propagator in the k-representation is given as

$$\tilde{G}(k, \bar{k}) = \frac{1}{k\bar{k} + m^2} . \tag{34}$$

Since, the propagator $\tilde{G}(k, \bar{k})$ behaves like $(k\bar{k})^{-1}$ for large $k\bar{k}$, the UV-divergences has to be expected. Indeed, let us consider the scalar field on the q-plane in a constant external field F, so that the interaction term has the simple form

$$V(\varphi, \varphi^\dagger) = F\varphi^\dagger(z, \bar{z})\varphi(z, \bar{z}) , \qquad F = \mathrm{const} .$$

Then the tadpole contribution is proportional to the divergent Jackson integral

$$\Gamma = \int_q d\bar{k}dk \frac{1}{k\bar{k} + m^2} = \infty . \tag{35}$$

In order to prove (35) we shall perform the evaluation of the tadpole contribution with the regularized propagator

$$\tilde{G}_s^{reg} = m^{2s} \int^q d\bar{z}dz\, e_{q^2}^{-m^2 \bar{z}z} e_{q^{-2}}^{-k\bar{k}z\bar{z}} (z\bar{z})^s \,. \tag{36}$$

For $s > -1$, this is a well-defined Jackson integral behaving asymptotically as $(k\bar{k})^{-1-s}$. For $s = 0$ it reduces to (34). Inserting (36) into the tadpole integral and interchanging the order of integrations, the regularized tadpole contribution can be calculated straightforwardly,

$$\Gamma_s^{reg} = \int^q d\bar{z}dz e_{q^2}^{-\bar{z}z} (z\bar{z})^{s-1} = \frac{1}{2}\Gamma_{q^2}(s) \,. \tag{37}$$

Here, $\Gamma_{q^2}(s)$ is a q-deformed Γ-function which is singular at $s = 0$, similarly to the usual Γ-function (repeating the indicated calculations in the standard case we recover (37) with the ordinary Γ-function). In the case of the external field $F(z, \bar{z}) = \sum_{n,m=0}^{\infty} a_{mn} \bar{z}^n z^m$ depending on coordinates, the tadpole contribution is given as

$$\int_q d\bar{k}dk \, \frac{g(k, \bar{k}, k, \bar{k})}{k\bar{k} + m^2} \,,$$

where $g(k_1, \bar{k}_1, k_2, \bar{k}_2) = \int^q d\bar{z}dz \, \Phi_{k_1}^{\dagger}(z, \bar{z}) F(z, \bar{z}) \Phi_{k_2}(z, \bar{z})$. In the commutative case the numerator $g(k, \bar{k}, k, \bar{k}) = \widetilde{F}(0, 0)$ is constant leading to the divergent tadpole contribution. In the noncommutative case we obtain even more divergent contributions which cannot cancel each other for a general field $F(z, \bar{z})$ (due to arbitrariness of the expansion coefficients a_{mn}). They are proportional to $\log q$ and disappear in the commutative limit.

We conclude that the q-deformation of a plane does not lead to the desired UV-regularization.

5 Conclusion

We have shown that transition to a noncommutative space-time does not necessarily lead to an ultraviolet regularization of the quantum field theory constructed in this space, at least in the most natural way of introducing noncommutativity as we have performed in this paper. In particular, QFT on noncommutative planes with the Heisenberg-like commutation relations for coordinates and the deformed plane with the quantum $E_q(2)$-symmetry still contain divergent tadpoles. However, in general, theories which have the same UV-behaviour on classical spaces may acquire essentially different properties after the quantization. The reason is that quantization procedure is highly sensitive to the topology of the manifold under consideration. Thus, while in the case of classical space-time the theories on a sphere, cylinder or plane have UV-divergences, in the case of noncommutative space-time the two-dimensional theories on the fuzzy sphere [8,9] and on the quantum cylinder do not have divergences at all. This can be traced to the compactness properties of the space-time in question:

- In the case of a fuzzy sphere, models contain a finite number of modes and thus all the usual integrations are replaced by final sums and, consequently, no UV-divergences can appear.
- In the case of a cylinder, a priori one can not claim whether the quantum field theory is finite. However, the non-commutativity of the space-time together with the compactness of the space (circle) lead to the intrinsic cut-off in the energy modes. This guarantees the removal of UV-divergences in the two-dimensional case.
- On a noncommutative plane (whose commutative limit is noncompact in both directions) with Heisenberg-like or even with deformed commutation relations, the noncommutativity of the space-time does not lead to an UV-regular theory.

We conclude with the following general picture:

- the noncommutativity itself does not guarantee the removal of UV-divergences;
- global topological restrictions are needed - namely, at most one dimension (time) is allowed to be noncompact, in order to achieve the removal of UV-divergences of a quantum field theory formulated in a noncommutative space-time of arbitrary dimensions.

It is of great interest to investigate the problem further in order to find out whether there exists any other, than the one presented in this paper, way of introducing noncommutativity so that it would remove the UV-divergences even in the case of fully noncompact space-times.

Acknowledgments

The financial support of the Academy of Finland under the Projects No. 163394 is greatly acknowledged. A.D.'s work was partially supported also by RFBR-98-02-16769 grant and P.P.'s work by VEGA project 1/4305/97.

References

1. A. Connes, *Noncommutative Geometry* (Academic Press, New York, 1994).
2. A. Connes and J.Lott, *Nucl. Phys. Suppl.* **B11**, 19 (1990).
3. H.-H. Chamseddin, G. Felder and J. Fröhlich, *Nucl. Phys.* **B395**, 672 (1993).
4. S. Doplicher, K. Fredenhagen and J. E. Roberts, *Phys. Lett.* **B331**, 39 (1994);
   S. Doplicher, K. Fredenhagen and J. E. Roberts, *Comm. Math. Phys.* **172**, 187 (1995).
5. N. Seiberg and E. Witten, preprint LASSNS-HEP-99-74; *hep-th/9908142*; *JHEP* **9909**, 032 (1999); to appear in *Nucl. Phys.* **B**.
6. M. Chaichian and A. Demichev, *Introduction to Quantum Groups*, World Scientific, Singapore, (1996).
7. F. A. Berezin, *Commun. Math. Phys.* **40**, 153 (1975).
8. J. Hoppe, *Elem. Part. Res. J.* **80**, 145 (1989);
   J. Madore, *Journ. Math. Phys.* **32**, 332 (1991).
9. H. Grosse, C. Klimčík and P. Prešnajder, *Int. Jour. Theor. Phys.* **35**, 231 (1996);
   H. Grosse, C. Klimčík and P. Prešnajder, *Commun. Math. Phys.* **185**, 155 (1997).
10. M. Chaichian, A. Demichev and P. Prešnajder, preprint HIP-1998-77/TH, *hep-th/9812180*, *Nucl. Phys.* **B567**, 360 (2000).
11. T. Filk, *Phys. Lett.* **B376**, 53 (1996).
12. A.A. Kirillov, *Elements of the Theory of Representations*, Springer, Berlin, (1976).
13. N.L. Balazs and B.K. Jennings, *Phys. Rep.* **104**, 347 (1984).
14. L.L. Vaksman and L.I. Korogodsky, *Dokl. Akad. Nauk SSSR* **304**, 1036 (1989).
15. H.T. Koelink, *Duke Math. J.* **76**, 483 (1994);
    S. Baaj, *C. R. Acad. Sci.* **314**, 1021 (1992).

INDUCED REPRESENTATIONS FOR DUALS OF TWO-PARAMETER GL(2) DEFORMATIONS[a]

V.K. DOBREV[1,2,3] AND S.G. MIHOV[3]
[1] *International Center for Theoretical Physics*
Strada Costiera 11, P.O. Box 586, 34100 Trieste, Italy

[2] *Arnold Sommerfeld Institute for Mathematical Physics*
Technical University Clausthal
Leibnizstr. 10, 38678 Clausthal-Zellerfeld, Germany[†]

[3] *Bulgarian Academy of Sciences*
Institute of Nuclear Research and Nuclear Energy
72 Tsarigradsko Chaussee, 1784 Sofia, Bulgaria[*]
[†]Current, resp., [*]permanent, address for V.K.D.

We construct induced infinite-dimensional representations of the two-parameter quantum algebras $U_{p,q}(gl(2))$ and $U_{g,h}(gl(2))$ which are in duality with the deformations $GL_{p,q}(2)$ and $GL_{g,h}(2)$, resp. The representations split into one-parameter representations of a one-generator central algebra and a three-generator quantum algebra, the latter in each case being a deformation of $U(sl(2))$. In both cases the representations can be mapped to representations in one complex variable, which are deformations of the standard one-parameter vector-field realization of $sl(2)$. The deformation in the case of $U_{g,h}(gl(2))$, actually of the Jordanian $U_{\tilde{g}}(sl(2))$, $\tilde{g} = g + h$, is a *new* deformation. We also obtain canonically finite-dimensional representations which can be restricted to the one-parameter three-generator subalgebras in both cases.

1 Introduction

In the last ten years there were many developments in the representation theory of those quantum groups introduced by Drinfeld [1] and Jimbo, [2] and others, [3–11] (see Dobrev [12] for a review). We call these objects quantum algebras. However, most of these developments were for weight modules over these quantum algebras. There were not many papers on induced representations of quantum algebras realized on the dual matrix quantum groups.[7,11] Later, in Dobrev [13] were developed induced representations generalizing the classical induction from parabolic subgroups. This was extended to the dual algebra of a matrix Lorentz group [14] and to $GL_q(n)$ and $SL_q(n)$ in Dobrev.[15] Note that since the earliest results of [13] were not published the reader should see Dobrev [15] from where the mentioned unpublished results may be recovered for $n = 2$. In the present paper we extend the approach of Dobrev [13] to the duals

[a]Talk given by V.K. Dobrev

of two-parametric deformations of $GL(2)$, namely of $GL_{p,q}(2)$ and $GL_{g,h}(2)$. The duals are the two-parameter quantum algebras $\mathcal{U}_{p,q} \equiv U_{p,q}(gl(2))$ and $\mathcal{U}_{g,h} \equiv U_{g,h}(gl(2))$, resp., and they were constructed in Dobrev [16] and Aneva et al.,[17] resp. They are deformations of $U(gl(2))$. We construct induced representations of $\mathcal{U}_{p,q}$ and $\mathcal{U}_{g,h}$. For $p = q = 1$, resp., $g = h = 0$ our representations coincide with the holomorphic representations induced from the lower, resp. upper, diagonal Borel subgroup $\mathcal{B}$ of $GL(2)$ and acting on the one-dimensional coset $G/\mathcal{B}$. The induced representations are infinite-dimensional and depend on two representation parameters, however, they split into one-parameter representations of a one-generator central algebra and a three-generator quantum algebra, the latter in each case being a deformation of $U(sl(2))$. In both cases the representations can be mapped to representations in one complex variable. These constitute deformations of standard one-parameter vector-field realization of $sl(2)$. The deformation in the case of $\mathcal{U}_{p,q}$ coincides with the one for the one-parameter Jimbo quantum algebra $U_t(sl(2))$, $t = \sqrt{pq}$. The deformation in the case of $\mathcal{U}_{g,h}$, actually of the Jordanian $U_{\tilde{g}}(sl(2))$, $\tilde{g} = g + h$, is a *new* deformation. The infinite-dimensional representations are reducible for some values of the representation parameters, and then we obtain canonically finite-dimensional representations which can be restricted to the one-parameter three-generator subalgebras in both cases.

2 The Case of $\mathbf{GL_{p,q}}$

2.1 The Matrix Quantum Group $GL_{\mathbf{p,q}}$ *and the Dual Quantum Algebra*

In this Section we construct induced representations of the quantum algebra $\mathcal{U}_{p,q}$, which was found [16] as the dual of the two-parameter matrix quantum group $GL_{p,q}(2)$ introduced in Demidov et al.[18] We follow (with some modifications) the similar construction for $U_q(sl(2))$ in Dobrev.[13] The representation spaces are built on formal power series of $GL_{p,q}(2)$, so we start by recalling it. The matrix quantum group $\mathcal{A}_{p,q} = GL_{p,q}(2)$ [18] is generated by four elements a, b, c, d, which may be collected for later use in a 2×2 matrix:

$$M = \begin{pmatrix} a & b \\ c & d \end{pmatrix} . \tag{1}$$

These elements obey the following commutation relations ($p, q \in \mathbb{C}$, $pq \neq 0$):

$$\begin{aligned} ab &= p^{-1}\, ba\,, & ac &= q^{-1}\, ca\,, & bd &= q^{-1}\, db\,, \\ cd &= p^{-1}\, dc\,, & q\, bc &= p\, cb\,, & da &= ad + (q - p^{-1})\, bc\,. \end{aligned} \tag{2}$$

There exists a multiplicative quantum determinant:

$$\mathcal{D} \equiv ad - p^{-1}bc = da - ts^{1/2}cb\,, \qquad t \equiv \sqrt{pq}, \quad s \equiv pq^{-1}\,, \tag{3}$$

which is not central (unless $p = q$) :

$$[\mathcal{D}, a] = [\mathcal{D}, d] = 0 , \qquad p\mathcal{D}b = qb\mathcal{D} , \qquad q\mathcal{D}c = pc\mathcal{D} . \tag{4}$$

For our purposes we shall suppose that d is invertible. Then from (2) we have $a = (\mathcal{D} + p^{-1}bc)d^{-1}$. Thus, we shall take the following basis of $\mathcal{A}_{p,q}$:

$$f = f_{mnk\ell} = b^m c^n d^\ell \mathcal{D}^k , \quad k, m, n \in \mathbb{Z}_+ , \ \ell \in \mathbb{Z} . \tag{5}$$

For the dual algebra $\mathcal{U}_{p,q}$ [16] we use the following generating elements: r, r^{-1}, h, h^{-1}, X^+, X^- and relations:

$$\begin{gathered} [r^\epsilon, h^\varepsilon] = 0, \quad [r^\epsilon, X^+] = 0 , \quad [r^\epsilon, X^-] = 0 , \quad \epsilon, \varepsilon = \pm 1, \\ rr^{-1} = 1_{\mathcal{U}} , \quad hh^{-1} = 1_{\mathcal{U}} , \\ h^\varepsilon X^+ = t^\varepsilon X^+ h^\varepsilon , \quad h^\varepsilon X^- = t^{-\varepsilon} X^- h^\varepsilon , \quad [X^+, X^-] = \frac{h^2 - h^{-2}}{t - t^{-1}} . \end{gathered} \tag{6}$$

Note that in Dobrev [16] instead of the parameters p, q were used $p^{1/2}, q^{1/2}$, and instead of the generators r, r^{-1}, h, h^{-1} were used H, K so that: $h^\varepsilon = t^{\varepsilon H/2}$, $r^\epsilon = s^{\epsilon K/2}$. Thus, strictly speaking, our algebra $\mathcal{U}_{p,q}$ is a subalgebra of the one in Dobrev. [16] To obtain from (6) the $U(gl(2))$ commutation relations one has to use the expansions: $h^\varepsilon \approx 1 + \varepsilon(\log t) H/2$, $r^\epsilon \approx 1 + \epsilon(\log s) K/2$, and then to set: $\log t \to 0$, $\log s \to 0$. Thus, one gets:

$$\begin{gathered} [X^+, X^-] = H , \quad [H, X^+] = 2X^+ , \quad [H, X^-] = -2X^- , \\ [K, X^+] = 0 , \quad [K, Y^-] = 0 , \quad [K, H] = 0 . \end{gathered} \tag{7}$$

The coalgebra relations are: [16]

$$\begin{gathered} \delta_{\mathcal{U}}(h^\varepsilon) = h^\varepsilon \otimes h^\varepsilon , \qquad \delta_{\mathcal{U}}(r^\epsilon) = r^\epsilon \otimes r^\epsilon , \\ \delta_{\mathcal{U}}(X^+) = X^+ \otimes rh + h^{-1} \otimes X^+ , \\ \delta_{\mathcal{U}}(X^-) = X^- \otimes r^{-1}h + h^{-1} \otimes X^- , \end{gathered} \tag{8}$$

$$\epsilon_{\mathcal{U}}(h^\varepsilon) = 1 , \quad \epsilon_{\mathcal{U}}(r^\epsilon) = 1 , \quad \epsilon_{\mathcal{U}}(X^+) = 0 , \quad \epsilon_{\mathcal{U}}(X^-) = 0 , \tag{9}$$

$$\begin{gathered} \gamma_{\mathcal{U}}(r^\epsilon) = r^{-\epsilon} , \quad \gamma_{\mathcal{U}}(h^\varepsilon) = h^{-\varepsilon} , \\ \gamma_{\mathcal{U}}(X^+) = -tX^+ r^{-1} , \quad \gamma_{\mathcal{U}}(X^-) = -t^{-1} X^- r . \end{gathered} \tag{10}$$

Recall that the two sets $r^\epsilon, h^\varepsilon, X^+$ and $r^\epsilon, h^\varepsilon, X^-$ generate conjugate Borel Hopf subalgebras of $\mathcal{U}_{p,q}$, and that it is not possible to decouple the r^ϵ generators. [16] The duality for the algebras $\mathcal{U}_{p,q}$ and $\mathcal{A}_{p,q}$ is given by the pairings:

$$\lg r , \begin{pmatrix} a & b \\ c & d \end{pmatrix} \rangle = \begin{pmatrix} s^{1/2} & 0 \\ 0 & s^{1/2} \end{pmatrix} ,$$

$$\lg h\,,\begin{pmatrix} a & b \\ c & d \end{pmatrix}\rangle = \begin{pmatrix} t^{1/2} & 0 \\ 0 & t^{-1/2} \end{pmatrix},$$
$$\lg X^{+}\,,\begin{pmatrix} a & b \\ c & d \end{pmatrix}\rangle = \begin{pmatrix} 0 & 1 \\ 0 & 0 \end{pmatrix},$$
$$\lg X^{-}\,,\begin{pmatrix} a & b \\ c & d \end{pmatrix}\rangle = \begin{pmatrix} 0 & 0 \\ 1 & 0 \end{pmatrix}. \tag{11}$$

The pairing between arbitrary elements of $\mathcal{U}_{p,q}$ and $\mathcal{A}_{p,q}$ follows then from the properties of the duality pairing. All this is given in Dobrev [16] and is not reproduced here since we shall not need these formulae.

2.2 *Left and Right Action of* $\mathcal{U}_{p,q}$

Now we introduce the left regular representation of $\mathcal{U}_{p,q}$ which in the classical case is the infinitesimal version of: $\pi_L(Y)M = Y^{-1}M,\, Y, M \in GL(2)$. Namely, we set:

$$\pi_L(X)\;M = \lg \gamma_{\mathcal{U}}(X), M\rangle\; M\;, \qquad X \in \mathcal{U}\,. \tag{12}$$

Note that in the first formulation of the approach, [13] which we follow in (12), and also in the papers treating $GL_q(n)$ and $SL_q(n)$, [15] was used the classical antipode $\gamma^0_{\mathcal{U}}$ (with deformation parameter(s) set to classical values) instead of $\gamma_{\mathcal{U}}$, since in these cases things differ only in some intermediate formulae by inessential $q^{\cdots}$ factors. This would be true also here for the $GL_{p,q}(2)$ case, but not for the $GL_{g,h}(2)$ case treated below. Thus, for uniformity we use $\gamma_{\mathcal{U}}$. Explicitly we get from (12) for the generators of $\mathcal{U}_{p,q}$:

$$\pi_L(r)\begin{pmatrix} a & b \\ c & d \end{pmatrix} = \begin{pmatrix} s^{-\frac{1}{2}}a & s^{-\frac{1}{2}}b, \\ s^{-\frac{1}{2}}c & s^{-\frac{1}{2}}d \end{pmatrix},$$
$$\pi_L(h)\begin{pmatrix} a & b \\ c & d \end{pmatrix} = \begin{pmatrix} t^{-\frac{1}{2}}a & t^{-\frac{1}{2}}b, \\ t^{\frac{1}{2}}c & t^{\frac{1}{2}}d \end{pmatrix},$$
$$\pi_L(X^{+})\begin{pmatrix} a & b \\ c & d \end{pmatrix} = \begin{pmatrix} -ts^{-\frac{1}{2}}c & -ts^{-\frac{1}{2}}d \\ 0 & 0 \end{pmatrix},$$
$$\pi_L(X^{-})\begin{pmatrix} a & b \\ c & d \end{pmatrix} = \begin{pmatrix} 0 & 0 \\ -t^{-1}s^{\frac{1}{2}}a & -t^{-1}s^{\frac{1}{2}}b \end{pmatrix}. \tag{13}$$

In order to derive the action of π on arbitrary elements of the basis we use the following twisted derivation rule consistent with the coproduct and the representation structure. Namely, we use: [13]

$$\pi_L(y)\varphi\psi = \hat{m}\left(\pi_L(\delta'_{\mathcal{U}}(y))(\varphi\otimes\psi)\right), \tag{14}$$

where $\hat{m}$ is the multiplication map: $\hat{m} : \mathcal{A} \oplus \mathcal{A} \longrightarrow \mathcal{A}$, $\hat{m}(f \oplus f') = ff'$; $\delta'_{\mathcal{U}} = \sigma \circ \delta_{\mathcal{U}}$ is the opposite coproduct, (σ is the permutation operator). Thus, in our concrete situation we have:

$$\begin{aligned} \pi_L(r)\varphi\psi &= \pi_L(r)\varphi \cdot \pi_L(r)\psi\,, \\ \pi_L(h)\varphi\psi &= \pi_L(h)\varphi \cdot \pi_L(h)\psi\,, \\ \pi_L(X^+)\varphi\psi &= \pi_L(rh)\varphi \cdot \pi_L(X^+)\psi \ + \ \pi_L(X^+)\varphi \cdot \pi_L(h^{-1})\psi\,, \\ \pi_L(X^-)\varphi\psi &= \pi_L(r^{-1}h)\varphi \cdot \pi_L(X^-)\psi \ + \ \pi_L(X^-)\varphi \cdot \pi_L(h^{-1})\psi\,, \\ \pi_L \begin{pmatrix} r & X^+ \\ X^- & h \end{pmatrix} 1_{\mathcal{A}} &= \begin{pmatrix} 1_{\mathcal{A}} & 0 \\ 0 & 1_{\mathcal{A}} \end{pmatrix}. \end{aligned} \tag{15}$$

Further, we obtain:

$$\begin{aligned} \pi_L \begin{pmatrix} r & X^+ \\ X^- & h \end{pmatrix} \mathcal{D} &= \begin{pmatrix} s^{-1}\mathcal{D} & 0 \\ 0 & \mathcal{D} \end{pmatrix}, \\ \pi_L(r) \begin{pmatrix} \mathcal{D}^n & b^n \\ c^n & d^n \end{pmatrix} &= \begin{pmatrix} s^{-n}\mathcal{D}^n & s^{-\frac{n}{2}} b^n \\ s^{-\frac{n}{2}} c^n & s^{-\frac{n}{2}} d^n \end{pmatrix}, \\ \pi_L(h) \begin{pmatrix} \mathcal{D}^n & b^n \\ c^n & d^n \end{pmatrix} &= \begin{pmatrix} \mathcal{D}^n & t^{-\frac{n}{2}} b^n \\ t^{\frac{n}{2}} c^n & t^{\frac{n}{2}} d^n \end{pmatrix}, \\ \pi_L(X^+) \begin{pmatrix} \mathcal{D}^n & b^n \\ c^n & d^n \end{pmatrix} &= -t^{\frac{n+1}{2}} s^{-\frac{n}{2}} [n]_t \begin{pmatrix} 0 & b^{n-1}d \\ 0 & 0 \end{pmatrix}, \\ \pi_L(X^-) \begin{pmatrix} \mathcal{D}^n & b^n \\ c^n & d^n \end{pmatrix} &= -t^{\frac{n-3}{2}} s^{\frac{1}{2}} [n]_t \begin{pmatrix} 0 & 0 \\ ac^{n-1} & bd^{n-1} \end{pmatrix}, \end{aligned} \tag{16}$$

where

$$[n]_t = \frac{t^n - t^{-n}}{t - t^{-1}} = \frac{t^n - t^{-n}}{\lambda}, \quad \lambda = t - t^{-1}\,.$$

Next we introduce the right regular representation $\pi_R(X)$ [13] (which is used also in Masuda et al.,[11] (though not given in this form) being called left action and denoted π_l):

$$\pi_R(X)\ M \ = \ M \ \lg X, M\rangle\,, \qquad X \in \mathcal{U}_{p,q}\,. \tag{17}$$

Of course, as in Dobrev [13] we shall use (17) as right action in order to reduce the left regular representation (and we could have also reversed the role of left and right). Explicitly we have:

$$\begin{aligned}
\pi_R(r)\begin{pmatrix} a & b \\ c & d \end{pmatrix} &= \begin{pmatrix} s^{1/2}a & s^{1/2}b \\ s^{1/2}c & s^{1/2}d \end{pmatrix}, \\
\pi_R(h)\begin{pmatrix} a & b \\ c & d \end{pmatrix} &= \begin{pmatrix} t^{1/2}a & t^{-1/2}b \\ t^{1/2}c & t^{-1/2}d \end{pmatrix}, \\
\pi_R(X^+)\begin{pmatrix} a & b \\ c & d \end{pmatrix} &= \begin{pmatrix} 0 & a \\ 0 & c \end{pmatrix}, \\
\pi_R(X^-)\begin{pmatrix} a & b \\ c & d \end{pmatrix} &= \begin{pmatrix} b & 0 \\ d & 0 \end{pmatrix}.
\end{aligned} \tag{18}$$

The twisted derivation rule (cf. Dobrev [13]) is now given by:

$$\pi_R(y)\varphi\psi = \hat{m}\left(\pi_R(\delta_{\mathcal{U}_g}(y))(\varphi\otimes\psi)\right), \tag{19}$$

i.e., in our concrete situation:

$$\begin{aligned}
\pi_R(r)\varphi\psi &= \pi_R(r)\varphi\cdot\pi_R(r)\psi, \\
\pi_R(h)\varphi\psi &= \pi_R(h)\varphi\cdot\pi_R(h)\psi, \\
\pi_R(X^+)\varphi\psi &= \pi_R(h^{-1})\varphi\cdot\pi_R(X^+)\psi + \pi_R(X^+)\varphi\cdot\pi_R(rh)\psi, \\
\pi_R(X^-)\varphi\psi &= \pi_R(h^{-1})\varphi\cdot\pi_R(X^-)\psi + \pi_L(X^-)\varphi\cdot\pi_L(r^{-1}h)\psi, \\
\pi_R\begin{pmatrix} r & X^+ \\ X^- & h \end{pmatrix} 1_{\mathcal{A}} &= \begin{pmatrix} 1_{\mathcal{A}} & 0 \\ 0 & 1_{\mathcal{A}} \end{pmatrix}.
\end{aligned} \tag{20}$$

Further, we obtain:

$$\begin{aligned}
\pi_R\begin{pmatrix} r & X^+ \\ X^- & h \end{pmatrix}\mathcal{D} &= \begin{pmatrix} s\mathcal{D} & 0 \\ 0 & \mathcal{D} \end{pmatrix}, \\
\pi_R(r)\begin{pmatrix} \mathcal{D}^n & b^n \\ c^n & d^n \end{pmatrix} &= \begin{pmatrix} s^n\mathcal{D}^n & s^{\frac{n}{2}}b^n \\ s^{\frac{n}{2}}c^n & s^{\frac{n}{2}}d^n \end{pmatrix}, \\
\pi_R(h)\begin{pmatrix} \mathcal{D}^n & b^n \\ c^n & d^n \end{pmatrix} &= \begin{pmatrix} \mathcal{D}^n & t^{-\frac{n}{2}}b^n \\ t^{\frac{n}{2}}c^n & t^{-\frac{n}{2}}d^n \end{pmatrix}, \\
\pi_R(X^+)\begin{pmatrix} \mathcal{D}^n & b^n \\ c^n & d^n \end{pmatrix} &= t^{\frac{n-1}{2}}s^{\frac{n-1}{2}}[n]_t\begin{pmatrix} 0 & ab^{n-1} \\ 0 & cd^{n-1} \end{pmatrix}, \\
\pi_R(X^-)\begin{pmatrix} \mathcal{D}^n & b^n \\ c^n & d^n \end{pmatrix} &= t^{\frac{n-1}{2}}[n]_t\begin{pmatrix} 0 & 0 \\ c^{n-1}d & 0 \end{pmatrix}.
\end{aligned} \tag{21}$$

2.3 Induced Representations of $\mathcal{U}_{p,q}$ and Intertwining Operators

Here we give the actual construction of the induced representations of $\mathcal{U}_{p,q}$. The induction in the deformed setting is performed by the imposition of the right covariance conditions, (cf. Dobrev [13]). (For the equivalence of this method and the usual one in the classical setting $p = q = 1$ we refer to Dobrev. [19]) Thus, we start with functions which are formal power series of the kind:

$$\varphi = \sum_{\substack{k,m,n \in \mathbb{Z}_+ \\ \ell \in \mathbb{Z}}} \mu_{k,\ell,m,n}\, b^m\, c^n\, d^\ell\, \mathcal{D}^k . \tag{22}$$

Then the right covariance conditions [19] with respect to X^-, h, r are:

$$\pi_R(X^-)\varphi = 0, \quad \pi_R(h)\varphi = t^{-\nu/2}\varphi, \quad \pi_R(r)\varphi = s^{\rho/2}\varphi . \tag{23}$$

Their implementation leads to the conditions:

$$n = 0, \quad \ell + m = \nu, \quad 2k + \ell + m = \rho , \tag{24}$$

from which follows that $\nu, \rho \in \mathbb{Z}$ and that $\mu_{k,\ell,m,n} \sim \delta_{n0}\delta_{\ell+m,\nu}\delta_{2k+\ell+m,\rho}$. Thus our reduced functions now are:

$$\varphi(b,d,\mathcal{D}) = \sum_{m\in\mathbb{Z}_+} \mu_m\, b^m\, d^{\nu-m}\, \mathcal{D}^{(\rho-\nu)/2} . \tag{25}$$

It is clear, also if we recall the following classical Gauss decomposition of $GL(2)$, which holds also here:

$$\begin{pmatrix} a & b \\ c & d \end{pmatrix} = \begin{pmatrix} 1 & bd^{-1} \\ 0 & 1 \end{pmatrix} \begin{pmatrix} d^{-1}\mathcal{D} & 0 \\ 0 & d \end{pmatrix} \begin{pmatrix} 1 & 0 \\ d^{-1}c & 1 \end{pmatrix} , \tag{26}$$

that the relevant variables are bd^{-1}, d, $\mathcal{D}$, since the variable c was already eliminated. Thus we shall also introduce the variable $\eta = bd^{-1}$. Then our functions become:

$$\tilde\varphi(\eta,d,\mathcal{D}) = \varphi(b,d,\mathcal{D}) = \sum_{m\in\mathbb{Z}_+} \hat\mu_m\, \eta^m\, d^\nu\, \mathcal{D}^{(\rho-\nu)/2} = \hat\varphi(\eta)\, d^\nu\, \mathcal{D}^{(\rho-\nu)/2} . \tag{27}$$

Now using formulae (16) we obtain:

$$\begin{aligned}
\pi_L(r)\, \eta^m d^\nu \mathcal{D}^{(\rho-\nu)/2} &= s^{-\rho/2}\, \eta^m\, d^\nu\, \mathcal{D}^{(\rho-\nu)/2} , \\
\pi_L(h)\, \eta^m d^\nu \mathcal{D}^{(\rho-\nu)/2} &= t^{-m+\nu/2}\, \eta^m\, d^\nu\, \mathcal{D}^{(\rho-\nu)/2} , \\
\pi_L(X^+)\, \eta^m d^\nu \mathcal{D}^{(\rho-\nu)/2} &= -t^{(3-\nu)/2} s^{-1/2} [m]_t\, \eta^{m-1}\, d^\nu\, \mathcal{D}^{(\rho-\nu)/2} , \\
\pi_L(X^-)\, \eta^m d^\nu \mathcal{D}^{(\rho-\nu)/2} &= t^{(\nu-3)/2} s^{1/2} [m-\nu]_t\, \eta^{m+1}\, d^\nu\, \mathcal{D}^{(\rho-\nu)/2} .
\end{aligned} \tag{28}$$

Then in terms of the functions $\tilde{\varphi}(\eta, d, \mathcal{D})$ we have:

$$\begin{array}{rcl}
\pi_L(r)\tilde{\varphi}(\eta,d,\mathcal{D}) & = & s^{-\rho/2}\,\tilde{\varphi}(\eta,d,\mathcal{D})\,, \\
\pi_L(h)\tilde{\varphi}(\eta,d,\mathcal{D}) & = & t^{\nu/2}\,T^{\eta}_{t^{-1}}\,\tilde{\varphi}(\eta,d,\mathcal{D})\,, \\
\pi_L(X^+)\tilde{\varphi}(\eta,d,\mathcal{D}) & = & -t^{(3-\nu)/2}s^{-1/2}\,D^{\eta}_t\,\tilde{\varphi}(\eta,d,\mathcal{D})\,, \\
\pi_L(X^-)\tilde{\varphi}(\eta,d,\mathcal{D}) & = & t^{(\nu-3)/2}s^{1/2}\,\frac{\eta}{\lambda}\left(t^{-\nu}T^{\eta}_t - t^{\nu}T^{\eta}_{t^{-1}}\right)\tilde{\varphi}(\eta,d,\mathcal{D})\,, \\
T^{\eta}_t f(\eta) = f(t\eta)\,, \qquad D^{\eta}_t f(\eta) & = & \frac{\eta^{-1}}{\lambda}\left(T^{\eta}_t f(\eta) - T^{\eta}_{t^{-1}} f(\eta)\right).
\end{array} \tag{29}$$

It is immediate to check that $\pi_L(r)$, $\pi_L(h)$, $\pi_L(X^+)$, $\pi_L(X^-)$ satisfy (6). It is also clear that if we redefine them by setting:

$$\begin{array}{rcl}
\pi_{\nu,\rho}(r) & = & \pi_L(r)\,, \\
\pi_{\nu,\rho}(h) & = & \pi_L(h)\,, \\
\pi_{\nu,\rho}(X^+) & = & t^{(\nu-3)/2}s^{1/2}\,\pi_L(X^+)\,, \\
\pi_{\nu,\rho}(X^-) & = & t^{(3-\nu)/2}s^{-1/2}\,\pi_L(X^-)\,,
\end{array} \tag{30}$$

then $\pi_{\nu,\rho}$ shall also satisfy (6). Explicitly, the representation is given by:

$$\begin{array}{c}
\pi_{\nu,\rho}(r)\,\tilde{\varphi} \;=\; s^{-\rho/2}\,\tilde{\varphi}\,, \qquad \pi_{\nu,\rho}(h)\,\tilde{\varphi} \;=\; t^{\nu/2}\,T^{\eta}_{t^{-1}}\,\tilde{\varphi}\,, \\
\pi_{\nu,\rho}(X^+)\,\tilde{\varphi} \;=\; -D^{\eta}_t\,\tilde{\varphi}\,, \pi_{\nu,\rho}(X^-)\,\tilde{\varphi} \;=\; \frac{\eta}{\lambda}\left(t^{-\nu}T^{\eta}_t - t^{\nu}T^{\eta}_{t^{-1}}\right)\tilde{\varphi}\,.
\end{array} \tag{31}$$

We denote the representation space of functions $\tilde{\varphi}(\eta, \delta, \mathcal{D})$ with covariance properties (23) and transformation laws (29) by $\mathcal{C}_{\nu,\rho}$. For $p = q = 1$ our representations coincide with the holomorphic representations induced from the lower diagonal Borel subgroup $\mathcal{B}$ of $GL(2)$ and acting on the one-dimensional coset $G/\mathcal{B}$. We notice that each representation decouples into a representation of the central subalgebra with generator r, and a representation of the Jimbo quantum algebra $U_t(sl(2))$. Further we discuss only generic t, i.e., which are not at nontrivial roots of unity. For generic t and $\nu \notin \mathbb{Z}_+$ the representations $\pi_{\nu,\rho}$ are irreducible. For generic t and $\nu \in \mathbb{Z}_+$ the representations $\pi_{\nu,\rho}$ are reducible. Indeed, it is easily seen from formulae (28) that the vectors $\eta^m d^{\nu}\mathcal{D}^{(\rho-\nu)/2}$ with $m = 0, 1, \ldots, \nu$ span an invariant subspace of $\mathcal{C}_{\nu,\rho}$, cf. (28). Let us denote these finite-dimensional invariant subspaces by $\mathcal{E}_{\nu,\rho}$. Thus, for fixed ρ and $\nu \in \mathbb{Z}_+$ there are two irreducible representations realized in the spaces $\mathcal{E}_{\nu,\rho}$ and $\mathcal{C}_{\nu,\rho}/\mathcal{E}_{\nu,\rho}$. Furthermore, for $\nu \in \mathbb{Z}_+$ the representations $\pi_{\nu,\rho}$ and $\pi_{-\nu-2,\rho}$ are partially equivalent. This partial equivalence should be realized by the operator:

$$\left(\pi_R(X^+)\right)^{\nu+1} \;:\; \mathcal{C}_{\nu,\rho} \longrightarrow \mathcal{C}_{-\nu-2,\rho}\,, \tag{32}$$

since the monomial $(X^+)^{\nu+1}$ is giving the singular vector $(X^+)^{\nu+1}\,v_0$ of the corresponding Verma module.[19,13] Indeed, we have (omitting a long calculation):

$$
\begin{aligned}
\left(\pi_R(X^+)\right)^{\nu+1} \eta^m\, d^\nu\, \mathcal{D}^{(\rho-\nu)/2} \;&=\; t^{-\frac{1}{2}(\nu+1)(2m-\nu-3)} s^{\frac{1}{2}(\nu+1)(\rho-1)} \\
&\times [m]_t [m-1]_t \ldots [m-\nu]_t\, \eta^{m-\nu-1}\, d^{-\nu-2}\, \mathcal{D}^{(\rho+\nu+2)/2}
\end{aligned}
\tag{33}
$$

or, finally, in terms of our functions:

$$
(\pi_R(X^+))^{\nu+1}\, \tilde{\varphi} \;=\; \left(t^{(\nu+3)/2}\, s^{(\rho-1)/2}\, D^\eta_t\, (T^\eta_t)^{-1}\right)^{\nu+1} \tilde{\varphi}\,. \tag{34}
$$

Further as in Dobrev[19] we introduce the restricted functions $\hat{\varphi}(\eta)$ by the formula which is prompted in (27):

$$
\hat{\varphi}(\eta) \;=\; (\hat{A}_{\nu,\rho}\, \tilde{\varphi})(\eta) \;\equiv\; \tilde{\varphi}(\eta, 1_{\mathcal{A}}, 1_{\mathcal{A}})\,. \tag{35}
$$

We denote the representation space of $\hat{\varphi}(\eta)$ by $\hat{\mathcal{C}}_{\nu,\rho}$ and the representation acting in $\hat{\mathcal{C}}_{\nu,\rho}$ by $\hat{\pi}_{\nu,\rho}$. Thus the operator $\hat{A}_{\nu,\rho}$ acts from $\mathcal{C}_{\nu,\rho}$ to $\hat{\mathcal{C}}_{\nu,\rho}$. We shall use also the inverse operator $\hat{A}^{-1}_{\nu,\rho}$ which is defined by:

$$
\tilde{\varphi}(\eta, d, \mathcal{D}) \;=\; (\hat{A}^{-1}_{\nu,\rho}\, \hat{\varphi})(\eta, d, \mathcal{D}) \;\equiv\; \hat{\varphi}(\eta)\, d^\nu\, \mathcal{D}^{(\rho-\nu)/2}\,. \tag{36}
$$

The properties of $\hat{\mathcal{C}}_{\nu,\rho}$ follow from the intertwining requirements:[19]

$$
\hat{\pi}_{\nu,\rho} \circ \hat{A}_{\nu,\rho} \;=\; \hat{A}_{\nu,\rho} \circ \pi_{\nu,\rho}\,, \qquad \pi_{\nu,\rho} \circ \hat{A}^{-1}_{\nu,\rho} \;=\; \hat{A}^{-1}_{\nu,\rho} \circ \hat{\pi}_{\nu,\rho}\,. \tag{37}
$$

In particular, the representation $\hat{\pi}_{\nu,\rho}$ is given by:

$$
\begin{aligned}
\hat{\pi}_{\nu,\rho}(r)\, \hat{\varphi}(\eta) \;&=\; s^{-\rho/2}\, \hat{\varphi}(\eta)\,, \\
\hat{\pi}_{\nu,\rho}(h)\, \hat{\varphi}(\eta) \;&=\; t^{\nu/2}\, T^\eta_{t^{-1}}\, \hat{\varphi}(\eta)\,, \\
\hat{\pi}_{\nu,\rho}(X^+)\, \hat{\varphi}(\eta) \;&=\; -D^\eta_t\, \hat{\varphi}(\eta)\,, \\
\hat{\pi}_{\nu,\rho}(X^-)\, \hat{\varphi}(\eta) \;&=\; \frac{\eta}{\lambda}\left(t^{-\nu} T^\eta_t - t^\nu T^\eta_{t^{-1}}\right) \hat{\varphi}(\eta)\,,
\end{aligned}
\tag{38}
$$

or, using the decomposition $\hat{\varphi}(\eta) \;=\; \sum_{m\in\mathbb{Z}_+} \tilde{\mu}_m\, \eta^m$ inherited from (27), we get the analogue of (28):

$$
\begin{aligned}
\hat{\pi}_{\nu,\rho}(r)\, \eta^m \;&=\; s^{-\rho/2}\, \eta^m\,, \\
\hat{\pi}_{\nu,\rho}(h)\, \eta^m \;&=\; t^{-m+\nu/2}\, \eta^m\,, \\
\hat{\pi}_{\nu,\rho}(X^+)\, \eta^m \;&=\; -[m]_t\, \eta^{m-1}\,, \\
\hat{\pi}_{\nu,\rho}(X^-)\, \eta^m \;&=\; [m-\nu]_t\, \eta^{m+1}\,.
\end{aligned}
\tag{39}
$$

At this moment we notice that we can consider (38),(39) for arbitrary complex ν, ρ. Actually the representation has decoupled into a representation of the central generator r, cf. the first formula of (39) and the well known representation of $U_t(sl(2))$, cf. the last three formulae of (39), which was obtained first in Ganchev and Petkova [20] by another method and then in Dobrev [13] by the method we follow here. We know that for generic $t, \nu \in \mathbb{C}$ the representations $\hat{\pi}_{\nu,\rho}$ are irreducible. For generic $t \in \mathbb{C}$ and $\nu \in \mathbb{Z}_+$ the representations $\hat{\pi}_{\nu,\rho}$ are reducible. Moreover, for $\nu \in \mathbb{Z}_+$ the representations $\hat{\pi}_{\nu,\rho}$ and $\hat{\pi}_{-\nu-2,\rho}$ are partially equivalent. The intertwining operators between these pairs is naturally obtained from the ones relating the pairs $\pi_{\nu,\rho}$ and $\pi_{-\nu-2,\rho}$, namely:

$$\begin{aligned} \mathcal{I}_\nu \ &: \ \hat{\mathcal{C}}_{\nu,\rho} \longrightarrow \hat{\mathcal{C}}_{-\nu-2,\rho}\,, \\ \mathcal{I}_\nu \ &\equiv \ \left(t^{-(\nu+3)/2}\, s^{(1-\rho)/2}\right)^{\nu+1} \hat{A}_{-\nu-2,\rho} \circ (\pi_R(X^+))^{\nu+1} \circ \hat{A}^{-1}_{\nu,\rho}\,, \\ \mathcal{I}_\nu\, \hat{\varphi} \ &= \ \left(D^\eta_t\, (T^\eta_t)^{-1}\right)^{\nu+1} \hat{\varphi}\,, \end{aligned} \tag{40}$$

where we have also made use of the fact that the intertwining operators are defined up to multiplicative factors. Formulae (40) (for $s = 1$) were obtained first in Dobrev.[13] The kernel $\hat{\mathcal{E}}_\nu$ of the operator (40) is an invariant subspace of $\hat{\mathcal{C}}_{\nu,\rho}$. It consists of polynomials of degree $\leq \nu$, $\dim \hat{\mathcal{E}}_\nu = \nu + 1$. The basis of $\hat{\mathcal{E}}_\nu$ may be taken as $1_\mathcal{A}, \eta, \ldots, \eta^\nu$, on which $\hat{\pi}_{\nu,\rho}$ acts as in (39). Thus $\hat{\mathcal{E}}_\nu$, $\nu \in \mathbb{Z}_+$, is a finite-dimensional representation space with highest weight vector $1_\mathcal{A}$ $(\hat{\pi}_{\nu,\rho}(X^+)1_\mathcal{A} = 0)$ and lowest weight vector η^ν $(\hat{\pi}_{\nu,\rho}(X^-)\eta^\nu = 0)$. Finally, we should note that since we have functions of one variable η we can treat it as complex variable z. In these terms we can also recover from (38) the classical vector-field representation of $gl(2)$ by setting (as noted above) $h^\varepsilon = t^{\varepsilon H/2}$, $r^\epsilon = s^{\epsilon K/2}$, expanding $h^\varepsilon \approx 1 + \varepsilon(\log t)\, H/2$, $r^\epsilon \approx 1 + \epsilon(\log s)\, K/2$, and taking the limit $\log t \to 0$, $\log s \to 0$. Thus, we get the standard realization which fulfills (7):

$$\begin{gathered} K \ = \ -\rho\,, \quad H \ = \ \nu - 2z\partial_z\,, \quad X^+ \ = \ -\partial_z\,, \\ X^- = z^2\partial_z - \nu z\,, \quad \partial_z \equiv \tfrac{d}{dz}\,. \end{gathered} \tag{41}$$

3 The Case of $\mathbf{GL_{g,h}}$

3.1 The Matrix Quantum Group $GL_{\mathbf{g,h}}$ and the Dual Quantum Algebra

In this Section we construct induced representations of the quantum algebra $\mathcal{U}_{g,h}$, which was found in Aneva et al. [17] as the dual of the Jordanian two-parameter matrix quantum group $GL_{g,h}(2)$ introduced in Aghamohammadi,[21] (and denoted $GL_{h,h'}$). We first recall $\mathcal{A}_{g,h} =$

$GL_{g,h}(2)$. One starts with a unital associative algebra generated by four elements a, b, c, d of a quantum matrix $M = \begin{pmatrix} a & b \\ c & d \end{pmatrix}$ with the following relations $(g, h \in \mathbb{C})$:

$$\begin{gathered}
[a,c] = gc^2 \,, \qquad [d,c] = hc^2 \,, \qquad [a,d] = gdc - hac \,, \\
[a,b] = h(\mathcal{D} - a^2) \,, \qquad [d,b] = g(\mathcal{D} - d^2) \,, \\
[b,c] = gdc + hac - ghc^2 , \\
\mathcal{D} = ad - bc + hac = da - cb - hac + ghc^2 = ad - cb - gdc + ghc^2 \,,
\end{gathered} \tag{42}$$

where $\mathcal{D}$ is a multiplicative quantum determinant which is not central (unless $g = h$) :

$$[a,\mathcal{D}] = [\mathcal{D},d] = (g-h)c\mathcal{D} \,, \qquad [b,\mathcal{D}] = (g-h)(\mathcal{D}d - a\mathcal{D}) \,, \qquad [c,\mathcal{D}] = 0 \,. \tag{43}$$

We shall take the following basis of $\mathcal{A}_{g,h}$:

$$f = f_{k,\ell,m,n} = b^m \, a^\ell \, c^n \, d^k \,, \qquad k,\ell,m,n \in \mathbb{Z}_+ \,. \tag{44}$$

The dual algebra $\mathcal{U}_{g,h}$ was found in Aneva et al. [17] There was used also the following change of generating elements and parameters:

$$\tilde{a} = \tfrac{1}{2}(a+d) \,, \quad \tilde{d} = \tfrac{1}{2}(a-d) \,, \quad \tilde{g} = \tfrac{1}{2}(g+h) \,, \quad \tilde{h} = \tfrac{1}{2}(g-h) \,, \tag{45}$$

with PBW basis:

$$f' = f'_{k,\ell,m,n} = \tilde{a}^k \, \tilde{d}^\ell \, c^n \, b^m \,, \qquad k,\ell,m,n \in \mathbb{Z}_+ \,. \tag{46}$$

The generating elements of $\mathcal{U}_{g,h}$ were introduced in Aneva et al. [17] as:

$$\begin{aligned}
\mathrm{lg}\ A \,,\ f' \,\rangle &= \varepsilon\left(\frac{\partial f'}{\partial \tilde{a}}\right) = k\delta_{\ell 0}\delta_{m0}\delta_{n0} \,, \\
\mathrm{lg}\ B \,,\ f' \,\rangle &= \varepsilon\left(\frac{\partial f'}{\partial b}\right) = \delta_{\ell 0}\delta_{m1}\delta_{n0} \,, \\
\mathrm{lg}\ C \,,\ f' \,\rangle &= \varepsilon\left(\frac{\partial f'}{\partial c}\right) = \delta_{\ell 0}\delta_{m0}\delta_{n1} \,, \\
\mathrm{lg}\ D \,,\ f' \,\rangle &= \varepsilon\left(\frac{\partial f'}{\partial \tilde{d}}\right) = \delta_{\ell 1}\delta_{m0}\delta_{n0} \,.
\end{aligned} \tag{47}$$

Actually, in Aneva et al. [17] was made a one-parameter change of basis from A, B, C, D to A, B, Y, H as follows:

$$\begin{aligned}
D &= e^{\mu B} \, H \, e^{\nu B} \,, \qquad \mu + \nu = \tilde{g} \,, \\
C &= e^{\mu B} \, Y \, e^{\nu B} - \tfrac{\tilde{g}}{2} \sinh(\tilde{g}B) e^{\tilde{g}B} - \tfrac{\tilde{g}}{2} A \,.
\end{aligned} \tag{48}$$

Moreover, similarly to the p, q case we shall explicitly work with a subalgebra of $\mathcal{U}_{g,h}$ with the basis : A, $K = K^+ = e^{\tilde{g}B}$, $K^{-1} = K^- = e^{-\tilde{g}B}$, Y, H. The point is that we can always extract the information for the generator B from $K^{\pm}$. It has the following commutation relations:[17]

$$\begin{aligned}
[K^{\pm}, Y] &= \pm \tilde{g} H K^{\pm} \pm \tfrac{\tilde{g}}{2}(1_{\mathcal{U}} - K^{\pm 2}), \qquad [H, K^{\pm}] = K^{\pm 2} - 1_{\mathcal{U}}, \\
[H, Y] &= -Y(K + K^{-1}) + \tfrac{\tilde{g}}{2} H (K^{-1} - K) + \tfrac{\tilde{g}}{4}(K^2 - K^{-2}), \\
K K^{-1} &= K^{-1} K = 1_{\mathcal{U}}, \\
[A, K] &= [A, K^{-1}] = 0, \qquad [A, Y] = 0, \qquad [A, H] = 0,
\end{aligned} \tag{49}$$

and coalgebra structure:

$$\begin{aligned}
\delta_{\mathcal{U}}(A) &= A \otimes 1_{\mathcal{U}} + 1_{\mathcal{U}} \otimes A, \qquad \delta_{\mathcal{U}}(K^{\pm}) = K^{\pm} \otimes K^{\pm}, \\
\delta_{\mathcal{U}}(H) &= H \otimes K^{-1} + K \otimes H + \tfrac{\tilde{h}}{\tilde{g}} (K^{-1} - K) \otimes A K^{-1}, \\
\delta_{\mathcal{U}}(Y) &= Y \otimes K^{-1} + K \otimes Y + \tfrac{\tilde{h}^2}{2\tilde{g}} (K^{-1} - K) \otimes A^2 K^{-1} \\
&\quad + \tilde{h} H \otimes A K^{-1},
\end{aligned} \tag{50}$$

$$\varepsilon_{\mathcal{U}}(Z) = 0, \quad Z = A, Y, H, \qquad \varepsilon_{\mathcal{U}}(Z) = 1, \quad Z = K, K^{-1}, \tag{51}$$

$$\begin{aligned}
\gamma_{\mathcal{U}}(A) &= -A, \\
\gamma_{\mathcal{U}}(K^{\pm}) &= K^{\mp}, \\
\gamma_{\mathcal{U}}(H) &= -K^{-1} H K + \tfrac{\tilde{h}}{\tilde{g}} (K^{-1} - K) A, \\
\gamma_{\mathcal{U}}(Y) &= -K^{-1} Y K + \tfrac{\tilde{h}^2}{2\tilde{g}} (K - K^{-1}) A^2 + \tilde{h} K^{-1} H A K.
\end{aligned} \tag{52}$$

Note that (49), (50), (51), (52) do not depend on the parameter μ (or ν). We recall from Aneva et al.[17] that A generates a central Hopf subalgebra $\mathcal{Z}$ of $\mathcal{U}_{g,h}$, and that $K^{\pm}, H, Y$ generate a subalgebra $\mathcal{U}'$ of $\mathcal{U}_{g,h}$, which is not a Hopf subalgebra since the generator A enters the co-algebra relations of H, Y. Note also that the generators $A, K^{\pm}, H$ generate a Borel Hopf subalgebra of $\mathcal{U}_{g,h}$, and that there is no conjugate Borel Hopf subalgebra as in the p, q case, since the Y relations involve also H. For further use we recall that to obtain from (49), (50), (51), (52) the classical $U(gl(2))$, one has to reintroduce the generator B by setting $K^{\pm} = e^{\pm \tilde{g} B}$, then to expand $K^{\pm} \approx 1_{\mathcal{U}} \pm \tilde{g} B$ and to take the limit $\tilde{g} \to 0$. Thus, we obtain from (49):

$$\begin{aligned}
[B, Y] &= H, \qquad [H, B] = 2B, \qquad [H, Y] = -2Y, \\
[A, B] &= 0, \qquad [A, Y] = 0, \qquad [A, H] = 0.
\end{aligned} \tag{53}$$

The duality for the algebras $\mathcal{U}_{g,h}$ and $\mathcal{A}_{g,h}$ is given by the pairings:[17]

$$
\begin{aligned}
\lg A \,, \begin{pmatrix} a & b \\ c & d \end{pmatrix} \rangle &= \begin{pmatrix} 1 & 0 \\ 0 & 1 \end{pmatrix}, \\
\lg K^{\pm} \,, \begin{pmatrix} a & b \\ c & d \end{pmatrix} \rangle &= \begin{pmatrix} 1 & \pm\tilde{g} \\ 0 & 1 \end{pmatrix}, \\
\lg H \,, \begin{pmatrix} a & b \\ c & d \end{pmatrix} \rangle &= \begin{pmatrix} 1 & \mu-\nu \\ 0 & -1 \end{pmatrix} = \begin{pmatrix} 1 & 2\mu-\tilde{g} \\ 0 & -1 \end{pmatrix}, \\
\lg Y \,, \begin{pmatrix} a & b \\ c & d \end{pmatrix} \rangle &= \begin{pmatrix} \frac{1}{2}\tilde{g}-\mu & \mu\nu \\ 1 & \frac{1}{2}\tilde{g}-\nu \end{pmatrix} = \begin{pmatrix} \frac{1}{2}\tilde{g}-\mu & \mu(\tilde{g}-\mu) \\ 1 & \mu-\frac{1}{2}\tilde{g} \end{pmatrix}.
\end{aligned}
\tag{54}
$$

The pairing between arbitrary elements of $\mathcal{U}_{g,h}$ and $\mathcal{A}_{g,h}$ follows then from the properties of the duality pairing, however, we shall not need these formulae.

3.2 Left and Right Action of $\mathcal{U}_{g,h}$

For the left regular representation we need the following formulae:

$$
\begin{aligned}
\lg \gamma_{\mathcal{U}}(A) \,, \begin{pmatrix} a & b \\ c & d \end{pmatrix} \rangle &= \begin{pmatrix} -1 & 0 \\ 0 & -1 \end{pmatrix} = -I_2 \,, \\
\lg \gamma_{\mathcal{U}}(K^{\pm}) \,, \begin{pmatrix} a & b \\ c & d \end{pmatrix} \rangle &= \begin{pmatrix} 1 & \mp\tilde{g} \\ 0 & 1 \end{pmatrix}, \\
\lg \gamma_{\mathcal{U}}(H) \,, \begin{pmatrix} a & b \\ c & d \end{pmatrix} \rangle &= \begin{pmatrix} -1 & -2\tilde{\mu} \\ 0 & 1 \end{pmatrix}, \qquad \tilde{\mu} \equiv \mu + \tilde{h} + \tfrac{1}{2}\tilde{g}\,, \\
\lg \gamma_{\mathcal{U}}(Y) \,, \begin{pmatrix} a & b \\ c & d \end{pmatrix} \rangle &= \begin{pmatrix} \tilde{\mu} & G \\ -1 & -\tilde{\mu} \end{pmatrix}, \qquad G \equiv \tilde{\mu}^2 - \tilde{g}^2/4\,.
\end{aligned}
\tag{55}
$$

The left regular representation of $\mathcal{U}_{g,h}$ is again given by (12). Explicitly we get:

$$
\begin{aligned}
\pi_L(A)\begin{pmatrix} a & b \\ c & d \end{pmatrix} &= \begin{pmatrix} -1 & 0 \\ 0 & -1 \end{pmatrix}\begin{pmatrix} a & b \\ c & d \end{pmatrix} = \begin{pmatrix} -a & -b \\ -c & -d \end{pmatrix}, \\
\pi_L(K^{\pm})\begin{pmatrix} a & b \\ c & d \end{pmatrix} &= \begin{pmatrix} 1 & \mp\tilde{g} \\ 0 & 1 \end{pmatrix}\begin{pmatrix} a & b \\ c & d \end{pmatrix} = \begin{pmatrix} a\mp\tilde{g}c & b\mp\tilde{g}d \\ c & d \end{pmatrix}, \\
\pi_L(H)\begin{pmatrix} a & b \\ c & d \end{pmatrix} &= \begin{pmatrix} -1 & -2\tilde{\mu} \\ 0 & 1 \end{pmatrix}\begin{pmatrix} a & b \\ c & d \end{pmatrix} = \begin{pmatrix} -a-2\tilde{\mu}c & -b-2\tilde{\mu}d \\ c & d \end{pmatrix}, \\
\pi_L(Y)\begin{pmatrix} a & b \\ c & d \end{pmatrix} &= \begin{pmatrix} \tilde{\mu} & G \\ -1 & -\tilde{\mu} \end{pmatrix}\begin{pmatrix} a & b \\ c & d \end{pmatrix} = \begin{pmatrix} \tilde{\mu}a+Gc & \tilde{\mu}b+Gd \\ -a-\tilde{\mu}c & -b-\tilde{\mu}d \end{pmatrix}.
\end{aligned}
\tag{56}
$$

In order to derive the action of π_L on arbitrary elements of the basis we use the following twisted derivation rule (14). In our situation here this gives:

$$\begin{aligned}
\pi_L(A)\,\varphi\psi &= \pi_L(A)\varphi\cdot\psi+\varphi\cdot\pi_L(A)\psi\,,\\
\pi_L(K^{\pm})\,\varphi\psi &= \pi_L(K^{\pm})\varphi\cdot\pi_L(K^{\pm})\psi\,,\\
\pi_L(H)\,\varphi\psi &= \pi_L(K^{-1})\varphi\cdot\pi_L(H)\psi\;+\;\pi_L(H)\varphi\cdot\pi_L(K)\psi\\
&\quad+\tfrac{\tilde h}{\tilde g}\,\pi_L(AK^{-1})\varphi\cdot\pi_L(K^{-1}-K)\psi\,,\\
\pi_L(Y)\,\varphi\psi &= \pi_L(K^{-1})\varphi\cdot\pi_L(Y)\psi\;+\;\pi_L(Y)\varphi\cdot\pi_L(K)\psi\\
&\quad+\tfrac{\tilde h^2}{2\tilde g}\,\pi_L(A^2K^{-1})\varphi\cdot\pi_L(K^{-1}-K)\psi\\
&\quad+\tilde h\,\pi_L(AK^{-1})\varphi\cdot\pi_L(H)\psi\,,\\
\pi_L\begin{pmatrix}A & H\\ K & Y\end{pmatrix}1_{\mathcal A} &= \begin{pmatrix}0 & 0\\ 1_{\mathcal A} & 0\end{pmatrix}.
\end{aligned}\tag{57}$$

Thus, we have also:

$$\pi_L\begin{pmatrix}A & H\\ K & Y\end{pmatrix}\mathcal D=\begin{pmatrix}-2\mathcal D & 0\\ \mathcal D & 0\end{pmatrix}.\tag{58}$$

Next we introduce the right regular representation $\pi_R(X)$ as in (17). Explicitly we get here:

$$\begin{aligned}
\pi_R(A)\begin{pmatrix}a & b\\ c & d\end{pmatrix} &= \begin{pmatrix}a & b\\ c & d\end{pmatrix}\begin{pmatrix}1 & 0\\ 0 & 1\end{pmatrix}=\begin{pmatrix}a & b\\ c & d\end{pmatrix},\\
\pi_R(K^{\pm})\begin{pmatrix}a & b\\ c & d\end{pmatrix} &= \begin{pmatrix}a & b\\ c & d\end{pmatrix}\begin{pmatrix}1 & \pm\tilde g\\ 0 & 1\end{pmatrix}=\begin{pmatrix}a & b\pm\tilde g a\\ c & d\pm\tilde g c\end{pmatrix},\\
\pi_R(H)\begin{pmatrix}a & b\\ c & d\end{pmatrix} &= \begin{pmatrix}a & b\\ c & d\end{pmatrix}\begin{pmatrix}1 & 2\mu-\tilde g\\ 0 & -1\end{pmatrix}=\begin{pmatrix}a & (2\mu-\tilde g)a-b\\ c & (2\mu-\tilde g)c-d\end{pmatrix},\\
\pi_R(Y)\begin{pmatrix}a & b\\ c & d\end{pmatrix} &= \begin{pmatrix}a & b\\ c & d\end{pmatrix}\begin{pmatrix}\frac12\tilde g-\mu & \mu(\tilde g-\mu)\\ 1 & \mu-\frac12\tilde g\end{pmatrix}\\
&= \begin{pmatrix}b+(\frac12\tilde g-\mu)a & \mu(\tilde g-\mu)a+(\mu-\frac12\tilde g)b\\ d+(\frac12\tilde g-\mu)c & \mu(\tilde g-\mu)c+(\mu-\frac12\tilde g)d\end{pmatrix}.
\end{aligned}\tag{59}$$

The twisted derivation rule is given again by (19). Thus, in our concrete situation:

$$\pi_R(A)\,\varphi\psi = \pi_R(A)\varphi\cdot\psi+\varphi\cdot\pi_R(A)\psi\,,$$

$$\begin{aligned}
\pi_R(K^\pm)\,\varphi\psi &= \pi_R(K^\pm)\varphi\cdot\pi_R(K^\pm)\psi\,,\\
\pi_R(H)\,\varphi\psi &= \pi_R(K)\varphi\cdot\pi_R(H)\psi + \pi_R(H)\varphi\cdot\pi_R(K^{-1})\psi\\
&\quad + \tfrac{\tilde h}{\tilde g}\,\pi_R(K^{-1}-K)\varphi\cdot\pi_R(AK^{-1})\psi\,,\\
\pi_R(Y)\,\varphi\psi &= \pi_R(K)\varphi\cdot\pi_R(Y)\psi + \pi_R(Y)\varphi\cdot\pi_R(K^{-1})\psi\\
&\quad + \tfrac{\tilde h^2}{2\tilde g}\,\pi_R(K^{-1}-K)\varphi\cdot\pi_R(A^2K^{-1})\psi\\
&\quad + \tilde h\,\pi_R(H)\varphi\cdot\pi_R(AK^{-1})\psi\,,\\
\pi_R\begin{pmatrix} A & H\\ K & Y\end{pmatrix} 1_{\mathcal A} &= \begin{pmatrix} 0 & 0\\ 1_{\mathcal A} & 0\end{pmatrix}. \qquad (60)
\end{aligned}$$

Thus, we have also:

$$\pi_R\begin{pmatrix} A & H\\ K & Y\end{pmatrix} \mathcal{D} = \begin{pmatrix} 2\mathcal{D} & 0\\ \mathcal{D} & 0\end{pmatrix}. \qquad (61)$$

3.3 Induced Representations of $\mathcal{U}_{g,h}$ and Intertwining Operators

As in the p,q case the construction of the induced representations of $\mathcal{U}_{g,h}$ is performed in the deformed setting by the imposition of the right covariance conditions, (cf. Dobrev [13]). We start with functions which are formal power series of the kind:

$$\varphi = \sum_{k,\ell,m,n\in\mathbb{Z}_+} \mu_{k,\ell,m,n}\, b^m\, a^\ell\, c^n\, d^k . \qquad (62)$$

Above we have defined left and right action of $\mathcal{U}_{g,h}$ on φ and as before we shall use the right action to reduce the left regular representation. Note that unlike the $\mathcal{U}_{p,q}$ case, where we could have used either of the two conjugate Borel subalgebras for the reduction, here we have only the 'upper diagonal' Borel subalgebra generated by $A, K^\pm, H$. (Note that this feature is true also in the one-parameter Jordanian deformation as used for its representation theory. [22]) Thus right covariance conditions will look as follows:

$$\pi_R(A)\,\varphi = \rho\,\varphi\,, \qquad (63)$$

$$\pi_R(K)\,\varphi = \varphi \quad\Leftrightarrow\quad \pi_R(B)\,\varphi = 0\,, \qquad (64)$$

$$\pi_R(H)\,\varphi = \nu\,\varphi\,. \qquad (65)$$

Note that from (64) follows: $\pi_R(K^-)\varphi = \varphi$.

Let us apply first (64). From (59) it is clear that only those functions which do not depend on b and d will satisfy (64). However, our functions may depend on b and d through the determinant $\mathcal{D}$ since the latter is preserved by the action of $\pi_R(K)$, cf. (61). Thus, instead of (22) we take for our functions:

$$\tilde{\varphi} = \sum_{k,\ell,n\in\mathbb{Z}_+} \mu_{k,\ell,n}\, a^\ell\, c^n\, \mathcal{D}^k\ , \tag{66}$$

with the same conditions (63),(64),(65) (with $\varphi \longrightarrow \tilde{\varphi}$). By construction (64) is fulfilled for all $\tilde{\varphi}$. To apply (63) we first obtain using (59) and (60):

$$\pi_R(A)\, a^\ell\, c^n\, \mathcal{D}^k \;=\; (2k+\ell+n)\, a^\ell\, c^n\, \mathcal{D}^k\ . \tag{67}$$

Combining (67) with (63) we obtain that: $2k+\ell+n \;=\; \rho$. Analogously, we obtain using (59) and (60):

$$\pi_R(H)\, a^\ell\, c^n\, \mathcal{D}^k \;=\; (\ell+n)\, a^\ell\, c^n\, \mathcal{D}^k\ . \tag{68}$$

Combining (68) with (65) we obtain that: $\ell+n \;=\; \nu$. Thus, we conclude that $\rho, \nu, \frac{1}{2}(\rho-\nu) \in \mathbb{Z}_+$, and that the functions in (66) reduce to:

$$\tilde{\varphi} = \sum_{n\in\mathbb{Z}_+} \mu_n\, a^n\, c^{\nu-n}\, \mathcal{D}^{(\rho-\nu)/2}\ . \tag{69}$$

Thus, our functions are given in the bases:

$$u_n \;=\; a^n\, c^{\nu-n}\, \mathcal{D}^{(\rho-\nu)/2}\ , \qquad n\in\mathbb{Z}_+\ . \tag{70}$$

Now if neither a or c has an inverse these bases will be finite-dimensional, in contrast to the undeformed case. However, these finite-dimensional representations we shall obtain also if we suppose that either a or c has an inverse (see below). Thus, further we shall suppose that c has an inverse. (The choice of a having an inverse would make the further formulae much more complicated, as can be anticipated from the comparison of the action of the algebra on a or c.) The transformation rules for the bases (70) are (with $\tilde{\mu}=0$):

$$\begin{aligned}
\pi_L(A)\, u_n \;&=\; -\rho\, u_n\ ,\\
\pi_L(K^\varepsilon)\, u_n \;&=\; (a-\varepsilon\tilde{g}c)^n\, c^{\nu-n}\, \mathcal{D}^{(\rho-\nu)/2} \;=\; \sum_{s=0}^{n} (-\varepsilon)^s \binom{n}{s}\, \Delta^\varepsilon_s\, u_{n-s}\ ,\\
&\qquad \Delta^\varepsilon_s \;=\; \prod_{j=0}^{s-1} (\tilde{g}+\varepsilon j g)\ , \qquad (\Delta^\varepsilon_0 \;=\; 1)\ , \qquad \varepsilon=\pm\ ,\\
\pi_L(H)\, u_n \;&=\; (\nu-n)\, (a+\tilde{g}c)^n\, c^{\nu-n}\, \mathcal{D}^{(\rho-\nu)/2}
\end{aligned}$$

$$
\begin{aligned}
& + \sum_{s=0}^{n} \tilde{\beta}_{ns}\ a^{n-s}\ c^{\nu-n+s}\ \mathcal{D}^{(\rho-\nu)/2} \\
&= \sum_{s=0}^{n} \binom{n}{s} \left((\nu - n)\ \Delta_s^- + \frac{(-1)^s}{s+1}\ \Delta_s^+\ \beta_{ns} \right) u_{n-s}\,, \\
\beta_{n0} &= -n, \\
\beta_{ns} &= 2(n-s)\tfrac{\tilde{h}}{\tilde{g}} + \sum_{\ell=1}^{s-1}(-1)^\ell \frac{\Delta_\ell^-}{\Delta_{\ell+1}^+} \\
& \times \left\{(n-s)(\tilde{g}+2\tilde{h}) - \ell(n+1)(\tilde{g}-\tilde{h})\right\}, \\
\pi_L(Y)u_n &= (n-\nu)u_{n+1} + \left[\binom{n}{2}\tilde{h} + \binom{\nu-n}{2}\tilde{g} + (n-\nu)n(\tilde{g}+\tilde{h})\right] u_n \\
& + \sum_{k=1}^{n} \left\{(-1)^k\gamma_{nk} + \binom{n}{k}\ \Delta_k^- \left[\binom{\nu-n}{2}\ \tilde{g}\right.\right. \\
& \left.\left. +(n-\nu)(n\tilde{h}+kg)\right] + \binom{n}{k+1}\ (n-\nu)\ \Delta_{k+1}^- \right\} u_{n-k}\,, \\
\gamma_{nk} &= \Delta_k^+ \left\{ \binom{n-2}{k-2}\tfrac{1}{4}\tilde{g} + \tfrac{1}{\tilde{g}}\left(\frac{(n+k-1)(n-2)!}{4(n-k-1)!k!}\tilde{g}^2 - \binom{n}{k+2}\tilde{h}^2 \right.\right. \\
& \left. - \binom{n+1}{k+2}\tilde{g}\tilde{h} \right) + \sum_{\ell=2}^{k}(-1)^\ell \frac{\Delta_\ell^-}{\Delta_\ell^+}\left[\ell\binom{n+1}{k+2} + \binom{n}{k+2}\right]\tilde{h} \\
& + \sum_{\ell=1}^{k-1}(-1)^\ell \frac{\Delta_\ell^-}{\Delta_{\ell+1}^+}\left[\binom{n}{k}\left(\tfrac{1}{4}\tilde{g}^2 + \ell^2\tilde{g}\tilde{h}\right)\right. \\
& \left.\left. + 2\binom{n}{k+1}\binom{\ell+1}{2}\tilde{g}\tilde{h} - \frac{n!(\ell+1)(\ell+n\ell+2n-k)}{(k+2)!(n-k-1)!}\tilde{h}^2\right]\right\}.
\end{aligned}
\tag{71}
$$

Thus, we have obtained infinite-dimensional representations of $\mathcal{U}_{g,h}$ parametrized by two integers. We shall denote by $\tilde{\mathcal{C}}_{\nu,\rho}$ the representation spaces of the functions $\tilde{\varphi}$ in (66). For $g = h = 0$ our representations coincide with the holomorphic representations induced from the upper diagonal Borel subgroup $\mathcal{B}$ of $GL(2)$ and acting on the one-dimensional coset $G/\mathcal{B}$. Let us comment that the two subalgebras $\mathcal{Z}$ and $\mathcal{U}'$ are represented separately with parameters ρ and ν, resp. First we notice from (71) that if $\nu \in \mathbb{Z}_+$ the representation space $\tilde{\mathcal{C}}_{\nu,\rho}$ is reducible since the vectors u_n, $n = 0, 1, \ldots, \nu$ span a finite-dimensional invariant subspace (see (71)). For further use we shall denote these representation spaces by $\mathcal{E}_{\nu,\rho}$, $\nu \in \mathbb{Z}_+$. Thus, if $\nu \in \mathbb{Z}_+$ we have two irreducible representations with representation spaces isomorphic to $\tilde{\mathcal{E}}_{\nu,\rho}$ and to $\tilde{\mathcal{C}}_{\nu,\rho}/\tilde{\mathcal{E}}_{\nu,\rho}$. We would like further to reduce the representation spaces. From the above we are prompted to use the variable $\chi \equiv ac^{-1}$. This is also related to the following classical Gauss decomposition of $GL(2)$:

$$\begin{pmatrix} a & b \\ c & d \end{pmatrix} = \begin{pmatrix} 1 & 0 \\ ca^{-1} & 1 \end{pmatrix} \begin{pmatrix} a & 0 \\ 0 & a^{-1}\mathcal{D} \end{pmatrix} \begin{pmatrix} 1 & a^{-1}b \\ 0 & 1 \end{pmatrix}, \qquad (72)$$

though instead of $\xi \equiv ca^{-1}$ we shall use χ as defined above. We note also the following connection between the two variable $\chi = \xi^{-1} - g$. The action on the new variable is:

$$\pi_L \begin{pmatrix} A & H \\ K^{\pm} & Y \end{pmatrix} \chi^n = \begin{pmatrix} 0 & -(2\chi + \tilde{g})\, \frac{(\chi+\tilde{g})^n - (\chi-\tilde{g})^n}{2\tilde{g}} \\ (\chi \mp \tilde{g})^n & (\chi^2 + \tilde{g}\chi - \frac{\tilde{g}^2}{4})\, \frac{(\chi+\tilde{g})^n - (\chi-\tilde{g})^n}{2\tilde{g}} \end{pmatrix}. \qquad (73)$$

Thus, our bases and functions shall be:

$$w_n = \chi^n\, c^\nu\, \mathcal{D}^{(\rho-\nu)/2}, \qquad n \in \mathbb{Z}_+, \qquad (74)$$

$$\phi = \phi(\chi, c, \mathcal{D}) = \sum_{n\in\mathbb{Z}_+} \mu_n\, \chi^n\, c^\nu\, \mathcal{D}^{(\rho-\nu)/2} = \hat{\phi}(\chi)\, c^\nu\, \mathcal{D}^{(\rho-\nu)/2}. \qquad (75)$$

We shall denote these representation spaces by $\mathcal{C}_{\nu,\rho}$, and the representation acting in these spaces by $\pi_{\nu,\rho}$. Explicitly the action is:

$$\begin{aligned}
\pi_{\nu,\rho}(A)\, w_n &= -\rho\, w_n, \\
\pi_{\nu,\rho}(K^{\pm})\, w_n &= (\chi \mp \tilde{g})^n\, c^\nu \mathcal{D}^{\frac{\rho-\nu}{2}} = \sum_{k=0}^{n} \binom{n}{k} (\mp\tilde{g})^k\, w_{n-k}, \\
\pi_{\nu,\rho}(H)\, w_n &= (\nu - \tfrac{1}{\tilde{g}}\chi - \tfrac{1}{2})\, (\chi + \tilde{g})^n\, c^\nu \mathcal{D}^{\frac{\rho-\nu}{2}} \\
&\quad + (\tfrac{1}{\tilde{g}}\chi + \tfrac{1}{2})(\chi - \tilde{g})^n\, c^\nu \mathcal{D}^{\frac{\rho-\nu}{2}} \\
&= (\nu - \tfrac{1}{\tilde{g}}\chi - \tfrac{1}{2})\, \pi_{\nu,\rho}(K^-)\, w_n + (\tfrac{1}{\tilde{g}}\chi + \tfrac{1}{2})\, \pi_{\nu,\rho}(K^+)\, w_n \\
&= (\nu - 2n)\, w_n + \sum_{k=0}^{n-1} \omega(n, k, \nu, \tilde{g})\, w_k, \\
\pi_{\nu,\rho}(Y)\, w_n &= \left\{\tfrac{1}{2\tilde{g}}\chi^2 + (\tfrac{1}{2} - \nu)\chi - \tfrac{\tilde{g}}{2}(\nu^2 + 3\nu + \tfrac{1}{4})\right\} (\chi + \tilde{g})^n\, c^\nu \mathcal{D}^{\frac{\rho-\nu}{2}} \\
&\quad - (\tfrac{1}{2\tilde{g}}\chi^2 + \tfrac{1}{2}\chi - \tfrac{\tilde{g}}{8})\, (\chi - \tilde{g})^n\, c^\nu \mathcal{D}^{\frac{\rho-\nu}{2}} \\
&= \left\{\tfrac{1}{2\tilde{g}}\chi^2 + (\tfrac{1}{2} - \nu)\chi - \tfrac{\tilde{g}}{2}(\nu^2 + 3\nu + \tfrac{1}{4})\right\} \pi_{\nu,\rho}(K^-)\, w_n \\
&\quad - (\tfrac{1}{2\tilde{g}}\chi^2 + \tfrac{1}{2}\chi - \tfrac{\tilde{g}}{8})\, \pi_{\nu,\rho}(K^+)\, w_n
\end{aligned}$$

$$= \quad (n-\nu)\, w_{n+1} \; + \; \sum_{k=0}^{n} \omega'(n,k,\nu,\tilde{g})\; w_k \,. \tag{76}$$

Also in these bases for $\nu \in \mathbb{Z}_+$ there exists a $(\nu+1)$-dimensional invariant subspace spanned here by the vectors w_k, $k \leq \nu$, cf. the last of formulae (76). Also, from the results of Dobrev [22] follows that for $\nu \in \mathbb{Z}_+$ the representations $\pi_{\nu,\rho}$ and $\pi_{-\nu-2,\rho}$ are partially equivalent. This partial equivalence should be realized by the operator:

$$Q_\nu(\pi_R(Y)) \; : \; \mathcal{C}_{\nu,\rho} \; \longrightarrow \; \mathcal{C}_{-\nu-2,\rho}\,, \tag{77}$$

where Q_ν is the polynomial in Y given in formula (37) of Dobrev, [22] (denoted there as $Q_p(p-1)$, $p = \Lambda(H_0)+1$, $\Lambda(H_0)$ corresponding to our ν here), which polynomial gives the singular vector of the reducible Verma module V^Λ, such that $\Lambda(H_0) = p-1 = \nu$. The general expression for the singular vectors from Dobrev [22] is:

$$v_s^p \;=\; \sum_{k=0}^{[p/2]} \alpha_{pk}\; \tilde{g}^{2k}\; Y^{p-2k}\; v_0\,, \qquad p = \nu+1\,, \tag{78}$$

where v_0 is the highest weight vector of V^Λ. The coefficients α_{pk} are known explicitly only for the low level cases. Thus, a few low level singular vectors are: [22]

$$\begin{aligned}
v_s^1 &= Y\, v_0\,, \qquad \nu = 0\,,\\
v_s^2 &= Y^2\, v_0\,, \qquad \nu = 1\,,\\
v_s^3 &= (Y^3 + \tilde{g}^2\, Y)\, v_0\,, \qquad \nu = 2\,,\\
v_s^4 &= (Y^4 + 9\tilde{g}^2\, Y^2)\, v_0\,, \qquad \nu = 3\,,\\
v_s^5 &= (Y^5 - 39\tilde{g}^2\, Y^3 + 36\tilde{g}^4\, Y)\, v_0\,, \qquad \nu = 4\,.
\end{aligned} \tag{79}$$

It is a matter of a tedious calculation to check explicitly the intertwining properties of $Q_\nu(\pi_R(Y))$, including also the fact that it annihilates the polynomials of degree up to ν (incl.), however, these properties follow also from the general theory. [19] Further we introduce the restricted functions $\hat{\phi}(\chi)$ by the formula which is prompted in (75):

$$\hat{\phi}(\chi) \;=\; (\hat{A}_{\nu,\rho}\,\tilde{\varphi})(\chi) \;\equiv\; \phi(\chi, 1_{\mathcal{A}}, 1_{\mathcal{A}})\,. \tag{80}$$

We denote the representation space of $\hat{\phi}(\chi)$ by $\hat{\mathcal{C}}_{\nu,\rho}$ and the representation acting in $\hat{\mathcal{C}}_{\nu,\rho}$ by $\hat{\pi}_{\nu,\rho}$. Thus the operator $\hat{A}_{\nu,\rho}$ acts from $\mathcal{C}_{\nu,\rho}$ to $\hat{\mathcal{C}}_{\nu,\rho}$. We shall use also the inverse operator $\hat{A}^{-1}_{\nu,\rho}$ which is defined by:

$$\tilde{\varphi}(\chi, c, \mathcal{D}) = (\hat{A}^{-1}_{\nu,\rho}\hat{\phi})(\chi, c, \mathcal{D}) \equiv \hat{\phi}(\chi)\, c^{\nu}\, \mathcal{D}^{(\rho-\nu)/2}. \tag{81}$$

The properties of $\hat{\mathcal{C}}_{\nu,\rho}$ follow from the intertwining requirements: [19]

$$\hat{\pi}_{\nu,\rho}\,\hat{A}_{\nu,\rho} = \hat{A}_{\nu,\rho}\,\pi_{\nu,\rho}, \qquad \pi_{\nu,\rho}\circ\hat{A}^{-1}_{\nu,\rho} = \hat{A}^{-1}_{\nu,\rho}\circ\hat{\pi}_{\nu,\rho}. \tag{82}$$

In particular, the representation $\hat{\pi}_{\nu,\rho}$ is given by:

$$\begin{aligned}
\hat{\pi}_{\nu,\rho}(A)\,\chi^n &= -\rho\,\chi^n, \\
\hat{\pi}_{\nu,\rho}(K^{\pm})\,\chi^n &= (\chi \mp \tilde{g})^n, \\
\hat{\pi}_{\nu,\rho}(H)\,\chi^n &= (\nu - \tfrac{1}{\tilde{g}}\chi - \tfrac{1}{2})\,(\chi+\tilde{g})^n + (\tfrac{1}{\tilde{g}}\chi + \tfrac{1}{2})(\chi-\tilde{g})^n \\
&= (\nu - \tfrac{1}{\tilde{g}}\chi - \tfrac{1}{2})\,\pi_{\nu,\rho}(K^-)\,\chi^n + (\tfrac{1}{\tilde{g}}\chi + \tfrac{1}{2})\,\pi_{\nu,\rho}(K^+)\,\chi^n \\
&= (\nu - 2n)\,\chi^n + \sum_{k=0}^{n-1}\omega(n,k,\nu,\tilde{g})\,\chi^k, \\
\hat{\pi}_{\nu,\rho}(Y)\,\chi^n &= \left\{\tfrac{1}{2\tilde{g}}\chi^2 + (\tfrac{1}{2}-\nu)\chi - \tfrac{\tilde{g}}{2}(\nu^2+3\nu+\tfrac{1}{4})\right\}(\chi+\tilde{g})^n \\
&\quad - (\tfrac{1}{2\tilde{g}}\chi^2 + \tfrac{1}{2}\chi - \tfrac{\tilde{g}}{8})(\chi-\tilde{g})^n \\
&= \left\{\tfrac{1}{2\tilde{g}}\chi^2 + (\tfrac{1}{2}-\nu)\chi - \tfrac{\tilde{g}}{2}(\nu^2+3\nu+\tfrac{1}{4})\right\}\hat{\pi}_{\nu,\rho}(K^-)\,\chi^n \\
&\quad - (\tfrac{1}{2\tilde{g}}\chi^2 + \tfrac{1}{2}\chi - \tfrac{\tilde{g}}{8})\,\hat{\pi}_{\nu,\rho}(K^+)\,\chi^n \\
&= (n-\nu)\,\chi^{n+1} + \sum_{k=0}^{n}\omega'(n,k,\nu,\tilde{g})\,\chi^k,
\end{aligned} \tag{83}$$

or in terms of the functions $\hat{\phi}$:

$$\begin{aligned}
\hat{\pi}_{\nu,\rho}(A)\,\hat{\phi}(\chi) &= -\rho\,\hat{\phi}(\chi), \\
\hat{\pi}_{\nu,\rho}(K^{\pm})\,\hat{\phi}(\chi) &= \hat{\phi}(\chi \mp \tilde{g}), \\
\hat{\pi}_{\nu,\rho}(H)\,\hat{\phi}(\chi) &= (\nu - \tfrac{1}{\tilde{g}}\chi - \tfrac{1}{2})\,\hat{\phi}(\chi+\tilde{g}) + (\tfrac{1}{\tilde{g}}\chi + \tfrac{1}{2})\hat{\phi}(\chi-\tilde{g}) \\
&= (\nu - \tfrac{1}{\tilde{g}}\chi - \tfrac{1}{2})\,\hat{\pi}_{\nu,\rho}(K^-)\,\hat{\phi}(\chi) + (\tfrac{1}{\tilde{g}}\chi + \tfrac{1}{2})\,\hat{\pi}_{\nu,\rho}(K^+)\,\hat{\phi}(\chi), \\
\hat{\pi}_{\nu,\rho}(Y)\,\hat{\phi}(\chi) &= \left\{\tfrac{1}{2\tilde{g}}\chi^2 + (\tfrac{1}{2}-\nu)\chi - \tfrac{\tilde{g}}{2}(\nu^2+3\nu+\tfrac{1}{4})\right\}\hat{\phi}(\chi+\tilde{g}) \\
&\quad - (\tfrac{1}{2\tilde{g}}\chi^2 + \tfrac{1}{2}\chi - \tfrac{\tilde{g}}{8})\,\hat{\phi}(\chi-\tilde{g}) \qquad (84) \\
&= \left\{\tfrac{1}{2\tilde{g}}\chi^2 + (\tfrac{1}{2}-\nu)\chi - \tfrac{\tilde{g}}{2}(\nu^2+3\nu+\tfrac{1}{4})\right\} \\
&\quad \times\hat{\pi}_{\nu,\rho}(K^-)\,\hat{\phi}(\chi) - (\tfrac{1}{2\tilde{g}}\chi^2 + \tfrac{1}{2}\chi - \tfrac{\tilde{g}}{8})\,\hat{\pi}_{\nu,\rho}(K^+)\,\hat{\phi}(\chi).
\end{aligned}$$

At this moment we notice that we can consider (83), (84) for arbitrary complex ν, ρ. Actually, the representation has decoupled into a representation of the central algebra with generator A, cf. the first formula of (83), or of (84), and the a *new* representation of the Jordanian $U_{\tilde{g}}(sl(2))$, cf. the last three formulae of (83), or of (84). Analogously to before for generic $\nu \in \mathbb{C}$ the representations $\hat{\pi}_{\nu,\rho}$ are irreducible. For $\nu \in \mathbb{Z}_+$ the representations $\hat{\pi}_{\nu,\rho}$ are reducible, since there is a $(\nu+1)$-dimensional invariant subspace of the polynomials in χ of degree up to ν, cf. (83). Also, from the results of Dobrev[22] follows that for $\nu \in \mathbb{Z}_+$ the representations $\hat{\pi}_{\nu,\rho}$ and $\hat{\pi}_{-\nu-2,\rho}$ are partially equivalent. The intertwining operators between these pairs is naturally obtained from the ones relating the pairs $\pi_{\nu,\rho}$ and $\pi_{-\nu-2,\rho}$, namely:

$$
\begin{aligned}
\mathcal{I}_\nu \ &: \ \hat{\mathcal{C}}_{\nu,\rho} \longrightarrow \hat{\mathcal{C}}_{-\nu-2,\rho}\,, \\
\mathcal{I}_\nu \ &\equiv \ \hat{A}_{-\nu-2,\rho} \circ Q_\nu(\pi_R(Y)) \circ \hat{A}^{-1}_{\nu,\rho}\,.
\end{aligned} \tag{85}
$$

Finally, we should note that since we have functions of one variable χ we can replace it with an ordinary complex variable z, and then the transformation properties (84) can be rewritten as follows:

$$
\begin{aligned}
\hat{\pi}_{\nu,\rho}(A)\,\hat{\phi}(z) \ &= \ -\rho\,\hat{\phi}(z)\,, \\
\hat{\pi}_{\nu,\rho}(K^\pm)\,\hat{\phi}(z) \ &= \ e^{\mp\tilde{g}\partial_z}\,\hat{\phi}(z)\,, \\
\hat{\pi}_{\nu,\rho}(H)\,\hat{\phi}(z) \ &= \ \left(-\tfrac{1}{\tilde{g}}\,z - \tfrac{1}{2} + \nu\right) e^{\tilde{g}\partial_z}\,\hat{\phi}(z) \ + \ \left(\tfrac{1}{\tilde{g}}\,z + \tfrac{1}{2}\right) e^{-\tilde{g}\partial_z}\,\hat{\phi}(z) \\
&= \ -\left(\tfrac{2}{\tilde{g}}\,z + 1\right)\,\sinh(\tilde{g}\partial_z)\,\hat{\phi}(z) \ + \ \nu\, e^{\tilde{g}\partial_z}\,\hat{\phi}(z)\,, \\
\hat{\pi}_{\nu,\rho}(Y)\,\hat{\phi}(z) \ &= \ \left\{\tfrac{1}{2\tilde{g}}\,z^2 \ + \ \left(\tfrac{1}{2} - \nu\right) z \ - \ \tfrac{\tilde{g}}{2}\left(\nu^2 + 3\nu + \tfrac{1}{4}\right)\right\} e^{\tilde{g}\partial_z}\,\hat{\phi}(z) \\
&\quad - \left(\tfrac{1}{2\tilde{g}}\,z^2 \ + \ \tfrac{1}{2}\,z - \ \tfrac{\tilde{g}}{8}\right) e^{-\tilde{g}\partial_z}\,\hat{\phi}(z) \\
&= \ \left(\tfrac{1}{\tilde{g}}\,z^2 \ + \ z \ - \ \tfrac{\tilde{g}}{4}\right)\,\sinh(\tilde{g}\partial_z)\,\hat{\phi}(z) \\
&\quad - \nu\left\{z \ + \ \tfrac{\tilde{g}}{2}\,(\nu + 3)\right\} e^{\tilde{g}\partial_z}\,\hat{\phi}(z)\,.
\end{aligned} \tag{86}
$$

In these terms we can also recover from (86) the classical vector-field representation of $gl(2)$ by setting (as noted above) $K^\pm \ = \ e^{\pm\tilde{g}B}$, expanding $K^\pm \approx 1_{\mathcal{U}} \pm \tilde{g}B$ and taking the limit $\tilde{g} \to 0$. Thus, we obtain standard realization which fulfills (53):

$$
A \ = \ -\rho\,, \quad H \ = \ \nu - 2z\partial_z\,, \quad B \ = \ -\partial_z\,, \quad Y = z^2\partial_z - \nu z\,. \tag{87}
$$

Thus, the representation (86), more precisely, the last three formulae of (86), give a *new* deformation of the classical vector-field realization of $sl(2)$.

Acknowledgments

The authors were supported in part by BNFR under contract Ph-643. V.K.D. is grateful for hospitality at the ICTP (where the first version of the paper was prepared), to the Deutsche Forschungsgemeinschaft for the financial support as Guest Professor at TU Clausthal, to Prof. Dr. H.-D. Doebner for hospitality at TU Clausthal, and to Prof. Dr. J. Lukierski for hospitality at the conference in Karpacz.

References

1. V.G. Drinfeld, *Soviet. Math. Dokl.* **32**, 254 (1985); in: Proceedings ICM, (MSRI, Berkeley, 1986) p. 798.
2. M. Jimbo, *Lett. Math. Phys.* **10**, 63 (1985) & **11**, 247 (1986).
3. P.P. Kulish and N.Yu. Reshetikhin, *Zap. Nauch. Semin. LOMI* **101**, 101 (1981), (in Russian); English translation: *J. Soviet. Math.* **23**, 2435 (1983).
4. A.N. Kirillov and N.Yu. Reshetikhin, LOMI Leningrad preprint E-9-88 (1988).
5. M. Rosso, *Comm. Math. Phys.* **117**, 581 (1987).
6. G. Lusztig, *Adv. Math.* **70**, 237 (1988); *Contemporary Math.* **82**, 59 (1988).
7. L.L. Vaksman and Ya.S. Soibelman, *Funkts. Anal. Prilozh.* **22**, No. 3, 1 (1988); English translation: *Funct. Anal. Appl.* **22**, 170 (1988).
8. L.C. Biedenharn, *J. Phys.* **A22**, L873 (1989).
9. A.J. MacFarlane, *J. Phys.* **A22**, 4581 (1989).
10. P. Roche and D. Arnaudon, *Lett. Math. Phys.* **17**, 295 (1989).
11. T. Masuda, K. Mimachi, Y. Nakagami, M. Noumi, Y. Sabuti and K. Ueno, *Lett. Math. Phys.* **19**, 187 (1990); *Lett. Math. Phys.* **19**, 195 (1990).
12. V.K. Dobrev,*Acta Appl. Math.* **44**,81 (1996).
13. V.K. Dobrev, unpublished, (May 1993); see [15] for $n = 2$.
14. L. Dabrowski, V.K. Dobrev and R. Floreanini, *J. Math. Phys.* **35**, 971 (1994).
15. V.K. Dobrev, *J. Phys.* **A27**, 4841& 6633 (1994).
16. V.K. Dobrev, *J. Math. Phys.* **33**, 3419 (1992).
17. B.L. Aneva, V.K. Dobrev and S.G. Mihov, *J. Phys.* **A30**, 6769 (1997).
18. E.E. Demidov, Yu.I. Manin, E.E. Mukhin and D.V. Zhdanovich, *Theor. Phys. Suppl.* **102**, 203 (1990).
19. V.K. Dobrev, *Rep. Math. Phys.* **25**, 159 (1988).
20. A.Ch. Ganchev and V.B. Petkova, *Phys. Lett.* **B233**, 374 (1989).

21. A. Aghamohammadi, *Mod. Phys. Lett.* **A8**, 2607 (1993).
22. V.K. Dobrev, Proc. X Intern. Conf. 'Problems of Quantum Field Theory', (Alushta, 1996), eds. D. Shirkov et. al. (Publ. Dept. JINR, Dubna, 1996) p. 104.

UNIFIED DESCRIPTION OF QUANTUM AFFINE (SUPER)ALGEBRAS $U_q(A_1^{(1)})$ AND $U_q(C(2)^{(2)})$[a]

S.M. KHOROSHKIN[1], J. LUKIERSKI[2], V.N. TOLSTOY[3]

[1] *Institute of Theoretical and and Experimental Physics 117259 Moscow, Russia (e-mail: khoroshkin@vxitep.itep.ru)*

[2] *Institute of Theoretical Physics, University of Wrocław 50-204 Wrocław & Poland (e-mail: lukier@ift.uni. wroc.pl)*

[3] *Institute of Nuclear Physics, Moscow State University 119899 Moscow, Russia (e-mail: tolstoy@nucl-th.npi.msu.su)*

We show that the quantum affine algebra $U_q(A_1^{(1)})$ and the quantum affine superalgebra $U_q(C(2)^{(2)})$ admit unified description. The difference between them consists in the phase factor which is equal to 1 for $U_q(A_1^{(1)})$ and is equal to -1 for $U_q(C(2)^{(2)})$. We present such a description for the construction of Cartan-Weyl generators and their commutation relations, as well for the universal R-matrices.

1 Introduction

Among variety of all affine Lie (super)algebras[b] (both quantized and non-quantized) the affine (super)algebras of rank 2 play a key role. In the first place, all affine series of the type $A(n|m)^{(1)}$, $B(n|m)^{(1)}$, $C(n)^{(1)}$, $D(n|m)^{(1)}$, $A(2n|2m{-}1)^{(2)}$, $A(2n{-}1|2m{-}1)^{(2)}$,$C(n)^{(2)}$, $D(n|m)^{(2)}$ and $A(2n|2m)^{(4)}$ begin with the affine (super)algebras of rank 2. Secondly, the contragredient Lie (super)algebras of rank 2 are basic structural blocks of any affine (super)algebras of arbitrary rank. This fact permits, for example, the reduction of the proofs of basic theorems for the extremal projector and the universal R-matrix to the proofs of such theorems for the (super)algebras of rank 2 (see Refs. [1], [11] - [13], [5] - [8]). Further, the representation theory of the affine (super)algebras (both quantized and non-quantized) contains some typical elements of the representation theory of the affine (super)algebras of rank 2. Besides, in applications of the affine (super)algebras, first of all the affine (super)algebras of rank 2 are used by virtue of their simplicity.

In this paper along the line of considerations presented in [9] for rank 2 affine superalgebra $U_q(B(0,1)^{(1)})$ we give detailed description of the quantum

[a]Talk given by V.N. Tolstoy

[b]We introduce the prefix "super" in brackets to stress that the Lie (super)algebras include the Lie algebras as well as the Lie superalgebras.

untwisted affine algebra $U_q(A_1^{(1)})$ ($\simeq U_q(\widehat{sl}(2))$) and the quantum twisted affine superalgebra $U_q(C(2)^{(2)})$. Moreover our goal is to show that these quantum (super)algebras are described in unified way. Namely, we present in unified way their defining relations, the construction of the Cartan-Weyl bases, the complete list of all commutation relations of the Cartan-Weyl generators corresponding to all root vectors and finally the unified form of their universal R-matrices. Difference between both considered quantum (super)algebras is only determined by a phase factor which is equal to 1 for $U_q(A_1^{(1)})$ and it is equal to -1 for $U_q(C(2)^{(2)})$. This situation is similar to the finite-dimensional case. Namely, in the paper [5] it was shown that all quantum (super)algebras $U_q(g)$, where g are the finite-dimensional contragredient Lie (super)algebras of rank 2, are divided into three classes. Each such class is characterized by the same Dynkin diagram and has the same reduced root system, provided that we neglect the type (colour) of the roots (white, grey or black). Consequently, all the (super)algebras of the same class have unified defining relations, unified construction of the Cartan-Weyl basis and its properties, as well as unified universal R-matrix. Difference between the (super)algebras of the same class is determined by some phase factor which takes values ± 1 depending on the colour of the nodes of the Dynkin diagram.

Basic information about the (super)algebras $A_1^{(1)}$ and $C(2)^{(2)}$ is presented in the Tables 1a and 1b (see Refs. [3,4,14]). In the Table 1a there are listed the standard and symmetric Cartan matrices A and A^{sym}, the corresponding extended symmetric matrices $\bar{A}^{sym}$ and their inverses $(\bar{A}^{sym})^{-1}$, and also the sets of odd roots (odd roots), the Dynkin diagrams (diagram), and the dimensions of these (super)algebras (dim). We remind some elementary definitions of the colour of the roots:

- All even roots are called white roots. A white root is pictured by the white node ○.
- An odd root γ is called a grey root if 2γ is not a root. Such an odd root is pictured by the grey node ⊗.
- An odd root γ is called a dark root if 2γ is a root. Such an odd root is pictured by the dark node ●.

We also remind the definition of the reduced system of the positive root system Δ_+ for any contragredient (super)algebras of finite growth.

- The system $\underline{\Delta}_+$ is called the reduced system if it is defined by the following way: $\underline{\Delta}_+ = \Delta_+ \backslash \{2\gamma \in \Delta_+ | \gamma \text{ is odd}\}$. That is the reduced system $\underline{\Delta}_+$ is obtained from the total system Δ_+ by removing of all doubled roots 2γ where γ is a dark odd root.

The total and reduced root systems of the (super)algebras $A_1^{(1)}$ and $C(2)^{(2)}$ are listed in the Table 1b. It is convenient to present the total $\Delta = \Delta_+ \bigcup (-\Delta_+)$ and reduced $\underline{\Delta} = \underline{\Delta}_+ \bigcup (-\underline{\Delta}_+)$ root systems by the pictures: Figs. 1, 2a, 2b. Comparing Fig. 1 and Fig. 2b we see that the reduced root systems of $A_1^{(1)}$ and $C(2)^{(2)}$ coincide if we neglect colour of the roots.

Table 1a

$g(A,\Upsilon)$	$A = A^{sym}$	$\bar{A}^{sym}$	$(\bar{A}^{sym})^{-1}$	odd	diagram
$A_1^{(1)}$	$\begin{pmatrix} 2 & -2 \\ -2 & 2 \end{pmatrix}$	$\begin{pmatrix} 0 & 1 & 0 \\ 1 & 2 & -2 \\ 0 & -2 & 2 \end{pmatrix}$	$\begin{pmatrix} 0 & 1 & 1 \\ 1 & 0 & 0 \\ 1 & 0 & \frac{1}{2} \end{pmatrix}$	$\emptyset$	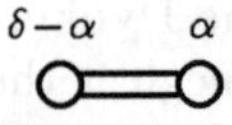
$C(2)^{(2)}$	$\begin{pmatrix} 2 & -2 \\ -2 & 2 \end{pmatrix}$	$\begin{pmatrix} 0 & 1 & 0 \\ 1 & 2 & -2 \\ 0 & -2 & 2 \end{pmatrix}$	$\begin{pmatrix} 0 & 1 & 1 \\ 1 & 0 & 0 \\ 1 & 0 & \frac{1}{2} \end{pmatrix}$	$\{\delta-\alpha, a\}$	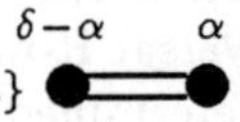

Table 1b

$g(A,\Upsilon)$	Δ_+	$\underline{\Delta}_+$
$A_1^{(1)}$	$\{\alpha,\ n\delta\pm\alpha,\ n\delta \mid n\in\mathbf{N}\}$	$\{\alpha,\ n\delta\pm\alpha,\ n\delta \mid n\in\mathbf{N}\}$
$C(2)^{(2)}$	$\{\alpha,\ 2\alpha,\ n\delta\pm\alpha,\ 2n\delta\pm2\alpha,\ n\delta \mid n\in\mathbf{N}\}$	$\{\alpha,\ n\delta\pm\alpha,\ n\delta \mid n\in\mathbf{N}\}$

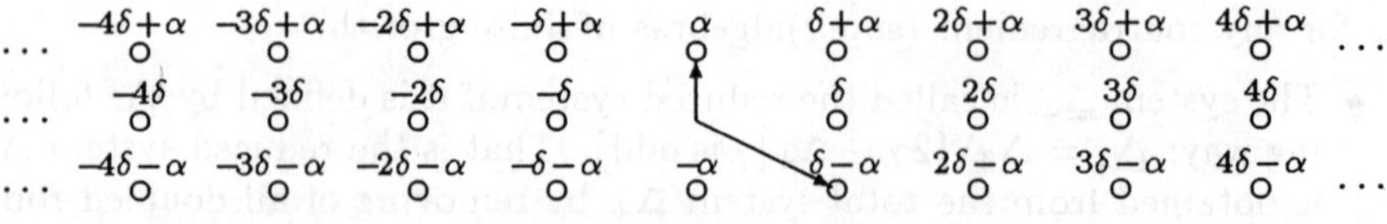

Fig. 1. The total and reduced root system ($\Delta = \underline{\Delta}$) of $A_1^{(1)} (\simeq \widehat{sl}_2)$.

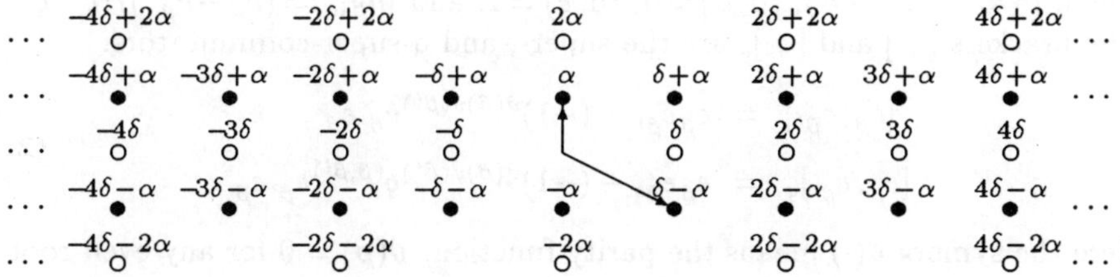

Fig. 2a. The total root system Δ of $C(2)^{(2)}$.

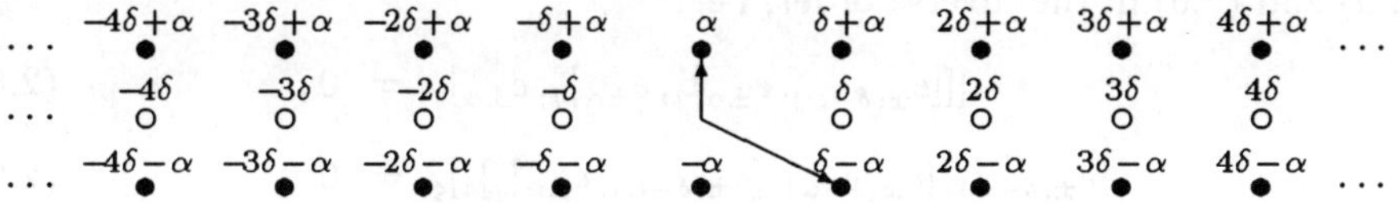

Fig. 2b. The reduced root system $\underline{\Delta}$ of $C(2)^{(2)}$.

2 Defining Relations of $U_q(A_1^{(1)})$ and $U_q(C(2)^{(2)})$

The quantum (q-deformed) affine (super)algebras $U_q(A_1^{(1)})$ and $U_q(C(2)^{(2)})$ are generated by the Chevalley elements $k_{\rm d}^{\pm 1} := q^{\pm h_{\rm d}}$, $k_{\alpha}^{\pm 1} := q^{\pm h_{\alpha}}$, $k_{\delta-\alpha}^{\pm 1} := q^{\pm h_{\delta-\alpha}}$, $e_{\pm\alpha}$, $e_{\pm(\delta-a)}$ with the defining relations:

$$k_{\gamma}k_{\gamma}^{-1} = k_{\gamma}^{-1}k_{\gamma} = 1\,, \qquad [k_{\gamma}^{\pm 1}, k_{\gamma'}^{\pm 1}] = 0\,, \tag{2.1}$$

$$k_{\gamma}e_{\pm\alpha}k_{\gamma}^{-1} = q^{\pm(\gamma,\alpha)}e_{\pm\alpha}\,, \qquad k_{\gamma}e_{\pm(\delta-\alpha)}k_{\gamma}^{-1} = q^{\pm(\gamma,\delta-\alpha)}e_{\pm(\delta-\alpha)}\,, \tag{2.2}$$

$$[e_{\alpha}, e_{-\delta+\alpha}] = 0\,, \qquad [e_{-\alpha}, e_{\delta-\alpha}] = 0\,, \tag{2.3}$$

$$[e_{\alpha}, e_{-\alpha}] = [h_{\alpha}]_q\,, \qquad [e_{\delta-\alpha}, e_{-\delta+\alpha}] = [h_{\delta-\alpha}]_q\,, \tag{2.4}$$

$$[e_{\pm\alpha}, [e_{\pm\alpha}, [e_{\pm\alpha}, e_{\pm(\delta-\alpha)}]_q]_q]_q = 0\,, \tag{2.5}$$

$$[[[e_{\pm\alpha}, e_{\pm(\delta-\alpha)}]_q, e_{\pm(\delta-\alpha)}]_q, e_{\pm(\delta-\alpha)}]_q = 0\,, \tag{2.6}$$

where $(\gamma = \mathrm{d}, \alpha, \delta-\alpha)$, $(\mathrm{d},\alpha)=0$, $(\mathrm{d},\delta)=1$, and $[h_\beta]_q := (k_\beta - k_\beta^{-1})/(q-q^{-1})$. The brackets $[\cdot,\cdot]$ and $[\cdot,\cdot]_q$ are the super-, and q-super-commutators:

$$\begin{aligned} [e_\beta, e_{\beta'}] &= e_\beta e_{\beta'} - (-1)^{\vartheta(\beta)\vartheta(\beta')} e_{\beta'} e_\beta \ , \\ [e_\beta, e_{\beta'}]_q &= e_\beta e_{\beta'} - (-1)^{\vartheta(\beta)\vartheta(\beta')} q^{(\beta,\beta')} e_{\beta'} e_\beta \ . \end{aligned} \tag{2.7}$$

Here the symbol $\vartheta(\cdot)$ means the parity function: $\vartheta(\beta)=0$ for any even root β and $\vartheta(\beta)=1$ for any odd root β.

Remark. The left-side sides of the relations (2.5) and (2.6) are invariant with respect to the replacement of q by q^{-1}. Indeed, if we remove the q-brackets we see that the left-hand of (2.5) and (2.6) contain the symmetric functions of q and q^{-1}. This property permits to write the q-commutators in (2.5) and (2.6) in the inverse order, i.e.

$$[[[e_{\pm(\delta-\alpha)}, e_{\pm\alpha}]_q, e_{\pm\alpha}]_q, e_{\pm\alpha}]_q = 0 \ , \tag{2.8}$$

$$[e_{\pm(\delta-\alpha)}, [e_{\pm(\delta-\alpha)}, [e_{\pm(\delta-\alpha)}, e_{\pm\alpha}]_q]_q]_q = 0 \ . \tag{2.9}$$

The standard Hopf structure of the quantum (super)algebras $U_q(A_1^{(1)})$ and $U_q(C(2)^{(2)})$ is given by the following formulas for the comultiplication Δ_q and antipode S_q:

$$\begin{aligned} \Delta_q(k_\gamma^{\pm1}) &= k_\gamma^{\pm1}\otimes k_\gamma^{\pm1} \ , & S_q(k_\gamma^{\pm1}) &= k_\gamma^{\mp1} \ , \\ \Delta_q(e_\beta) &= e_\beta\otimes 1 + k_\beta^{-1}\otimes e_\beta \ , & S_q(e_\beta) &= -k_\beta e_\beta \ , \\ \Delta_q(e_{-\beta}) &= e_{-\beta}\otimes k_\beta + 1\otimes e_{-\beta} \ , & S_q(e_{-\beta}) &= -e_{-\beta}k_\beta^{-1} \ , \end{aligned} \tag{2.10}$$

where $\beta=\alpha,\ \delta-\alpha$; $\gamma=\mathrm{d},\ \beta$. It is not hard to verify by direct calculations for the defining relations (2.1)-(2.6) that the quantum affine (super)algebras $U_q(A_1^{(1)})$ and $U_q(C(2)^{(2)})$ have the following simple involutive (anti)automorphisms:

(i) The non-graded antilinear antiinvolution or conjugation "":*

$$\begin{aligned} (q^{\pm1})^* &= q^{\mp1} \ , & (k_\gamma^{\pm1})^* &= k_\gamma^{\mp1} \ , \\ e_\beta^* &= e_{-\beta} \ , & e_{-\beta}^* &= e_\beta \end{aligned} \tag{2.11}$$

($(xy)^* = y^*x^*$ for $\forall\ x,y\in U_q(g)$).

(ii) The graded antilinear antiinvolution or graded conjugation "‡":

$$\begin{aligned} (q^{\pm1})^\ddagger &= q^{\mp1} \ , & (k_\gamma^{\pm1})^\ddagger &= k_\gamma^{\mp1} \ , \\ e_\beta^\ddagger &= (-1)^{\vartheta(\beta)} e_{-\beta} \ , & e_{-\beta}^\ddagger &= e_\beta \end{aligned} \tag{2.12}$$

($(xy)^{\rm deg} = (-1)^{\deg x \deg y} y^{\ddagger} x^{\ddagger}$ for any homogeneous elements $x, y \in U_q(g)$).
(iii) The Chevalley graded involution ω:

$$\begin{aligned} \omega(q^{\pm 1}) &= q^{\mp 1}\,, & \omega(k_{\gamma}^{\pm 1}) &= k_{\gamma}^{\pm 1}\,, \\ \omega(e_{\beta}) &= -e_{-\beta}\,, & \omega(e_{-\beta}) &= -(-1)^{\theta(\beta)} e_{\beta}\,. \end{aligned} \tag{2.13}$$

(iv) The Dynkin involution τ which is associated with the automorphism of the Dynkin diagrams of the (super)algebras $A_1^{(1)}$ and $C(2)^{(2)}$:

$$\begin{aligned} \tau(q^{\pm 1}) &= q^{\pm 1}\,, & \tau(k_{\rm d}^{\pm 1}) &= k_{\rm d}^{\pm 1}\,, \\ \tau(k_{\beta}^{\pm 1}) &= k_{\delta-\beta}^{\pm 1}\,, & \tau(k_{-\beta}^{\pm 1}) &= k_{-\delta+\beta}^{\pm 1}\,, \\ \tau(e_{\beta}) &= e_{\delta-\beta}\,, & \tau(e_{-\beta}) &= e_{-\delta+\beta}\,. \end{aligned} \tag{2.14}$$

Here in (2.11)-(2.14) $\beta = \alpha, \delta - \alpha$; $\gamma = {\rm d}, \beta$.

It should be noted that the graded conjugation "$\ddagger$" and the Chevalley graded involution ω are involutive (anti)automorphism of the fourth order, i.e., for example, $(\omega)^4 = {\rm id}$. Note also that the Dynkin involution τ commutes with all other three involutions, i.e. $\tau(x^*) = (\tau(x))^*$, $\tau(x^{\ddagger}) = (\tau(x))^{\ddagger}$ and $\omega\tau(x) = \tau\omega(x)$ for any element $x \in U_q(g)$ ($g = A_1^{(1)}, C(2,0)^{(2)}$).

In the next Section we construct the Cartan-Weyl basis and describe its properties in detail.

3 Cartan-Weyl Basis for $U_q(A_1^{(1)})$ and $U_q(C(2)^{(2)})$

A general scheme for construction of a Cartan-Weyl basis for quantized Lie algebras and superalgebras was proposed in Ref. [11] The scheme was applied in detail at first for quantized finite-dimensional Lie (super)algebras [5] and then to quantized untwisted affine algebras [13]. This procedure is based on a notion of "normal ordering" for the reduced positive root system. For affine Lie (super)algebras this notation was introduced in Ref. [12] (see also Refs. [11,6,7]).

In our case the reduced positive system has only two normal orderings:

$$\alpha, \delta+\alpha, 2\delta+\alpha, \ldots, \infty\delta+\alpha, \delta, 2\delta, 3\delta, \ldots, \infty\delta, \infty\delta-\alpha, \ldots, 2\delta-\alpha, \delta-\alpha, \tag{3.1}$$

$$\delta-\alpha, 2\delta-\alpha, \ldots, \infty\delta-\alpha, \delta, 2\delta, 3\delta, \ldots, \infty\delta, \infty\delta+\alpha, \ldots, 2\delta+\alpha, \delta+\alpha, \alpha. \tag{3.2}$$

The first normal ordering (3.1) corresponds to "clockwise" ordering for positive roots in Fig. 1, 2b if we start from root α to root $\delta - \alpha$. The inverse normal ordering (3.2) corresponds to "anticlockwise" ordering for the positive roots when we move from $\delta-\alpha$ to α. In accordance with the normal ordering (3.1) we set

$$e_{\delta} := [e_{\alpha}, e_{\delta-\alpha}]_q , \qquad e_{-\delta} := [e_{-\delta+\alpha}, e_{-\alpha}]_{q^{-1}} , \tag{3.3}$$

$$e_{n\delta+\alpha} := \frac{1}{a}[e_{(n-1)\delta+\alpha}, e_{\delta}] , \qquad e_{-n\delta-\alpha} := \frac{1}{a}[e_{-\delta}, e_{-(n-1)\delta-\alpha}] , \tag{3.4}$$

$$e_{(n+1)\delta-\alpha} := \frac{1}{a}[e_{\delta}, e_{n\delta-\alpha}] , \qquad e_{-(n+1)\delta+\alpha} := \frac{1}{a}[e_{-n\delta+\alpha}, e_{-\delta}] , \tag{3.5}$$

$$e'_{n\delta} := [e_{\alpha}, e_{n\delta-\alpha}]_q , \qquad e'_{-n\delta} := [e_{-n\delta+\alpha}, e_{-\alpha}]_{q^{-1}} , \tag{3.6}$$

where $n = 1, 2, \ldots$, and a is given by the formula:

$$a := [(\alpha, \alpha)]_q = \frac{q^{(\alpha,\alpha)}-q^{-(\alpha,\alpha)}}{q-q^{-1}} . \tag{3.7}$$

Analogously for the inverse normal ordering (3.2) we set

$$\tilde{e}_{\delta} := [e_{\delta-\alpha}, e_{\alpha}]_q , \qquad \tilde{e}_{-\delta} := [e_{-\alpha}, e_{-\delta+\alpha}]_{q^{-1}}, \tag{3.8}$$

$$\tilde{e}_{(n+1)\delta-\alpha} := \frac{1}{a}[\tilde{e}_{n\delta-\alpha}, \tilde{e}_{\delta}] , \qquad \tilde{e}_{-(n+1)\delta+\alpha} := \frac{1}{a}[\tilde{e}_{-\delta}, \tilde{e}_{-n\delta+\alpha}] , \tag{3.9}$$

$$\tilde{e}_{n\delta+\alpha} := \frac{1}{a}[\tilde{e}_{\delta}, \tilde{e}_{(n-1)\delta+\alpha}] , \qquad \tilde{e}_{-n\delta-\alpha} := \frac{1}{a}[\tilde{e}_{-(n-1)\delta-\alpha}, \tilde{e}_{-\delta}] , \tag{3.10}$$

$$\tilde{e}'_{n\delta} := [e_{\delta-\alpha}, \tilde{e}_{(n-1)\delta+\alpha}]_q , \qquad \tilde{e}'_{-n\delta} := [e_{-\delta+\alpha}, \tilde{e}_{-(n-1)\delta-\alpha}]_{q^{-1}}, \tag{3.11}$$

where $n = 1, 2, \ldots$. Thus, we have two systems of the Cartan-Weyl generators: 'direct' and 'inverse'. Each such system together with the Cartan generators $k_{\alpha}^{\pm 1}$, $k_{\delta-\alpha}^{\pm 1}$ $e_{\pm\alpha}$ and $e_{\pm(\delta-a)}$ are called the q-analog of the Cartan-Weyl basis (or simply the Cartan-Weyl basis) for the quantum (super)algebras $U_q(A_1^{(1)})$ and $U_q(C(2)^{(2)})$.

Now we consider some properties of these bases. First of all, the explicit construction of the Cartan-Weyl generators (3.3)-(3.6) (or (3.8)-(3.11)) permits easily to find their properties with respect to the (anti)involutions (2.11)-(2.13). For example, it is evident that

$$(e_{\pm\gamma})^* = e_{\mp\gamma} , \qquad \forall\gamma \in \underline{\Delta}_+ . \tag{3.12}$$

Further, it is easy to see that the 'direct' and 'inverse' Cartan-Weyl generators (3.3)-(3.6) and (3.8)-(3.11) have very simple connection with the Dynkin involution τ:

$$\begin{aligned}
\tau(e_{n\delta+\alpha}) &= \tilde{e}_{(n+1)\delta-\alpha}, & \tau(\tilde{e}_{n\delta+\alpha}) &= e_{(n+1)\delta-\alpha} & (n \in \mathbb{Z}), \\
\tau(e_{n\delta-\alpha}) &= \tilde{e}_{(n-1)\delta+\alpha}, & \tau(\tilde{e}_{n\delta-\alpha}) &= e_{(n-1)\delta+\alpha} & (n \in \mathbb{Z}), \\
\tau(e_{n\delta}) &= \tilde{e}_{n\delta}, & \tau(\tilde{e}_{n\delta}) &= e_{n\delta} & (n \neq 0).
\end{aligned} \tag{3.13}$$

Proposition 3.1 *The root vectors (3.3)-(3.6) satisfy the following permutation relations:*

$$\begin{aligned}
k_{\rm d} e_{n\delta\pm\alpha} k_{\rm d}^{-1} &= q^{n({\rm d},\delta)} e_{n\delta\pm\alpha}, & k_{\rm d} e'_{n\delta} k_{\rm d}^{-1} &= q^{n({\rm d},\delta)} e'_{n\delta}, \\
k_{\gamma} e_{n\delta\pm\alpha} k_{\gamma} &= q^{\pm(\gamma,\alpha)} e_{n\delta\pm\alpha}, & k_{\gamma} e'_{n\delta} k_{\gamma}^{-1} &= e'_{n\delta},
\end{aligned} \tag{3.14}$$

for any $n \in \mathbb{Z}$ and any $\gamma \in \underline{\Delta}_{+}$, and also

$$[e_{n\delta+\alpha}, e_{-n\delta-\alpha}] = (-1)^{n\theta(\alpha)} \frac{k_{n\delta+\alpha} - k_{n\delta+\alpha}^{-1}}{q - q^{-1}} \qquad (n \geq 0), \tag{3.15}$$

$$[e_{n\delta-\alpha}, e_{-n\delta+\alpha}] = (-1)^{(n-1)\theta(\alpha)} \frac{k_{n\delta-\alpha} - k_{n\delta-\alpha}^{-1}}{q - q^{-1}} \qquad (n > 0), \tag{3.16}$$

$$[e_{n\delta+\alpha}, e_{(n+2m-1)\delta+\alpha}]_q = (q_{\alpha}^2 - 1) \sum_{l=1}^{m-1} q_{\alpha}^{-l} e_{(n+l)\delta+\alpha} e_{(n+2m-1-l)\delta+\alpha}, \tag{3.17}$$

$$\begin{aligned}
[e_{n\delta+\alpha}, e_{(n+2m)\delta+\alpha}]_q &= (q_{\alpha} - 1) q_{\alpha}^{-m+1} e_{(n+m)\delta+\alpha}^2 \\
&\quad + (q_{\alpha}^2 - 1) \sum_{l=1}^{m-1} q_{\alpha}^{-l} e_{(n+l)\delta+\alpha} e_{(n+2m-l)\delta+\alpha},
\end{aligned} \tag{3.18}$$

for any integers $n \geq 0$, $m > 0$;

$$[e_{(n+2m-1)\delta-\alpha}, e_{n\delta-\alpha}]_q = -(q_{\alpha}^2 - 1) \sum_{l=1}^{m-1} q_{\alpha}^{-l} e_{(n+2m-1-l)\delta-\alpha} e_{(n+l)\delta-\alpha}, \tag{3.19}$$

$$\begin{aligned}
[e_{(n+2m)\delta-\alpha}, e_{n\delta-\alpha}]_q &= -(q_{\alpha} - 1) q_{\alpha}^{-m+1} e_{(n+m)\delta-\alpha}^2 \\
&\quad - (q_{\alpha}^2 - 1) \sum_{l=1}^{m-1} q_{\alpha}^{-l} e_{(n+l)\delta-\alpha} e_{(n+2m-l)\delta-\alpha},
\end{aligned} \tag{3.20}$$

for any integers n, m > 0;

$$[e_{-n\delta+\alpha}, e_{(n+2m-1)\delta+\alpha}] = -(-1)^{(n-1)\theta(\alpha)}(q_\alpha^2-1)$$
$$\times \sum_{l=n}^{n+m-1} q_\alpha^{-l} k_{n\delta-\alpha} e_{(l-n)\delta+\alpha} e_{(n+2m-1-l)\delta+\alpha}$$
$$+(q_\alpha^2-1)\sum_{l=1}^{n-1}(-1)^{l\theta(\alpha)} q_\alpha^{-l} k_\delta^l e_{(-n+l)\delta+\alpha} e_{(n+2m-1-l)\delta+\alpha}, \quad (3.21)$$

$$[e_{-n\delta+\alpha}, e_{(n+2m)\delta+\alpha}] = -(-1)^{(n-1)\theta(\alpha)}(q_\alpha^2-1)$$
$$\times \sum_{l=n}^{n+m-1} q_\alpha^{-l} k_{n\delta-\alpha} e_{(l-n)\delta+\alpha} e_{(n+2m-l)\delta+\alpha}$$
$$+(q_\alpha^2-1)\sum_{l=1}^{n-1}(-1)^{l\theta(\alpha)} q_\alpha^{-l} k_\delta^l e_{(-n+l)\delta+\alpha} e_{(n+2m-l)\delta+\alpha}$$
$$-(-1)^{(n-1)\theta(\alpha)}(q_\alpha-1) q_\alpha^{-m-n+1} k_{n\delta-\alpha} e_{m\delta+\alpha}^2, \quad (3.22)$$

for any integers n, m > 0;

$$[e_{(n+2m-1)\delta-\alpha}, e_{-n\delta-\alpha}] = (-1)^{(n+1)\theta(\alpha)}(q_\alpha^2-1)$$
$$\times \sum_{l=n+1}^{n+m-1} q_\alpha^{-l} e_{(n+2m-1-l)\delta-\alpha} e_{(l-n)\delta-\alpha} k_{n\delta+\alpha}^{-1}$$
$$-(q_\alpha^2-1)\sum_{l=1}^{n-1}(-1)^{l\theta(\alpha)} q_\alpha^{-l} e_{(n+2m-1-l)\delta-\alpha} e_{(-n+l)\delta-\alpha} k_\delta^{-l}, \quad (3.23)$$

$$[e_{(n+2m)\delta-\alpha}, e_{-n\delta-\alpha}] = (-1)^{(n+1)\theta(\alpha)}(q_\alpha^2-1)$$
$$\times \sum_{l=n}^{n+m-1} q_\alpha^{-l} e_{(n+2m-l)\delta-\alpha} e_{(l-n)\delta-\alpha} k_{n\delta+\alpha}^{-1}$$
$$-(q_\alpha^2-1)\sum_{l=1}^{n-1}(-1)^{l\theta(\alpha)} q_\alpha^{-l} e_{(n+2m-l)\delta-\alpha} e_{(-n+l)\delta-\alpha} k_\delta^{-l}$$
$$+(-1)^{(n+1)\theta(\alpha)}(q_\alpha-1) q_\alpha^{-m-n+1} e_{m\delta-\alpha}^2 k_{n\delta+\alpha}^{-1}, \quad (3.24)$$

for any integers $n \geq 0$, $m > 0$;

$$[e_{n\delta+\alpha}, e_{m\delta-\alpha}]_q = e'_{(n+m)\delta} \qquad (n \geq 0,\ m > 0), \quad (3.25)$$

$$[e_{n\delta+\alpha}, e_{-m\delta-\alpha}] = -(-1)^{(m+1)\theta(\alpha)} e'_{(n-m)\delta} k^{-1}_{m\delta+\alpha} \quad (n > m \geq 0), \tag{3.26}$$

$$[e_{-m\delta+\alpha}, e_{n\delta-\alpha}] = -(-1)^{(m-1)\theta(\alpha)} k_{m\delta-\alpha} e'_{(n-m)\delta} \quad (n > m > 0), \tag{3.27}$$

$$[e'_{n\delta}, e'_{m\delta}] = [e'_{-n\delta}, e'_{-m\delta}] = 0 \quad (n > 0,\ m > 0), \tag{3.28}$$

$$[e_{n\delta+\alpha}, e'_{m\delta}] = q_\alpha^{-m+1} a e_{(n+m)\delta+\alpha} + (q_\alpha^2 - 1) \sum_{l=1}^{m-1} q_\alpha^{-l} e_{(n+l)\delta+\alpha} e'_{(m-l)\delta}, \tag{3.29}$$

for any integers $n \geq 0$, $m > 0$*;*

$$[e'_{m\delta}, e_{n\delta-\alpha}] = q_\alpha^{-m+1} a e_{(n+m)\delta-\alpha} + (q_\alpha^2 - 1) \sum_{l=1}^{m-1} q_\alpha^{-l} e'_{(m-l)\delta} e_{(n+l)\delta-\alpha}, \tag{3.30}$$

for any integers n, $m > 0$*;*

$$\begin{aligned}[e_{-n\delta+\alpha}, e'_{m\delta}] = & -(-1)^{(n-1)\theta(\alpha)} q_\alpha^{-m+1} a k_{n\delta-\alpha} e_{(m-n)\delta+\alpha} \\ & -(-1)^{(n-1)\theta(\alpha)} (q_\alpha^2 - 1) k_{n\delta-\alpha} \sum_{l=n}^{m-1} q_\alpha^{-l} e_{(l-n)\delta+\alpha} e'_{(m-l)\delta} \\ & +(q_\alpha^2 - 1) \sum_{l=1}^{n-1} (-1)^{-l\theta(\alpha)} q_\alpha^{-l} k_\delta^l e_{(-n+l)\delta+\alpha} e'_{(m-l)\delta},\end{aligned} \tag{3.31}$$

for any integers $m \geq n > 0$*;*

$$\begin{aligned}[e_{-n\delta+\alpha}, e'_{m\delta}] = & (-1)^{m\theta(\alpha)} q_\alpha^{-m+1} a k_\delta^m e_{(-n+m)\delta+\alpha} \\ & +(q_\alpha^2 - 1) \sum_{l=1}^{m-1} (-1)^{-l\theta(\alpha)} q_\alpha^{-l} k_\delta^l e_{(-n+l)\delta+\alpha} e_{(m-l)\delta},\end{aligned} \tag{3.32}$$

for any integers $n > m > 0$*;*

$$\begin{aligned}[e'_{m\delta} e_{-n\delta-\alpha}] = & -(-1)^{(n+1)\theta(\alpha)} q_\alpha^{-m+1} a e_{(m-n)\delta-\alpha} k^{-1}_{n\delta+\alpha} \\ & -(-1)^{(n+1)\theta(\alpha)} (q_\alpha^2 - 1) \sum_{l=n+}^{m-1} q_\alpha^{-l} e'_{(m-l)\delta} e_{(l-n)\delta-\alpha} k^{-1}_{n\delta+\alpha} \\ & +(q_\alpha^2 - 1) \sum_{l=1}^{n} (-1)^{-l\theta(\alpha)} q_\alpha^{-l} e_{(m-l)\delta} e_{(-n+l)\delta-\alpha} k_\delta^{-l},\end{aligned} \tag{3.33}$$

for any integers $m > n \geq 0$*;*

$$\begin{aligned}[e'_{m\delta} e_{-n\delta-\alpha}] = & (-1)^{m\theta(\alpha)} q_\alpha^{-m+1} a e_{(-n+m)\delta-\alpha} k_\delta^{-m} \\ & +(q_\alpha^2 - 1) \sum_{l=1}^{m-1} (-1)^{-l\theta(\alpha)} q_\alpha^{-l} e_{(m-l)\delta} e_{(-n+l)\delta-\alpha} k_\delta^{-l},\end{aligned} \tag{3.34}$$

for any integers $n \geq m > 0$.

Here in the relations (3.17)-(3.34) and in what follows we denote $q_\alpha := (-1)^{\theta(\alpha)}q^{(\alpha,\alpha)}$. The imaginary root vectors $e'_{n\delta}$ do not satisfy the relations of the type (3.15) and therefore we introduce new imaginary roots vectors $e_{\pm n\delta}$ by the following (Schur) relations:

$$e'_{n\delta} = \sum_{p_1+2p_2+\ldots+np_n=n} \frac{\left((-1)^{\theta(\alpha)}(q-q^{-1})\right)^{\sum p_i-1}}{p_1!\cdots p_n!} e_\delta^{p_1}\cdots e_{n\delta}^{p_n}. \tag{3.35}$$

In terms of generating functions

$$\mathcal{E}'(u) := (q-q^{-1})\sum_{n\geq 1} e'_{n\delta}u^{-n} , \tag{3.36}$$

$$\mathcal{E}(u) = (q-q^{-1})\sum_{n\geq 1} e_{n\delta}u^{-n} \tag{3.37}$$

the relation (3.35) may be rewritten in the form

$$\mathcal{E}'(u) = -1 + \exp\mathcal{E}(u) \tag{3.38}$$

or

$$\mathcal{E}(u) = \ln(1+\mathcal{E}'(u)) . \tag{3.39}$$

This provides a formula inverse to (3.35)

$$e_{n\delta} = \sum_{p_1+2p_2+\ldots+np_n=n} \frac{\left((-1)^{\theta(\alpha)}(q^{-1}-q)\right)^{\sum p_i-1}(\sum_{i=1}^n p_i-1)!}{p_1!\cdots p_n!}(e'_\delta)^{p_1}\cdots(e'_{n\delta})^{p_n}. \tag{3.40}$$

The new root vectors corresponding to negative roots are obtained by the Cartan conjugation (*):

$$e_{-n\delta} = (e_{n\delta})^* . \tag{3.41}$$

Proposition 3.2 *The new root vectors $e_{\pm n\delta}$ satisfy the following commutation relations:*

$$[e_{n\delta+\alpha}, e_{m\delta}] = (-1)^{(m-1)\theta(\alpha)}a(m)e_{(n+m)\delta+\alpha} \qquad (n\geq 0,\, m>0), \tag{3.42}$$

$$[e_{m\delta}, e_{n\delta-\alpha}] = (-1)^{(m-1)\theta(\alpha)}a(m)e_{(n+m)\delta-\alpha} \qquad (n,\, m>0), \tag{3.43}$$

$$[e_{-n\delta+\alpha}, e_{m\delta}] = -(-1)^{(n+m)\theta(\alpha)}a(m)k_{n\delta-\alpha}e_{(m-n)\delta+\alpha} \quad (m \geq n > 0), \tag{3.44}$$

$$[e_{-n\delta+\alpha}, e_{m\delta}] = (-1)^{\theta(\alpha)}a(m)k_{\delta}^{m}e_{(-n+m)\delta+\alpha} \quad (n > m > 0), \tag{3.45}$$

$$[e_{m\delta}, e_{-n\delta-\alpha}] = -(-1)^{(n+m)\theta(\alpha)}a(m)e_{(m-n)\delta-\alpha}k_{n\delta+\alpha}^{-1} \quad (m > n \geq 0), \tag{3.46}$$

$$[e_{n\delta-\alpha}, e_{-m\delta}] = (-1)^{\theta(\alpha)}a(m)e_{(-n+m)\delta-\alpha}k_{\delta}^{-m} \quad (n \geq m > 0), \tag{3.47}$$

$$[e_{n\delta}, e_{-m\delta}] = \delta_{nm}\, a(m)\, \frac{k_{\delta}^{m}-k_{\delta}^{-m}}{q-q^{-1}} \quad (n, m > 0), \tag{3.48}$$

where

$$a(n) := \frac{q^{n(\alpha,\alpha)}-q^{-n(\alpha,\alpha)}}{n(q-q^{-1})}\ . \tag{3.49}$$

All the relations of Propositions (3.1), (3.2) together with the ones obtained from them by the conjugation describe complete list of the permutation relations of the Cartan-Weyl bases corresponding to the 'direct' normal ordering (3.1). Applying to these relations the Dynkin involution τ, it is easy to obtain these results for the 'inverse' normal ordering (3.2).

4 Universal R-matrix for $U_q(A_1^{(1)})$ and $U_q(C(2)^{(2)})$

Any quantum (super)algebra $U_q(g)$ is a non-cocommutative Hopf (super)algebra which has the intertwining operator called the universal R-matrix. By definition [2], the universal R-matrix for the Hopf (super)algebra $U_q(g)$ is an invertible element R of some extension $U_q(g) \otimes U_q(g)$, satisfying the equations

$$\tilde{\Delta}_q(a) = R\Delta_q(a)R^{-1} \qquad \forall\, a \in U_q(g)\ , \tag{4.1}$$

$$(\Delta_q \otimes \mathrm{id})R = R^{13}R^{23}\ , \qquad (\mathrm{id} \otimes \Delta_q)R = R^{13}R^{12}, \tag{4.2}$$

where $\tilde{\Delta}_q$ is the opposite comultiplication: $\tilde{\Delta}_q = \sigma\Delta_q$, $\sigma(a \otimes b) = (-1)^{\deg a \deg b} b \otimes a$ for all homogeneous elements a, $b \in U_q(g)$. In the relation (4.2) we use the standard notations $R^{12} = \sum a_i \otimes b_i \otimes \mathrm{id}$, $R^{13} = \sum a_i \otimes \mathrm{id} \otimes b_i$, $R^{23} = \sum \mathrm{id} \otimes a_i \otimes b_i$ if R has the form $R = \sum a_i \otimes b_i$. We employ the following standard notation for the q-exponential:

$$\exp_q(x) := 1 + x + \frac{x^2}{(2)_q!} + \ldots + \frac{x^n}{(n)_q!} + \ldots = \sum_{n\geq 0} \frac{x^n}{(n)_q!}\ , \tag{4.3}$$

where

$$(n)_q := \frac{q^n-1}{q-1}\ . \tag{4.4}$$

An explicit expression of the universal R-matrix R for our case $U_q(g)$ can be presented as follows:

$$R = R_+ R_0 R_- K \ . \tag{4.5}$$

Here the factors K and $R_\pm$ have the following form

$$K \ = \ q^{\frac{1}{(\alpha,\alpha)}h_\alpha \otimes h_\alpha + h_\delta \otimes h_d + h_d \otimes h_\delta} \ , \tag{4.6}$$

$$R_+ \ = \ \prod_{n\geq 0}^{\rightarrow} \mathcal{R}_{n\delta+\alpha}, \qquad R_- = \prod_{n\geq 1}^{\leftarrow} \mathcal{R}_{n\delta-\alpha}. \tag{4.7}$$

were the elements $\mathcal{R}_\gamma$ are given by the formula

$$\mathcal{R}_\gamma = \exp_{q_\gamma^{-1}}\Big(A(\gamma)(q-q^{-1})(e_\gamma \otimes e_{-\gamma})\Big) \ , \tag{4.8}$$

where

$$q_\gamma \ = \ (-1)^{\vartheta(\gamma)} q^{(\gamma,\gamma)} \ , \tag{4.9}$$

$$A(\gamma) \ = \ \left\{ \begin{array}{ll} (-1)^{n\theta(\alpha)} & \text{if } \gamma = n\delta + \alpha \ , \\ (-1)^{(n-1)\theta(\alpha)} & \text{if } \gamma = n\delta - \alpha \ . \end{array} \right. \tag{4.10}$$

Finally, the factor R_0 is defined as follows

$$R_0 = \exp\Big((q-q^{-1}) \sum_{n>0} d(n) e_{n\delta} \otimes e_{-n\delta}\Big) \ , \tag{4.11}$$

where $d(n)$ is the inverse to $a(n)$, i.e.

$$d(n) = \frac{n(q-q^{-1})}{q^{n(\alpha,\alpha)} - q^{-n(\alpha,\alpha)}} \ . \tag{4.12}$$

Acknowledgments

The third author is grateful to Institute of Theoretical Physics, University of Wrocław and Foundation "Kasa Mianowskiego" for the support of his visit on the Symposium. S.M. Khoroshkin and V.N. Tolstoy would like to thank for the support by the Russian Foundation for Fundamental Research (grant No. 98-01-00303) as well as by the program of French-Russian scientific cooperation (CNRS grant PICS-608 and grant RFBR-98-01-22033) and J. Lukierski for the support by KBN grant 2PO3B13012.

References

1. R.M. Asherova, Yu.F. Smirnov and V.N. Tolstoy, A description of certain class of projection operators for complex semisimple Lie algebras, (in Russian) *Matem. Zametki* **26**, no. 3, 15–25 (1979).
2. V. G. Drinfeld, Quantum groups, Proc. ICM-86 (Berkely USA), *Amer. Math. Soc.* **1**, 798–820 (1987).
3. V.G. Kac, Lie superalgebras, *Adv. Math.* **26**, 8–96 (1977).
4. V.G. Kac, Infinite dimensional Lie algebras, (Cambridge University Press, 1990).
5. S.M. Khoroshkin and V.N. Tolstoy, Universal R-matrix for quantized (super)algebras, *Commun. Math. Phys.* **141**, no. 3, 599–617 (1991).
6. S.M. Khoroshkin and V.N. Tolstoy, The Cartan-Weyl basis and the universal R-matrix for quantum Kac-Moody algebras and superalgebras, *Quantum Symmetries (Clausthal 1991)*. 336–351, (World Scientific Publishing, River Edge, NJ, 1993).
7. S.M. Khoroshkin and V.N. Tolstoy, The uniqueness theorem for the universal R-matrix, *Lett. Math. Phys.* **24**, no. 3, 231–244 (1992).
8. S.M. Khoroshkin and V.N. Tolstoy, Extremal projector and universal R-matrix for quantum contragredient Lie (super)algebras, Quantum groups and related topics (Wroclaw, 1991), *Math. Phys. Stud.*, **13**, 23–32 (Kluwer Acad. Publ., Dordrecht, 1992).
9. J. Lukierski and V.N. Tolstoy, Cartan-Weyl basis for quantum affine superalgebra $U_q(\widehat{\text{osp}}(1|2))$, *Czech. Journ. Phys.* **47**, no. 12, 1231–1239 (1997).
10. G. Lusztig, Canonical bases arising from quantized enveloping algebras, *J. Amer. Math. Soc.* **3**, 447–498 (1990).
11. V.N. Tolstoy, Extremal projectors for quantized Kac-Moody superalgebras and some of their applications, Quantum Groups (Clausthal, 1989), *Lectures Notes in Phys.* **370**, 118–125 (Springer, Berlin, 1990).
12. V.N. Tolstoy, Extremal projectors for contragredient Lie algebras and superalgebras of finite growth, (in Russian) *Uspekhi Math. Nauk* **44**, no.1(265), 211–212 (1989); translation in Russian *Math. Serveys* **44**, no. 1, 257–258 (1989).
13. V.N. Tolstoy and S.M. Khoroshkin, The Universal R-matrix for quantum nontwisted affine Lie algebras, (in Russian) *Funktsional. Anal. i Prilozhen.* **26**, no. 1, 85–88 (1992); translation in *Functional Anal. Appl.* **26**, no. 1, 69–71 (1992).
14. J.W. Van der Leur, Contragredient Lie superalgebras of finite grouth, *Utrecht thesis*, (1985).

TWIST DEFORMATIONS FOR GENERALIZED HEISENBERG ALGEBRAS

V.D. LYAKHOVSKY

Theoretical Department, St. Petersburg State University,
198904, St. Petersburg, Russia
E-mail address: lyakhovs@snoopy.phys.spbu.ru

Multidimensional Heisenberg algebras, whose creation a^+ and annihilation a^- operators are the n-dimensional vectors, can be injected into simple Lie algebras g. It is demonstrated that the spectrum of their deformations can be investigated using chains of extended Jordanian twists applied to $U(g)$. In the case of $U(sl(N))$ (for $N > 5$) the two-dimensional Heisenberg subalgebras $\widetilde{\mathcal{H}}$ have nine deformed costructures connected by four "internal" and "external" twists composing the commutative diagram.

1 Introduction

A Hopf algebra $\mathcal{A}(m, \Delta, \epsilon, S)$ can be transformed [1] by an invertible **twisting element** $\mathcal{F} \in \mathcal{A} \otimes \mathcal{A}$, into a **twisted algebra** $\mathcal{A}_{\mathcal{F}}(m, \Delta_{\mathcal{F}}, \epsilon, S_{\mathcal{F}})$, that has the same multiplication and counit but the twisted coproduct and the antipode given by

$$\Delta_{\mathcal{F}}(a) = \mathcal{F}\Delta(a)\mathcal{F}^{-1}, S_{\mathcal{F}}(a) = vS(a)v^{-1}, v = \sum f_i^{(1)} S(f_i^{(2)}), a \in \mathcal{A}. \quad (1)$$

The twisting element has to satisfy the equations

$$\mathcal{F}_{12}(\Delta \otimes id)(\mathcal{F}) = \mathcal{F}_{23}(id \otimes \Delta)(\mathcal{F}), \quad (\epsilon \otimes id)(\mathcal{F}) = (id \otimes \epsilon)(\mathcal{F}) = 1. \quad (2)$$

Let $\mathcal{A}$ and $\mathcal{B}$ be the universal enveloping algebras: $\mathcal{A} = U(l) \subset \mathcal{B} = U(g)$ with $l \subset g$. If $U(l)$ is the minimal subalgebra on which $\mathcal{F}$ is completely defined as $\mathcal{F} \in U(l) \otimes U(l)$ then l is called the **carrier** algebra for $\mathcal{F}$.

The first and well known example of the twist that was written in the explicit form [2] corresponds to the carrier subalgebra $B(2)$ with generators H and E, $[H, E] = E$. It is called the **Jordanian twist** and has the twisting element

$$\Phi_{\mathcal{J}} = e^{H \otimes \sigma}, \qquad \sigma = \ln(1 + E). \quad (3)$$

For the **extended Jordanian twists** suggested in [3] the carrier subalgebra **L** is four-dimensional:

$$\begin{aligned} &[H, E] = E, \quad [H, A] = \alpha A, \quad [H, B] = \beta B, \\ &[A, B] = E, \quad [E, A] = [E, B] = 0, \qquad \alpha + \beta = 1. \end{aligned} \quad (4)$$

The corresponding twisting element contains the **Jordanian factor** $\Phi_{\mathcal{J}}$:

$$\mathcal{F}_{\mathcal{E}(\alpha,\beta)} = \Phi_{\mathcal{E}(\alpha,\beta)}\Phi_{\mathcal{J}} \tag{5}$$

and the **extension**

$$\Phi_{\mathcal{E}(\alpha,\beta)} = \exp\{A \otimes Be^{-\beta\sigma}\}\,. \tag{6}$$

This twist defines the deformed Hopf algebras $\mathbf{L}_{\mathcal{E}(\alpha,\beta)}$ with the costructure

$$\begin{array}{rcl} \Delta_{\mathcal{E}(\alpha,\beta)}(H) & = & H \otimes e^{-\sigma} + 1 \otimes H - A \otimes Be^{-(\beta+1)\sigma}\,, \\ \Delta_{\mathcal{E}(\alpha,\beta)}(A) & = & A \otimes e^{-\beta\sigma} + 1 \otimes A\,, \\ \Delta_{\mathcal{E}(\alpha,\beta)}(B) & = & B \otimes e^{\beta\sigma} + e^{\sigma} \otimes B\,, \\ \Delta_{\mathcal{E}(\alpha,\beta)}(E) & = & E \otimes e^{\sigma} + 1 \otimes E\,. \end{array} \tag{7}$$

In general the composition of two twists is not a twist. But there are some important examples of the opposite behavior. When $\mathbf{L}$ is a subalgebra $\mathbf{L} \subset \mathrm{g}$ there may exist several pairs of generators of the type (A, B) arranged so that the Jordanian twist can acquire several similar extensions. [3] This demonstrates that some twistings can be applied successively to the initial Hopf algebra even in the case when their carrier subalgebras are nontrivially linked. In the universal enveloping algebras for classical Lie algebras there exists the possibility to construct systematically the special sequences of twists called **chains**:[4]

$$\mathcal{F}_{\mathcal{B}_{p\prec 0}} \equiv \mathcal{F}_{\mathcal{B}_p}\mathcal{F}_{\mathcal{B}_{p-1}}\ldots\mathcal{F}_{\mathcal{B}_0}\,. \tag{8}$$

The factors $\mathcal{F}_{\mathcal{B}_k} = \Phi_{\mathcal{E}_k}\Phi_{\mathcal{J}_k}$ of the chain are the twisting elements of the extended Jordanian twists for the initial Hopf algebra $\mathcal{A}_0$. Here the extensions $\{\Phi_{\mathcal{E}_k}, k = 0, \ldots p-1\}$ contain the fixed set of normalized factors $\Phi_{\mathcal{E}(\alpha,\beta)} = \exp\{A \otimes Be^{-\beta\sigma}\}$, the so called **full set**. It was proved that in the classical Lie algebras that conserve symmetric invariant forms such chains can be made maximal and proper. This means that for the algebras $U(A_n)$, $U(B_n)$ and $U(D_n)$ there exist chains $\mathcal{F}_{\mathcal{B}_{p\prec 0}}$ that cannot be reduced to a chain for a simple subalgebra and their full sets of extensions are the maximal sets in the sense described below.

To construct a maximal proper chain for a classical Lie algebra the sequence $\mathcal{A} \equiv \mathcal{A}_0 \supset \mathcal{A}_1 \supset \ldots \supset \mathcal{A}_{p-1} \supset \mathcal{A}_p$ of Hopf subalgebras is to be fixed. For example in the case of $U(sl(N))$ the corresponding sequence is: $U(sl(N)) \supset U(sl(N-2)) \supset \ldots \supset U(sl(N-2k))\ldots$. In each element of the sequence the **initial root** λ_0^k must be chosen. Here $\lambda_0^k = e_1 - e_2$ for each $sl(M)$ (the roots are written in the standard e-basis). For the initial root let us form the set π_k of **constituent roots**,

$$\pi_k = \{\lambda', \lambda'' | \lambda' + \lambda'' = \lambda_0^k; \quad \lambda' + \lambda_0^k, \lambda'' + \lambda_0^k \notin \Lambda_{\mathcal{A}}\} \tag{9}$$

(here $\Lambda_{\mathcal{A}}$ is the root system of $\mathcal{A}_0$). For each element $\lambda' \in \pi_k$ one can choose such an element $\lambda'' \in \pi_k$ that $\lambda' + \lambda'' = \lambda_0^k$. So, π_k is naturally decomposed as

$$\pi_k = \pi_k' \cup \pi_k'', \qquad \pi_k' = \{\lambda'\}, \quad \pi_k'' = \{\lambda''\}. \tag{10}$$

In these terms the factors $\mathcal{F}_{\mathcal{B}_k}$ of the chain (8) are fixed as follows:

$$\mathcal{F}_{\mathcal{B}_k} = \Phi_{\mathcal{E}_k} \Phi_{\mathcal{J}_k} \tag{11}$$

with

$$\Phi_{\mathcal{J}_k} = \exp\{H_{\lambda_0^k} \otimes \sigma_0^k\}, \qquad \sigma_0^k = \ln(1 + L_{\lambda_0^k}); \tag{12}$$

$$\Phi_{\mathcal{E}_k} = \prod_{\lambda' \in \pi_k'} \Phi_{\mathcal{E}_{\lambda'}} = \prod_{\lambda' \in \pi_k'} \exp\{L_{\lambda'} \otimes L_{\lambda_0^k - \lambda'} e^{-\frac{1}{2}\sigma_0^k}\} \tag{13}$$

(here L_λ is the generator associated to the root λ).

2 Deformed Heisenberg Algebras

One of the characteristic features of the extension $\Phi_{\mathcal{E}}$ in the ordinary extended Jordanian twists is that it connects the Heisenberg subalgebras $\mathcal{H}_{\mathcal{J}}$ and $\mathcal{H}_{\mathcal{E}\mathcal{J}}$ with $[A, B] = E$, $\alpha + \beta = 1$ and different deformed costructures (see Ref. [5]):

$$\begin{array}{c} \Phi_{\mathcal{E}} : \left\{ \begin{array}{lcl} \Delta_{\mathcal{J}}(A) & = & A \otimes e^{\alpha\sigma} + 1 \otimes A, \\ \Delta_{\mathcal{J}}(B) & = & B \otimes e^{\beta\sigma} + 1 \otimes B, \\ \Delta_{\mathcal{J}}(E) & = & E \otimes e^{\sigma} + 1 \otimes E, \end{array} \right\} \longrightarrow \\ \left\{ \begin{array}{lcl} \Delta_{\mathcal{E}\mathcal{J}}(A) & = & A \otimes e^{-\beta\sigma} + 1 \otimes A, \\ \Delta_{\mathcal{E}\mathcal{J}}(B) & = & B \otimes e^{\beta\sigma} + e^{\sigma} \otimes B, \\ \Delta_{\mathcal{E}\mathcal{J}}(E) & = & E \otimes e^{\sigma} + 1 \otimes E. \end{array} \right\} \end{array} \tag{14}$$

A chain of twists (8) $\mathcal{F}_{\mathcal{B}_{p\prec 0}} \equiv \mathcal{F}_{\mathcal{B}_p} \mathcal{F}_{\mathcal{B}_{p-1}} \ldots \mathcal{F}_{\mathcal{B}_0} : \mathcal{A} \longrightarrow \mathcal{A}_{\mathcal{B}_{p\prec 0}}$ contains $p+1$ Jordanian factors $\Phi_{\mathcal{J}_k}$. Their product $\Phi_{\mathcal{J}_{p\prec 0}} \equiv \prod_k \Phi_{\mathcal{J}_{k-1}}$ is also a twisting element for the universal enveloping algebra $\mathcal{A}$, it produces the **multijordanian deformation** $\Phi_{\mathcal{J}_{p\prec 0}} : \mathcal{A} \longrightarrow \mathcal{A}_{\mathcal{J}_{p\prec 0}}$. This means that the product $\widehat{\Phi}_{\mathcal{E}_{p\prec 0}} \equiv \prod_k \widehat{\Phi}_{\mathcal{E}_k}$ with $\widehat{\Phi}_{\mathcal{E}_k} = (\prod_{m>k} \Phi_{\mathcal{J}_m}) \Phi_{\mathcal{E}_k} (\prod_{m>k} \Phi_{\mathcal{J}_m})^{-1}$ is in turn a twisting element for $\mathcal{A}_{\mathcal{J}_{p\prec 0}}$, $\widehat{\Phi}_{\mathcal{E}_{p\prec 0}} : \mathcal{A}_{\mathcal{J}_{p\prec 0}} \longrightarrow \mathcal{A}_{\mathcal{B}_{p\prec 0}}$.

In Ref. [6] it was shown that $\widehat{\Phi}_{\mathcal{E}_{p\prec 0}}$ plays the role of an extension for the multijordanian twists $\Phi_{\mathcal{J}_{p\prec 0}}$. It was proved that there are the subalgebras in $\mathcal{A}$ that are carriers for $\widehat{\Phi}_{\mathcal{E}_{p\prec 0}}$ and whose costructures are shifted from one possible deformed "state" to the other by the twists $\widehat{\Phi}_{\mathcal{E}_{p\prec 0}}$.

Here we consider a certain type of such subalgebras, namely the multidimensional Heisenberg subalgebras $\widetilde{\mathcal{H}}(2, N-4)$ in $U(sl(N))$ generated by the 2×2-block of central and $2 \times (N-4)$ pairs of creation and annihilation operators:

$$\boxed{\begin{array}{cccc} E_{13} & E_{14} & \dots & E_{1,N-2} \\ E_{23} & E_{24} & \dots & E_{2,N-2} \end{array}} \quad \begin{array}{c} \boxed{\begin{array}{cc} E_{1,N-1} & E_{1N} \\ E_{2,N-1} & E_{2N} \end{array}} \\ \boxed{\begin{array}{cc} E_{3,N-1} & E_{3N} \\ \vdots & \vdots \\ E_{N-3,N-1} & E_{N-3,N} \\ E_{N-2,N-1} & E_{N-2,N} \end{array}} \end{array} \tag{15}$$

We shall normalize the Cartan elements as $H_{i,k} = 1/2(E_{ii} - E_{kk})$, use the standard $gl(N)$-basis $\{E_{ij}\}_{i,j=1,\dots N}$, and $\sigma = \ln(1+E)$.

First we shall study the properties of $\widetilde{\mathcal{H}}_{\mathcal{F}}(2, N-4)$ twisted by the factors of the 2-chain:

$$\mathcal{F}_{\mathcal{B}_{1 \prec 0}} = \Phi_{\mathcal{E}_1} \Phi_{\mathcal{J}_1} \Phi_{\mathcal{E}_0} \Phi_{\mathcal{J}_0} \tag{16}$$

with

$$\begin{array}{rcl} \Phi_{\mathcal{J}_{k-1}} & = & \exp(H_{k,N-k+1} \otimes \sigma_{k,N-k+1}), \\ \Phi_{\mathcal{E}_{k-1}} & = & \exp\left(\sum_{s=k+1}^{N-k} E_{k,s} \otimes E_{s,N-k+1} e^{-\frac{1}{2}\sigma_{k,N-k+1}}\right) \\ & = & \prod_{r=k+1}^{N-k} \Phi_{\mathcal{E}_{k-1}(r)} \, . \end{array} \tag{17}$$

To visualize the resulting deformations the following notations will be used for the costructures:

$$\begin{array}{lcl} P^0(L) & \equiv & L \otimes 1 + 1 \otimes L, \\ P_i^{\pm}(L) & \equiv & L \otimes e^{\pm\frac{1}{2}\sigma_{i,N+1-i}} + 1 \otimes L, \\ R_i(L) & \equiv & L \otimes e^{\frac{1}{2}\sigma_{i,N+1-i}} + e^{\sigma_{i,N+1-i}} \otimes L, \\ T_i(L) & \equiv & L \otimes e^{\sigma_{i,N+1-i}} + 1 \otimes L, \\ T^{++}(L) & \equiv & L \otimes e^{\frac{1}{2}\sigma_{1,N} + \frac{1}{2}\sigma_{2,N-1}} + 1 \otimes L, \\ T^{\mp\pm}(L) & \equiv & L \otimes e^{\mp\frac{1}{2}\sigma_{1,N} \pm \frac{1}{2}\sigma_{2,N-1}} + 1 \otimes L, \\ T_{R_i}(L) & \equiv & L \otimes e^{\frac{1}{2}\sigma_{1,N} + \frac{1}{2}\sigma_{2,N-1}} + e^{\sigma_{i,N+1-i}} \otimes L, \\ S_2^- & \equiv & -E_{2,r} \otimes E_{1,N-1} e^{-\frac{1}{2}\sigma_{2,N-1}}, \\ S_2^+ & \equiv & E_{2,N} \otimes E_{r,N-1} e^{\frac{1}{2}\sigma_{1,N} - \frac{1}{2}\sigma_{2,N-1}}, \\ S_1^- & \equiv & -E_{1,r} \otimes E_{2,N} e^{-\frac{1}{2}\sigma_{1,N}}, \\ S_1^+ & \equiv & E_{1,N-1} \otimes E_{r,N} e^{-\frac{1}{2}\sigma_{1,N} + \frac{1}{2}\sigma_{2,N-1}}, \\ & & i = 1, 2; \quad r = 3, \dots, N-2. \end{array} \tag{18}$$

In these terms the transformations performed in $\mathcal{H}$ by the factors of the canonical extended Jordanian twist (use the formulas (4)-(7) with $\alpha = \beta = 1/2$)

can be written as

$$\Phi_{\mathcal{J}}:\left\{\begin{array}{cc} P^0(A) & P^0(E) \\ & P^0(B) \end{array}\right\} \longrightarrow \left\{\begin{array}{cc} P^+(A) & T(E) \\ & P^+(B) \end{array}\right\}. \tag{19}$$

$$\Phi_{\mathcal{E}}:\left\{\begin{array}{cc} P^+(A) & T(E) \\ & P^+(B) \end{array}\right\} \longrightarrow \left\{\begin{array}{cc} P^-(A) & T(E) \\ & R(B) \end{array}\right\}. \tag{20}$$

Let us perform the 2-Jordanian twisting in $\widetilde{\mathcal{H}}(2, N-4)$, $\Phi_{\mathcal{J}_1\mathcal{J}_2} : \widetilde{\mathcal{H}}(2, N-4) \longrightarrow \widetilde{\mathcal{H}}_{\mathcal{J}_1\mathcal{J}_2}(2, N-4)$. Here the deformed coproducts will be presented by the schemes analogous to (19) and (20). In terms of (18) the costructure of $\widetilde{\mathcal{H}}_{\mathcal{J}_1\mathcal{J}_2}(2, N-4)$ will acquire the following form:

$$\begin{array}{|ccc|}\hline P_1^+(E_{13}) & \ldots & P_1^+(E_{1,N-2}) \\ P_2^+(E_{23}) & \ldots & P_2^+(E_{2,N-2}) \\ \hline \end{array} \quad \begin{array}{|cc|}\hline T^{++}(E_{1,N-1}) & T_1(E_{1N}) \\ T_2(E_{2,N-1}) & T^{++}(E_{2N}) \\ \hline P_2^+(E_{3,N-1}) & P_1^+(E_{3N}) \\ \vdots & \vdots \\ P_2^+(E_{N-2,N-1}) & P_1^+(E_{N-2,N}) \\ \hline \end{array} \tag{21}$$

As we have mentioned above any Heisenberg subalgebra with the costructure $\{P^+, T, P^+\}$ can be twisted by the corresponding extension $\Phi_{\mathcal{E}}$. In (21) such are the triples $\{P_i^+(E_{ir}),\ T_i(E_{i,N-i+1}),\ P_i^+(E_{r,N-i+1})\}$. Obviously the deformations performed by the extensions $\Phi_{\mathcal{E}_{i-1}(r)} = \exp\left(E_{i,r} \otimes E_{r,N-i+1} e^{-\frac{1}{2}\sigma_{i,N-i+1}}\right)$ in these triples do not touch the generators other than $\{E_{1,r}, E_{2,r}\ , E_{r,N-1}, E_{r,N}\}$. Thus to study the deformations induced on $\widetilde{\mathcal{H}}(2, N-4)$ by the chain of twists (16) it is sufficient to focus the attention on the behavior of the subalgebra $\widetilde{\mathcal{H}} \equiv \widetilde{\mathcal{H}}(2, 1)$,

$$\begin{array}{|c|}\hline E_{1r} \\ E_{2r} \\ \hline \end{array} \quad \begin{array}{|cc|}\hline E_{1,N-1} & E_{1N} \\ E_{2,N-1} & E_{2N} \\ \hline E_{r,N-1} & E_{rN} \\ \hline \end{array} \tag{22}$$

Starting with its 2-Jordanian deformation $\widetilde{\mathcal{H}}_{\mathcal{J}_1\mathcal{J}_0}$ one can apply any of the two extensions: $\Phi_{\mathcal{E}_{i-1}(r)} = \exp\left(E_{i,r} \otimes E_{r,N-i+1} e^{-\frac{1}{2}\sigma_{i,N-i+1}}\right)$ $(i = 1, 2)$. Notice that in the chain (16) the extension $\Phi_{\mathcal{E}_1(r)}$ commutes not only with all the other extension factors $\Phi_{\mathcal{E}_{i-1}(r)}$ but also with $\Phi_{\mathcal{J}_1}$. We get three twisted "states":

$$\widetilde{\mathcal{H}}_{\mathcal{E}_0\mathcal{J}_1\mathcal{J}_0} \overset{\Phi_{\mathcal{E}_0(r)}}{\longleftarrow} \widetilde{\mathcal{H}}_{\mathcal{J}_1\mathcal{J}_0} \overset{\Phi_{\mathcal{E}_1(r)}}{\longrightarrow} \widetilde{\mathcal{H}}_{\mathcal{E}_1\mathcal{J}_1\mathcal{J}_0}$$

The corresponding coalgebras are transformed as follows

$$
\begin{array}{l}
\left\{\begin{array}{c}\boxed{\begin{array}{c}P_1^-\\ P_2^+ + S_1^-\end{array}}\end{array}\ \begin{array}{c}\boxed{\begin{array}{cc}T^{++} & T_1\\ T_2 & T^{++}\end{array}}\\ \boxed{\begin{array}{cc}P_2^+ + S_1^+ & R_1\end{array}}\end{array}\right\} \longleftarrow \left\{\begin{array}{c}\boxed{\begin{array}{c}P_1^+\\ P_2^+\end{array}}\end{array}\ \begin{array}{c}\boxed{\begin{array}{cc}T^{++} & T_1\\ T_2 & T^{++}\end{array}}\\ \boxed{\begin{array}{cc}P_2^+ & P_1^+\end{array}}\end{array}\right\}\\
\qquad\qquad \longrightarrow \left\{\begin{array}{c}\boxed{\begin{array}{c}P_1^+ + S_2^-\\ P_2^-\end{array}}\end{array}\ \begin{array}{c}\boxed{\begin{array}{cc}T^{++} & T_1\\ T_2 & T^{++}\end{array}}\\ \boxed{\begin{array}{cc}R_2 & P_1^+ + S_2^+\end{array}}\end{array}\right\}
\end{array}
\tag{23}
$$

Now we shall show that among the twists for $U\left(sl(N)\right)$ one can find those that perform further deformations of the states (23). Their carrier subalgebras are not included in the $\widetilde{\mathcal{H}}$, so with respect to $\widetilde{\mathcal{H}}$ these twists must be treated as external. Consider again the chain (16). According to the "matreshka" effect (see Ref. [4]) after the first two twists $\Phi_{\mathcal{J}_0}$ and $\Phi_{\mathcal{E}_0}$ the generators in the subalgebra $U_{\mathcal{E}_0\mathcal{J}_0}\left(sl(N-2)\right)$ will acquire trivial coproducts. Consequently when the first three factors $\Phi_{\mathcal{J}_1}\Phi_{\mathcal{E}_0}\Phi_{\mathcal{J}_0}$ are applied the generators $\{E_{2,r}, E_{r,N-1}\}$ return to the state P_2^+. The twisting factor that performs this transition appears when we drag the Jordanian twisting element $\Phi_{\mathcal{J}_1}$ to the right:

$$
\begin{aligned}
\Phi_{\mathcal{J}_1}\Phi_{\mathcal{E}_0}\Phi_{\mathcal{J}_0} &= \Phi_{\mathcal{J}_1}\Phi_{\mathcal{E}_0(2)}\Phi_{\mathcal{E}_0(N-1)}\Phi_{\mathcal{E}_0(N-2\prec 3)}\Phi_{\mathcal{J}_0}\\
&= \Phi_{\mathcal{J}_1}\Phi_{\mathcal{E}_0(2)}\Phi_{\mathcal{E}_0(N-1)}\Phi_{\mathcal{J}_1}^{-1}\Phi_{\mathcal{J}_1}\Phi_{\mathcal{E}_0(N-2\prec 3)}\Phi_{\mathcal{J}_0}\\
&= \widetilde{\Phi}_{\mathcal{E}_0(2,N-1)}\Phi_{\mathcal{E}_0(N-2\prec 3)}\Phi_{\mathcal{J}_1\mathcal{J}_0}\,.
\end{aligned}
\tag{24}
$$

The factor $\Phi_{\mathcal{E}_0(N-2\prec 3)}$ is a twisting element for $U_{\mathcal{J}_1\mathcal{J}_0}$. The relation (25) signifies that

$$
\begin{aligned}
\widetilde{\Phi}_{\mathcal{E}_0(2,N-1)} &= \exp\Big(\big(E_{1,2}+\tfrac{1}{2}E_{1,N-1}+E_{1,N-1}H_{2,N-1}\big)\\
&\quad \otimes E_{2,N}e^{-\frac{1}{2}(\sigma_{1,N}+\sigma_{2,N-1})}\\
&\quad + E_{1,N-1}\otimes E_{N-1,N}e^{-\frac{1}{2}(\sigma_{1,N}-\sigma_{2,N-1})}\Big)\,,
\end{aligned}
\tag{25}
$$

twists the costructure of $U_{\mathcal{E}_0(N-2\prec 3)\mathcal{J}_1\mathcal{J}_0}$. It is important that for the generators in (25) the coproducts in $U_{\mathcal{J}_1\mathcal{J}_0}$ and in $U_{\mathcal{E}_0(N-2\prec 3)\mathcal{J}_1\mathcal{J}_0}$ are the same. This means that the factor $\widetilde{\Phi}_{\mathcal{E}_0(2,N-1)}$ can twist directly the algebra $U_{\mathcal{J}_1\mathcal{J}_0}$ and the subalgebra $\widetilde{\mathcal{H}}_{\mathcal{J}_1\mathcal{J}_0}$ in it. The structure constants of this subalgebra are invariant with respect to the renumbering $(1 \rightleftharpoons 2, N-1 \rightleftharpoons N)$ of the indices of generators. Thus there exists the second external twisting factor

$$\begin{aligned} \widetilde{\Phi}_{\mathcal{E}_1(2,N-1)} &= \exp\Big(\big(E_{2,1} + \tfrac{1}{2}E_{2,N} + E_{2,N}H_{1,N}\big) \\ &\quad \otimes E_{1,N-1}e^{-\frac{1}{2}(\sigma_{1,N}+\sigma_{2,N-1})} \\ &\quad + E_{2,N} \otimes E_{N,N-1}e^{+\frac{1}{2}(\sigma_{1,N}-\sigma_{2,N-1})}\Big) \end{aligned} \qquad (26)$$

that can be applied to $\widetilde{\mathcal{H}}_{\mathcal{J}_1\mathcal{J}_0}$. One can find this twisting element directly considering the chain

$$\mathcal{F}'_{\mathcal{B}_{0\prec 1}} = \Phi'_{\mathcal{E}_0}\Phi_{\mathcal{J}_0}\Phi'_{\mathcal{E}_1}\Phi_{\mathcal{J}_1}\,, \qquad (27)$$

where the maximal set of the constituent roots is used in the extension $\Phi'_{\mathcal{E}_1}$ (that is for the root $(e_2 - e_{N-1})$). The algebra $U_{\mathcal{J}_1\mathcal{J}_0} = U_{\mathcal{J}_0\mathcal{J}_1}$ can be considered as an intermediate deformed object for both chains (16) and (27).

When any of the external factors $\widetilde{\Phi}_{\mathcal{E}_0(2,N-1)}$ and $\widetilde{\Phi}_{\mathcal{E}_1(2,N-1)}$ are applied to $\widetilde{\mathcal{H}}_{\mathcal{J}_1\mathcal{J}_0}$ one of the pairs of creation-annihilation generators retains the costructure P^+:

$$\left\{ \boxed{\begin{array}{c} P_1^+ \\ P_2^+ - S_1^- \end{array}} \; \begin{array}{c} \boxed{\begin{array}{cc} T^{-+} & T_1 \\ T_2 & T_{R_1} \end{array}} \\ \boxed{P_2^+ - S_1^+ \quad P_1^+} \end{array} \right\} \leftarrow \left\{ \boxed{\begin{array}{c} P_1^+ \\ P_2^+ \end{array}} \; \begin{array}{c} \boxed{\begin{array}{cc} T^{++} & T_1 \\ T_2 & T^{++} \end{array}} \\ \boxed{P_2^+ \quad P_1^+} \end{array} \right\}$$

$$\longrightarrow \left\{ \boxed{\begin{array}{c} P_1^+ - S_2^- \\ P_2^+ \end{array}} \; \begin{array}{c} \boxed{\begin{array}{cc} T_{R_2} & T_1 \\ T_2 & T^{+-} \end{array}} \\ \boxed{P_2^+ \quad P_1^+ - S_2^+} \end{array} \right\} \qquad (28)$$

This shows that to any of them the corresponding extension $\Phi_{\mathcal{E}_{i-1}}$ can be applied. Returning to the sequence (28) we see that these extensions remove the summands of the form $S_i^{\pm}$ so that the alternative extension also becomes applicable. As a result we get for the Heisenberg algebra $\widetilde{\mathcal{H}}$ nine deformed costructures

$$\Delta_{\mathcal{J}_1\mathcal{J}_0}\left(\widetilde{\mathcal{H}}\right) = \left\{ \begin{array}{ccc} P_1^+ & T^{++} & T_1 \\ P_2^+ & T_2 & T^{++} \\ & P_2^+ & P_1^+ \end{array} \right\},$$

$$\Delta_{\widetilde{\mathcal{E}}_0\mathcal{J}_1\mathcal{J}_0}\left(\widetilde{\mathcal{H}}\right) = \left\{ \begin{array}{ccc} P_1^+ & T^{-+} & T_1 \\ P_2^+ - S_1^- & T_2 & T_{R_1} \\ & P_2^+ - S_1^+ & P_1^+ \end{array} \right\},$$

$$\Delta_{\widetilde{\mathcal{E}}_1\mathcal{J}_1\mathcal{J}_0}\left(\widetilde{\mathcal{H}}\right) = \left\{ \begin{array}{ccc} P_1^+ - S_2^- & T_{R_2} & T_1 \\ P_2^+ & T_2 & T^{+-} \\ & P_2^+ & P_1^+ - S_2^+ \end{array} \right\},$$

$$\Delta_{\mathcal{E}_0\mathcal{J}_1\mathcal{J}_0}\left(\widetilde{\mathcal{H}}\right) = \left\{ \begin{array}{ccc} P_1^- & T^{++} & T_1 \\ P_2^+ + S_1^- & T_2 & T^{++} \\ & P_2^+ + S_1^+ & R_1 \end{array} \right\},$$

$$\Delta_{\widetilde{\mathcal{E}}_0\mathcal{E}_0\mathcal{J}_1\mathcal{J}_0}\left(\widetilde{\mathcal{H}}\right) = \left\{ \begin{array}{ccc} P_1^- & T^{-+} & T_1 \\ P_2^+ & T_2 & T_{R_1} \\ & P_2^+ & R_1 \end{array} \right\},$$

$$\Delta_{\mathcal{E}_1\mathcal{E}_0\widetilde{\mathcal{E}}_1\mathcal{J}_1\mathcal{J}_0}\left(\widetilde{\mathcal{H}}\right) = \left\{ \begin{array}{ccc} P_1^- & T_{R_2} & T_1 \\ P_2^- + S_1^- & T_2 & T^{+-} \\ & R_2 + S_1^+ & R_1 \end{array} \right\},$$

$$\Delta_{\mathcal{E}_1\mathcal{J}_1\mathcal{J}_0}\left(\widetilde{\mathcal{H}}\right) = \left\{ \begin{array}{ccc} P_1^+ + S_2^- & T^{++} & T_1 \\ P_2^- & T_2 & T^{++} \\ & R_2 & P_1^+ + S_2^+ \end{array} \right\},$$

$$\Delta_{\mathcal{E}_1\mathcal{E}_0\widetilde{\mathcal{E}}_0\mathcal{J}_1\mathcal{J}_0}\left(\widetilde{\mathcal{H}}\right) = \left\{ \begin{array}{ccc} P_1^- + S_2^- & T^{-+} & T_1 \\ P_2^- & T_2 & T_{R_1} \\ & R_2 & R_1 + S_2^+ \end{array} \right\},$$

$$\Delta_{\mathcal{E}_1\widetilde{\mathcal{E}}_1\mathcal{J}_1\mathcal{J}_0}\left(\widetilde{\mathcal{H}}\right) = \left\{ \begin{array}{ccc} P_1^+ & T_{R_2} & T_1 \\ P_2^- & T_2 & T^{+-} \\ & R_2 & P_1^+ \end{array} \right\}.$$

Here is how these costructures are connected by the external and internal extension twists:

$$
\begin{array}{ccccc}
\widetilde{\mathcal{H}}_{\mathcal{E}_1\mathcal{E}_0\widetilde{\mathcal{E}}_0\mathcal{J}_1\mathcal{J}_0} & & & & \\
\uparrow \Phi_{\mathcal{E}_1} & & & & \\
\widetilde{\mathcal{H}}_{\mathcal{E}_0\widetilde{\mathcal{E}}_0\mathcal{J}_1\mathcal{J}_0} & \stackrel{\widetilde{\Phi}_{\mathcal{E}_0}}{\longleftarrow} & \widetilde{\mathcal{H}}_{\mathcal{E}_0\mathcal{J}_1\mathcal{J}_0} & & \\
\uparrow \Phi_{\mathcal{E}_0} & & \uparrow \Phi_{\mathcal{E}_0} & & \\
\widetilde{\mathcal{H}}_{\widetilde{\mathcal{E}}_0\mathcal{J}_1\mathcal{J}_0} & \stackrel{\widetilde{\Phi}_{\mathcal{E}_0}}{\longleftarrow} & \widetilde{\mathcal{H}}_{\mathcal{J}_1\mathcal{J}_0} & \stackrel{\widetilde{\Phi}_{\mathcal{E}_1}}{\longrightarrow} & \widetilde{\mathcal{H}}_{\widetilde{\mathcal{E}}_1\mathcal{J}_1\mathcal{J}_0} \\
 & & \downarrow \Phi_{\mathcal{E}_1} & & \downarrow \Phi_{\mathcal{E}_1} \\
 & & \widetilde{\mathcal{H}}_{\mathcal{E}_1\mathcal{J}_1\mathcal{J}_0} & \stackrel{\widetilde{\Phi}_{\mathcal{E}_1}}{\longrightarrow} & \widetilde{\mathcal{H}}_{\mathcal{E}_1\widetilde{\mathcal{E}}_1\mathcal{J}_1\mathcal{J}_0} \\
 & & & & \downarrow \Phi_{\mathcal{E}_0} \\
 & & & & \widetilde{\mathcal{H}}_{\mathcal{E}_1\mathcal{E}_0\widetilde{\mathcal{E}}_1\mathcal{J}_1\mathcal{J}_0} .
\end{array}
\tag{29}
$$

The vertical arrows are the ordinary extension twists for Heisenberg subalgebras. The horizontal arrows correspond to the external twists borrowed from the chains for $U(sl(N))$. The squares of the diagram are commutative. Its asymmetry is justified by the fact that $\Phi_{\mathcal{E}_i}$ and $\widetilde{\Phi}_{\mathcal{E}_j}$ commute only for $i = j$. At the same time in the columns of the diagram (29) both $\Phi_{\mathcal{E}_i}$ are applicable, one – as a legal extension for the $\mathcal{H}_{\mathcal{J}}$ and the other due to the corresponding "matreshka" effect.

3 Conclusions

We have shown that the multidimensional Heisenberg algebras (of the type (15)) have the fixed spectrum of deformed costructures. The properties of this spectrum is tightly connected with different chains of extended twists that can be applied to the initial universal enveloping algebra. It is obvious that such Heisenberg algebras can be included in universal enveloping algebras other than $U(sl(N))$ and there they can be treated analogously. It is equally obvious that increasing the number of initial Jordanian twists (and enlarging correspondingly the rows and the columns of creation and annihilation operators in the multidimensional Heisenberg subalgebra) one can get more complicated costructures. The main feature here is the appearance of the $S_i^{\pm}$ summands that mix the creation (annihilation) operators from different rows (columns).

Acknowledgments

The author is heartily grateful to the organizers of the XIV-th Max Born Symposium for their warm hospitality.

This work has been partially supported by the Russian Foundation for Fundamental Research under the grant N 00-01-00500.

References

1. V.G. Drinfeld, *Leningrad Math. J.*, **1**, 1419 (1990).
2. O.V. Ogievetsky, *Suppl. Rendiconti Cir. Math.* Palermo, Serie II **37**, 185 (1993).
3. P.P. Kulish, V.D. Lyakhovsky, A.I. Mudrov, "Extended jordanian twists for Lie algebras", *J. Math. Phys.*, **40**, 4569 (1999); (math.QA/9806014).
4. P.P. Kulish, V.D. Lyakhovsky, M.A. del Olmo, *Journ. Phys. A: Math. Gen.*, **32**, 8671 (1999); (math.QA/9908061).
5. V.D. Lyakhovsky, M.A. del Olmo, *Journ. Phys. A: Math. Gen.*, **32**, 4541 (1999); (math.QA/9811153).
6. V.D. Lyakhovsky, M.A. del Olmo, "Chains of twists and induced deformations", *Czech. Journ. Phys.*, **50**, 129 (2000).

A REMARK ON KAPPA-DEFORMED SELF-DUAL PHASE-SPACE[a]

A. NOWICKI
Institute of Physics, Pedagogical University
pl. Słowiański 6, 65-029 Zielona Góra, Poland

We define the phase space as a cross product algebra extending κ-deformed self-dual momentum and position spaces.

1 Introduction

Deformations of space-time symmetries are extensively investigated in last years. In this approach the notion of symmetries is generalized to quantum groups [1] i.e. Hopf algebras and in consequence we deal with noncommuting space-time. Having a deformed configuration space, using the notion of duality for Hopf algebras we can build dual momentum space, which has also a Hopf algebra structure.
For a given configuration space one can define a momentum space using the concept of duality. It appears that such a momentum space has also a Hopf algebra structure.
It is known that from a pair of dual Hopf algebras describing the quantum symmetry group and its dual quantum Lie algebra one can construct *cross product algebra*, [1] which is not a Hopf algebra, however it is still associative algebra.

This construction was presented for the κ-deformed space-time symmetry in Lukierski and Nowicki. [2] In this case, the commuting momentum space and noncommuting configuration space are not self-dual as in the nondeformed case.
In this note, we extend both κ-deformed momentum and configuration spaces in a self-dual manner and then, using cross product algebra construction we define self-dual phase space.

2 Self–Dual Configuration and Momentum Spaces

One can extend the four-dimensional κ-deformed momentum and position (*space-time*) spaces in such a way, that both spaces are self-dual under the following duality relation (x_μ and p_ν are position and momentum generators):

$$< x_\mu, p_\nu > = \; -i\hbar g_{\mu\nu} \,, \qquad g_{\mu\nu} = (-1,1,1,1) \,, \tag{1}$$

[a]The paper is supported by KBN grant 2P03B130.12

where: $\hbar$ is the Planck's constant.
Introducing the four real parameters $\alpha, \tilde{\alpha}, \beta, \tilde{\beta} \in \mathbb{R}$, we can describe these spaces as the Hopf algebras by the relations:

$\mathcal{X}$-configuration space:

$$[x_0, x_k] = i\hbar(\tilde{\beta} - \tilde{\alpha})\, x_k\,, \tag{2a}$$

$$\Delta(x_0) = x_0 \otimes 1 + 1 \otimes x_0\,, \tag{2b}$$

$$\Delta(x_k) = x_k \otimes e^{\alpha x_0} + e^{\beta x_0} \otimes x_k\,, \tag{2c}$$

$\mathcal{P}$-momentum space:

$$[p_0, p_k] = i\hbar(\beta - \alpha)\, p_k\,, \tag{3a}$$

$$\Delta(p_0) = p_0 \otimes 1 + 1 \otimes p_0\,, \tag{3b}$$

$$\Delta(p_k) = p_k \otimes e^{\tilde{\alpha} p_0} + e^{\tilde{\beta} p_0} \otimes p_k\,. \tag{3c}$$

It is obvious that choosing the parameters properly we obtain momentum and position spaces of κ-Poincaré deformation. For example, using the deformation parameter κ with the mass dimension and the light velocity c we get:

(a) κ - deformed configuration and momentum spaces in the bicrossproduct basis [3]

$$\alpha = \beta = \tilde{\alpha} = 0, \quad \tilde{\beta} = \frac{1}{\kappa c}, \tag{4a}$$

(b) κ - deformed configuration and momentum spaces in the standard basis [4]

$$\alpha = \beta = 0, \quad \tilde{\alpha} = -\tilde{\beta} = \frac{1}{2\kappa c}, \tag{4b}$$

(c) trivially deformed momentum space with commuting, standard configuration space

$$\alpha = \beta = 0, \quad \tilde{\alpha} = \tilde{\beta} = \frac{1}{2\kappa c}. \tag{4c}$$

3 Phase Space as a Cross Product Algebra

In order to construct a deformed phase space one has to extend the commutation relations (2) and (3) by adding a cross commutators between $\mathcal{X}$ and $\mathcal{P}$. In this way a phase space $\Pi \sim \mathcal{X} \otimes \mathcal{P}$ as a vector space and becomes an associative algebra.
We find the cross commutators using the notion of a left (right) *cross product algebra.*

It is obvious that the algebra $\mathcal{P}$ is a left $\mathcal{X}$-module if we introduce the following *left action* (we use the Sweedler's notation)

$$\triangleright : \mathcal{P} \otimes \mathcal{X} \to \mathcal{X} : p \otimes x \to p \triangleright x = < p, x_{(2)} > x_{(1)} \,. \tag{5}$$

We recall the following definition [1]

Let $\mathcal{P}$ be a Hopf algebra and $\mathcal{X}$ a left $\mathcal{P}$-module algebra. Left *cross product algebra* $\mathcal{X} \rtimes \mathcal{P}$ is a vector space $\mathcal{X} \otimes \mathcal{P}$ with product

$$(x \otimes p)(\tilde{x} \otimes \tilde{p}) = x(p_{(1)} \triangleright \tilde{x}) \otimes p_{(2)}\tilde{p} \qquad (left\ cross\ product)\,, \tag{6}$$

with unit element $1 \otimes 1$, where $x, \tilde{x} \in \mathcal{X}$ and $p, \tilde{p} \in \mathcal{P}$.

The obvious isomorphism $\mathcal{X} \sim \mathcal{X} \otimes 1$, $\mathcal{P} \sim 1 \otimes \mathcal{P}$ gives us the following cross relations between the position and momentum space

$$p \circ x = (p_{(1)} \triangleright x) \circ p_{(2)} \qquad (left\ cross\ product)\,, \tag{7}$$

which one can rewrite as a cross commutation relations $[x, p] = x \circ p - p \circ x$.

From the relations (2), (3) and the cross relations (6) it follows the commutation relations for the phase space $\Pi \sim \mathcal{X} \rtimes \mathcal{P}$

$$[p_0, p_k] = i\hbar(\beta - \alpha)p_k\,, \tag{8a}$$

$$[x_0, x_k] = i\hbar(\tilde{\beta} - \tilde{\alpha})x_k\,, \tag{8b}$$

$$[x_k, p_l] = i\hbar\delta_{kl}e^{\beta x_0}e^{\tilde{\alpha}p_0} + \left(1 - e^{i\hbar\alpha\tilde{\beta}}\right) x_k p_l\,, \tag{8c}$$

$$[p_k, x_0] = i\hbar\tilde{\beta}\, p_k\,, \tag{8d}$$

$$[x_k, p_0] = -i\hbar\alpha\, x_k\,, \tag{8e}$$

$$[x_0, p_0] = -i\hbar\,. \tag{8f}$$

Denoting the deformed commutator as $[a, b]_q = ab - qba$ the relation (8c) takes the form

$$[x_k, p_l]_q = i\hbar\delta_{kl}qe^{\beta x_0}e^{\tilde{\alpha}p_0}\,, \qquad q = e^{-i\hbar\alpha\tilde{\beta}}\,. \tag{9}$$

Introducing the dilatation generator

$$D = x_i p_i\,, \tag{10a}$$

we can rewrite (8c) as follows

$$[x_k, p_k] = i\hbar \left\{ e^{\beta x_0} e^{\tilde{\alpha} p_0} + \left(1 - e^{i\hbar\alpha\tilde{\beta}}\right) D \right\}, \qquad \text{no summation}, \qquad (10b)$$

$$x_k p_l = e^{-i\hbar\alpha\tilde{\beta}} p_l x_k , \qquad k \neq l . \qquad (10c)$$

We see that depending on the parameters $\alpha, \tilde{\beta}$ the dilatation generator D can or not appear in phase space structure.
Further we consider two possibilities $\alpha = 0$ or $\tilde{\beta} = 0$ i.e. the generator D disappears.

Introducing additional time parameter τ, we distinguish the following *self-dual* phase spaces:

(a) *standard form of nondeformed covariant phase space*

$$\alpha = \tilde{\alpha} = \beta = \tilde{\beta} = 0 , \qquad (11)$$

(b) *deformed phase spaces with noncommuting position and momentum*

the case (i)

$$\alpha = \tilde{\beta} = 0, \quad \tilde{\alpha} = \pm\frac{1}{\kappa c}, \quad \beta = \pm\frac{1}{\tau c} , \qquad (12a)$$

defined by the relations ($x_0 = ct, \quad p_0 = E/c$):

- *configuration space*

$$[t, x_k] = \mp\frac{i\hbar}{\kappa c^2} x_k \quad \Delta(t) = t \otimes 1 + 1 \otimes t \quad \Delta(x_k) = x_k \otimes 1 + e^{\pm\frac{t}{\tau}} \otimes x_k , \quad (12b)$$

- *momentum space*

$$[E, p_k] = \pm\frac{i\hbar}{\tau} p_k \quad \Delta(E) = E \otimes 1 + 1 \otimes E \quad \Delta(p_k) = p_k \otimes e^{\pm\frac{E}{\kappa c^2}} + 1 \otimes p_k , \quad (12c)$$

- *cross relations*

$$[x_k, p_l] = i\hbar\delta_{kl} e^{\pm\frac{t}{\tau}} e^{\pm\frac{E}{\kappa c^2}} \qquad [t, E] = -i\hbar , \qquad (12d)$$

the case (ii)

$$\alpha = \tilde{\alpha} = 0, \qquad \beta = \pm\frac{1}{\tau c}, \qquad \tilde{\beta} = \pm\frac{1}{\kappa c} , \qquad (13a)$$

defined by the relations:

- *configuration space*

$$[t, x_k] = \pm\frac{i\hbar}{\kappa c^2} x_k \quad \Delta(t) = t \otimes 1 + 1 \otimes t \quad \Delta(x_k) = x_k \otimes 1 + e^{\pm\frac{t}{\tau}} \otimes x_k , \quad (13b)$$

- *momentum space*

$$[E, p_k] = \pm\frac{i\hbar}{\tau}p_k \quad \Delta(E) = E\otimes 1 + 1\otimes E \quad \Delta(p_k) = p_k\otimes 1 + e^{\pm\frac{E}{\kappa c^2}}\otimes p_k\,, \tag{13c}$$

- *cross relations*

$$[x_k, p_l] = i\hbar\delta_{kl}e^{\pm\frac{t}{\tau}} \quad [p_k, t] = \pm\frac{i\hbar}{\kappa c^2}p_k \qquad [t, E] = -i\hbar\,, \tag{13d}$$

the case (iii)

$$\beta = \tilde{\beta} = 0, \qquad \alpha = \pm\frac{1}{\tau c}, \qquad \tilde{\alpha} = \pm\frac{1}{\kappa c}\,, \tag{14a}$$

defined by the relations:

- *configuration space*

$$[t, x_k] = \mp\frac{i\hbar}{\kappa c^2}x_k \quad \Delta(t) = t\otimes 1 + 1\otimes t \quad \Delta(x_k) = x_k\otimes e^{\pm\frac{t}{\tau}} + 1\otimes x_k\,, \tag{14b}$$

- *momentum space*

$$[E, p_k] = \mp\frac{i\hbar}{\tau}p_k \quad \Delta(E) = E\otimes 1 + 1\otimes E \quad \Delta(p_k) = p_k\otimes e^{\pm\frac{E}{\kappa c^2}} + 1\otimes p_k\,, \tag{14c}$$

- *cross relations*

$$[x_k, p_l] = i\hbar\delta_{kl}e^{\pm\frac{E}{\kappa c^2}} \quad [x_k, E] = \mp\frac{i\hbar}{\tau c^2}x_k \qquad [t, E] = -i\hbar\,. \tag{14d}$$

It is worthwhile to notice that the construction of self-dual phase space requires the introduction of two dimensionful parameters κ and τ. Therefore, one parameter deformed phase spaces can not contain a self-dual configuration and momentum spaces.

Using κ parameter ($\tau = \infty$) we get the following well known realizations:

(i) κ - deformed phase space in the bicrossproduct basis

$$\alpha = \beta = \tilde{\alpha} = 0, \quad \tilde{\beta} = \frac{1}{\kappa c}\,, \tag{15a}$$

(ii) κ - deformed phase space in the standard basis

$$\alpha = \beta = 0, \quad \tilde{\alpha} = -\tilde{\beta} = \frac{1}{2\kappa c}\,, \tag{15b}$$

(iii) κ - trivially deformed phase space with commuting configuration space

$$\alpha = \beta = 0, \quad \tilde{\alpha} = \tilde{\beta} = \frac{1}{2\kappa c}. \tag{15c}$$

If we use τ parameter ($\kappa = \infty$), one can introduce the analog of (i) - τ - *deformed phase space*

$$\tilde{\alpha} = \tilde{\beta} = \alpha = 0, \quad \beta = \pm\frac{1}{\tau c}, \tag{16a}$$

defined by the relations:

- *configuration space*

$$[t, x_k] = 0 \quad \Delta(t) = t \otimes 1 + 1 \otimes t \quad \Delta(x_k) = x_k \otimes 1 + e^{\pm\frac{t}{\tau}} \otimes x_k\,, \tag{16b}$$

- *momentum space*

$$[E, p_k] = \mp\frac{i\hbar}{\tau}p_k \quad \Delta(E) = E \otimes 1 + 1 \otimes E \quad \Delta(p_k) = p_k \otimes 1 + 1 \otimes p_k\,, \tag{16c}$$

- *cross relations*

$$[x_k, p_l] = i\hbar\delta_{kl}e^{\pm\frac{t}{\tau}} \quad [x_0, p_0] = -i\hbar\,. \tag{16d}$$

The next step of our investigations is to consider the quantum Fourier transform between noncommuting, self-dual configuration and momentum spaces $\mathcal{X}$ and $\mathcal{P}$. Similar problem, in the case of self-dual, twodimensional $\mathcal{X}$ and $\mathcal{P}$ spaces was described in Majid and Oeckl [5] as the quantum Fourier transform on the Planck scale Hopf algebra. This extension of quantum Fourier transform for our algebras (2) and (3) is now under consideration.

References

1. S. Majid, "*Foundations of Quantum Group Theory*", Cambridge University Press, Cambridge 1995.
2. J. Lukierski, A. Nowicki, "Heisenberg Double Description of κ-Poincaré Algebra and κ-deformed Phase Space" in: *Proceedings of XXI International Colloquium on Group Theoretical Methods in Physics*, Goslar 1996.
3. S. Majid, H. Ruegg, *Phys. Lett.* **B 334**, 348 (1994).
4. J. Lukierski, A. Nowicki, H. Ruegg, *Phys. Lett.* **B 293**, 419 (1993).
5. S. Majid, R. Oeckl, "Twisting of Quantum Differentials and the Planck Scale Hopf Algebra", Preprint DAMTP/98-118, math.QA/9811054.

Particles, Strings, Superbranes and their SUSY Extensions

ON ACTION FUNCTIONALS FOR INTERACTING BRANE SYSTEMS[a]

I. BANDOS
Institute for Theoretical Physics,
NSC Kharkov Institute of Physics and Technology,
310108, Kharkov, Ukraine
e-mail: bandos@hep.itp.tuwien.ac.at
bandos@kipt.kharkov.ua

W. KUMMER
Institut für Theoretische Physik,
Technische Universität Wien,
Wiedner Hauptstrasse 8-10, A-1040 Wien
e-mail: wkummer@tph.tuwien.ac.at

We present an action functional and derive equations of motion for a coupled system of a bosonic Dp–brane and an open string ending on the Dp-brane. With this example we address the key issues of the recently proposed method [1,2] for the construction of manifestly supersymmetric action functionals for interacting superbrane systems. We clarify, in particular, how the arbitrariness in sources localized on the intersection [1,2] is related to the standard description of the flat D-branes as rigid planes where the string endpoints 'live'.

Introduction

Recently an approach to obtain a supersymmetric action functional for a coupled system of interacting superbranes has been proposed. [1,2] The aim of the study [1,2] was to construct a basis for a future search for new solitonic solutions of worldvolume equations of superbranes and of supergravity equations with a complicated system of brane sources.

The system of open fundamental superstring ending on a super–Dp-brane[3] was considered [1,2] as a generic case of interacting brane system in the frame of the 'brane democracy' concept. [4] On the other hand, the manifestly supersymmetric description of such system can be useful in a search for a quasiclassical effective action of the system of coincident branes which are conjectured to carry non-Abelian gauge fields [5] (non-Abelian Dirac–Born–Infeld action [6]).

In this contribution we would like to discuss the key issues of our approach [1,2] by studying a pure bosonic limit of the interacting brane system. We elaborate the case of open bosonic string with endpoints living on the *dynamical* bosonic Dp–brane and discuss the relation of the approach [1,2] with the

[a]Talk given by I. Bandos

standard description [3] where the Dp-branes are treated as rigid p-dimensional hyperplanes.

1 Standard Description of Open String

To clarify the problem and to establish the notations let us begin with the standard consideration of the open bosonic string in flat D–dimensional space–time $\underline{\mathcal{M}}^D$. The action can be written as [7]

$$S_{str} = \frac{1}{2} \int_{\mathcal{M}^2} d^2\xi \sqrt{-g} g^{mn}(\xi) \partial_m \hat{X}^{\underline{m}}(\xi) \partial_n \hat{X}^{\underline{n}}(\xi) \eta_{\underline{mn}} \, . \tag{1}$$

Here $\{(\xi^m)\} = \{(\tau, \sigma)\}$ $(m = 0, 1)$ are local coordinates on the string worldsheet $\mathcal{M}^2$, $\hat{X}^{\underline{m}}(\xi)$ $(\underline{m} = 0, \ldots, (D-1))$ are coordinate (embedding) functions which define (locally) an embedding of the worldsheet $\mathcal{M}^2$ into the D–dimensional space–time (target) $\underline{\mathcal{M}}^D$

$$X^{\underline{m}} = \hat{X}^{\underline{m}}(\xi) : \quad \mathcal{M}^2 \quad \rightarrow \underline{\mathcal{M}}^D \, . \tag{2}$$

The flat metric of $\underline{\mathcal{M}}^D$ is chosen to be 'mostly minus' $\eta_{\underline{mn}} = diag(+, -, \ldots, -)$.

When the worldsheet has a boundary $\partial\mathcal{M}^2$, the variation of the action (1) acquires the form ($d\xi^m \wedge d\xi^n = \varepsilon^{mn} d^2\xi$, $(d\xi)^{\wedge 2} = 1/2 \varepsilon_{mn} d\xi^m \wedge d\xi^n = -d^2\xi$)

$$\begin{aligned} \delta S_{str} \;=\; & \frac{1}{2} \int_{\mathcal{M}^2} d^2\xi \sqrt{-g} \delta g^{mn}(\xi) \left(\partial_m \hat{X}^{\underline{m}} \partial_n \hat{X}_{\underline{m}} - \frac{1}{2} g_{mn} g^{kl} \partial_k \hat{X}^{\underline{m}} \partial_l \hat{X}_{\underline{m}} \right) \\ & - \int_{\mathcal{M}^2} d^2\xi \delta \hat{X}_{\underline{m}} \partial_m \left(\sqrt{-g} g^{mn} \partial_n \hat{X}^{\underline{m}} \right) \\ & - \int_{\partial\mathcal{M}^2} d\xi^m \varepsilon_{mn} \sqrt{-g} g^{nk} \partial_k \hat{X}^{\underline{m}} \delta \hat{X}_{\underline{m}} \, . \end{aligned} \tag{3}$$

The variation with respect to the auxiliary intrinsic metric $g_{mn}(\xi)$ results in the equation

$$\partial_m \hat{X}^{\underline{m}}(\xi) \partial_n \hat{X}_{\underline{m}}(\xi) = \frac{1}{2} g_{mn}(\xi) g^{pq}(\xi) \partial_p \hat{X}^{\underline{m}}(\xi) \partial_q \hat{X}_{\underline{m}}(\xi) \, . \tag{4}$$

One easily finds that the trace part of Eq. (4) is satisfied identically. This is the Noether identity reflecting the gauge Weyl symmetry of the action (1)

$$g_{mn}(\xi) \rightarrow g'_{mn}(\xi) = e^{\Lambda} g_{mn}(\xi) \, , \qquad X^{\underline{m}}(\xi) \rightarrow X^{\underline{m}}(\xi) \, . \tag{5}$$

With this symmetry one can fix the gauge

$$g^{mn}(\xi)\partial_m \hat{X}^{\underline{m}}(\xi)\partial_n \hat{X}_{\underline{m}}(\xi) = 2\,. \tag{6}$$

Then Eq. (4) becomes

$$g_{mn}(\xi) = \partial_m \hat{X}^{\underline{m}}(\xi)\partial_n \hat{X}_{\underline{m}}(\xi)\,, \tag{7}$$

and implies that the metric is induced by the embedding.

The *usual way* to deal with the coordinate variation of the action (1) (presented in the second line of Eq. (3)) is to assume that the *bulk and the boundary inputs should vanish separately*. Then one arrives at the well known result that

- proper equations of motion have the form

$$\partial_m \left(\sqrt{-g} g^{mn}(\xi)\partial_n \hat{X}^{\underline{m}}(\xi)\right) = 0\,; \tag{8}$$

- the boundary conditions should be chosen in a way which provides

$$d\xi^m \varepsilon_{mn} \sqrt{-g} g^{nk} \partial_k \hat{X}^{\underline{m}} \delta \hat{X}_{\underline{m}}|_{\partial\mathcal{M}^2:\ \xi^m=\tilde{\xi}^m(\tau)} = 0\,, \tag{9}$$

  where the functions $\tilde{\xi}^m(\tau)$ define parametrically the 'embedding' of (a connected piece of) the world sheet boundary $\partial\mathcal{M}^2$ into the worldsheet

$$\xi^m = \tilde{\xi}^m(\tau): \qquad \partial\mathcal{M}^2 \quad \rightarrow \mathcal{M}^2\,. \tag{10}$$

To satisfy Eq. (9) in a simple way one can assume that for some part $\hat{X}^i(\xi)$ of the embedding functions

$$\hat{X}^{\underline{m}} = (\hat{X}^a, \hat{X}^i), \quad a = 0, \ldots p, \quad i = (p+1), \ldots, (D-1)\,, \tag{11}$$

the Neumann boundary conditions are imposed

$$d\xi^m \varepsilon_{mn} \sqrt{-g} g^{nk} \partial_k \hat{X}^a|_{\xi^m=\tilde{\xi}^m(\tau)} = 0, \quad \Leftrightarrow \quad \partial_\perp \hat{X}^a|_{\partial\mathcal{M}^2} = 0\,, \tag{12}$$

while the remaining equations from the set (9) are satisfied due to the Dirichlet boundary conditions. The latter can be treated as ones imposed *on the variation* $\delta\hat{X}^i(\tilde{\xi}^m(\tau)) = 0$ or, equivalently, as the boundary conditions imposed on the coordinate functions $\hat{X}^i(\xi)$ directly

$$\hat{X}^i(\tilde{\xi}^m(\tau)) = \tilde{X}^i(= const) \qquad \Leftrightarrow \qquad \delta\hat{X}^i(\tilde{\xi}^m(\tau)) = 0\,. \tag{13}$$

Note that *for the Dirichlet boundary problem the equations (8) appear formally only on the proper open subset of the worldsheet, i.e. only outside the boundary*, because the variation on the boundary vanishes just due to Eq. (13). In the Neumann problem we can regard Eq. (8) as valid on the whole worldsheet. But in this case we shall assume that the boundary conditions (12) *are imposed after the variation of the action has been performed* (as we did not restrict the variations by the condition (12) when derived the equations of motion).

The boundary conditions (12), (13) certainly break Lorentz invariance for $p \neq (D-1)$. Such a breaking reflects the existence of some $d = (p+1)$-dimensional defects where the string endpoints move. These are worldvolumes of Dp-branes [3] considered as flat hyperplanes.

2 Extended Variational Problem

However, we can proceed in a different manner which we will call *extended variational problem* or *extended variational approach*.

Let us introduce the current density distribution with support on the boundary of the worldsheet ($d\xi^m \wedge d\xi^n = d^2\xi\epsilon^{mn}$)

$$j_1 = d\xi^m \varepsilon_{mn} \int_{\partial\mathcal{M}^2} d\tilde{\xi}^n(\tau)\delta^2\left(\xi - \tilde{\xi}(\tau)\right) = \varepsilon_{mn} d\xi^m j^n \, . \tag{14}$$

Then one easily find that

$$\int_{\mathcal{M}^2} j_1 \wedge \hat{\mathcal{A}}_1 = \int_{\partial\mathcal{M}^2} \tilde{\hat{\mathcal{A}}}_1 \, , \tag{15}$$

holds for any one-form $\hat{\mathcal{A}}_1 = d\xi^m \mathcal{A}_m(\xi)$ defined on the worldsheet, $\tilde{\hat{\mathcal{A}}}_1 = d\tilde{\xi}^m(\tau)\mathcal{A}_m(\tilde{\xi}(\tau))$.

With the use of j_1 the coordinate variation of the string action can be written in the form

$$\begin{aligned} \delta S_{str} &= -\int_{\mathcal{M}^2} \left(d^2\xi\partial_m \left(\sqrt{-g}g^{mn}\partial_n \hat{X}^{\underline{m}}\right) \right. \\ &\quad \left. - j_1 \wedge d\xi^m \varepsilon_{mn}\sqrt{-g}g^{nk}\partial_k \hat{X}^{\underline{m}}\right) \delta\hat{X}_{\underline{m}} \, . \end{aligned} \tag{16}$$

The equations of motion are (7) and

$$d^2\xi\partial_m\left(\sqrt{-g}g^{mn}(\xi)\partial_n\hat{X}^{\underline{m}}(\xi)\right) = j_1 \wedge d\xi^m \varepsilon_{mn}\sqrt{-g}g^{nk}\partial_k\hat{X}^{\underline{m}} \tag{17}$$

$$\Leftrightarrow \quad \partial_m \left(\sqrt{-g} g^{mn} \partial_n \hat{X}^{\underline{m}}\right) = - \int\limits_{\partial \mathcal{M}^2} d\tilde{\xi}^m \varepsilon_{mn} \sqrt{-g} g^{nk} \partial_k \hat{X}^{\underline{m}} \delta^2 \left(\xi - \tilde{\xi}(\tau)\right) .$$

Thus no boundary conditions appear, but the equations of motion acquire a source localized at the boundary of the worldsheet. The result (17) of such an *'extended variational problem'* looks different from the standard one (8), (12), (13).

To clarify the situation let us, for the moment, fix the conformal gauge and accept the parametrization where the string endpoints correspond to the values $\sigma = 0$ and $\sigma = \pi$. Then the standard approach produces the free (Laplace) equations with boundary conditions

$$(\partial_\tau^2 - \partial_\sigma^2)\hat{X}^a(\tau,\sigma) = 0 , \qquad \partial_\sigma \hat{X}^a|_{\mathcal{M}^2} = 0 , \tag{18}$$

$$(\partial_\tau^2 - \partial_\sigma^2)\hat{X}^i(\tau,\sigma) = 0 , \quad \hat{X}^i(\tau,0) = \tilde{X}^i_0 , \quad \hat{X}^i(\tau,\pi) = \tilde{X}^i_\pi , \tag{19}$$

while the extended variational problem results in the equations with sources for both types of the coordinate functions

$$(\partial_\tau^2 - \partial_\sigma^2)\hat{X}^a(\tau,\sigma) = \partial_\sigma \hat{X}^a \left(\delta(\sigma) - \delta(\sigma - \pi)\right) , \tag{20}$$

$$(\partial_\tau^2 - \partial_\sigma^2)\hat{X}^i(\tau,\sigma) = \partial_\sigma \hat{X}^i \left(\delta(\sigma) - \delta(\sigma - \pi)\right) . \tag{21}$$

It is evident that the problem (18) can be obtained as a particular case of (20) because, after imposing an additional condition $\partial_\sigma \hat{X}^a|_{\mathcal{M}^2} = 0$, the r.h.s. of Eq. (20) vanishes. As in such a way we arrive at the free equations on the whole worldsheet, this corresponds just to the result of Neumann variation problem with the supposition that the boundary conditions are imposed *after* the variation has been performed (i.e. imposed 'by hand').

This does *not* hold for (19), because $\hat{X}^i|_{\mathcal{M}^2} = const$ does not imply $\partial_\sigma \hat{X}^i|_{\mathcal{M}^2} = 0$. Thus the value of Laplace operator acting on the coordinate function *is not indefinite* on the boundary, as it is in the classical Dirichlet problem, but is determined by the value of the derivative $\partial_\sigma \hat{X}^i|_{\mathcal{M}^2}$.

The fact that the straightforward application of the extended variational approach does not produce the Dirichlet boundary problem even as a particular case does not look not so surprising if one remembers that it assumes that the variations of the coordinate functions are unrestricted everywhere, while the Dirichlet boundary conditions (13) come just as restrictions on the variations.

3 Lagrange Multiplier Approach for Dirichlet Problem

If one would like to reproduce the Dirichlet problem (19) from the extended variational approach, one can try to incorporate the boundary conditions (13) with Lagrange multiplier one-form $P_1^i \equiv d\tau \mathcal{P}^i(\tau)$ into the action written in the conformal gauge $g_{mn} = \partial_m X^{\underline{m}} \partial_n X_{\underline{m}} = \eta_{mn} \equiv diag(+1,-1)$ (cf. Ref.[1])

$$\begin{aligned} S_{str} &= \frac{1}{2} \int_{\mathcal{M}^2} d^2\xi \left(\partial_m \hat{X}^a \partial^m \hat{X}_a - \partial_m \hat{X}^i \partial^m \hat{X}^i \right) \\ &+ \int_{\partial\mathcal{M}^2} d\tau \mathcal{P}_i \left(\hat{X}^i(\xi(\tau)) - \tilde{X}^i \right) . \end{aligned} \tag{22}$$

Then the Dirichlet boundary conditions are produced by the variation with respect to Lagrange multiplier $\mathcal{P}^i(\tau)$ or, more precisely, two Lagrange multipliers $\mathcal{P}_i^0(\tau)$ and $\mathcal{P}_i^\pi(\tau)$ (each living on the worldline of the corresponding endpoint of the string), while the dynamical equations for the coordinate functions $\hat{X}^i$ read

$$(\partial_\tau^2 - \partial_\sigma^2)\hat{X}^i(\tau,\sigma) = \left(\partial_\sigma \hat{X}^i - \mathcal{P}_i^0\right)\delta(\sigma) - \left(\partial_\sigma \hat{X}^i - \mathcal{P}_i^\pi\right)\delta(\sigma - \pi) . \tag{23}$$

What one can see from the very beginning is the arbitrariness in the Lagrangian multipliers $\mathcal{P}_i^{0,\pi}(\tau)$ which cannot be fixed by any gauge symmetry of the action. So one can conclude that the Lagrange multipliers carry some degrees of freedom and some doubts may arise concerning the applicability of the Lagrangian multiplier method to the problem.

However if one notes that
i) Eq. (23) becomes free when considered outside the boundary (i.e. on the open proper subset of the string worldsheet),
ii) the boundary conditions (13) are produced as equations of motion for Lagrange multipliers,
iii) the action of the Laplace operator on the coordinate functions X^i is *indefinite* on the boundary just due to the *arbitrariness* of the Lagrange multiplier, one easily concludes that the Dirichlet boundary problem is indeed reproduced.

When one can find a solution for it (as it can be done in the case of string), then, substituting it into the equations, one can fix the value of the Lagrange multiplier $\mathcal{P}^i(\tau)$.

This way to reproduce the Dirichlet boundary problem could be regarded as an artificial one. However it just provides the possibility to describe the system of open (super)string and *dynamical* (super-)Dp–brane and, more generally, of open (super–)brane(s) ending on closed 'host' (super–)brane(s) on the level of (quasi)classical action functional.[1]

4 Dynamical String with the Ends on Dynamical D–Brane

The action for free Dp-brane [3,8] has the form

$$S_{Dp} = \int_{\mathcal{M}^{p+1}} d^{p+1}\zeta \sqrt{|G|}, \qquad |G| = (-)^p det\,(G_{\tilde{m}\tilde{n}})\,, \tag{24}$$

where

$$G_{\tilde{m}\tilde{n}} \equiv \partial_{\tilde{m}} \tilde{X}^{\underline{m}}(\zeta) \partial_{\tilde{n}} \tilde{X}_{\underline{m}}(\zeta) - \mathcal{F}_{\tilde{m}\tilde{n}} \tag{25}$$

is the non-symmetric 'open string induced metric', [9] the coordinate functions $\hat{X}^{\underline{m}}(\zeta^{\tilde{m}})$ define (locally) an embedding of the $d = (p+1)$-dimensional world-volume $\mathcal{M}^{p+1} = \{(\zeta^{\tilde{m}})\}$ ($\tilde{m} = 0, \ldots, p$) of the Dp-brane into the D(=10)–dimensional space–time $\underline{\mathcal{M}}^D$

$$X^{\underline{m}} = \tilde{X}^{\underline{m}}(\zeta): \qquad \mathcal{M}^{p+1} \quad \Rightarrow \quad \underline{\mathcal{M}}^D, \tag{26}$$

$$\mathcal{F} \equiv dA = \frac{1}{2} d\zeta^{\tilde{m}} d\zeta^{\tilde{n}} \mathcal{F}_{\tilde{n}\tilde{m}} \tag{27}$$

is the field strength of the worldvolume gauge field of the Dp–brane $A = d\zeta^{\tilde{m}} A_{\tilde{m}}(\zeta)$.

With this notation the variation of the *free* Dp-brane action (24) can be written in the form

$$\begin{aligned}\delta S_{Dp} \;\; = \;\; & + \int_{\mathcal{M}^{p+1}} d^{p+1}\zeta \partial_{\tilde{m}} \left(\sqrt{|G|} G^{-1[\tilde{m}\tilde{n}]} \right) \delta A_{\tilde{n}}(\zeta) \\ & - \int_{\mathcal{M}^{p+1}} d^{p+1}\zeta \partial_{\tilde{m}} \left(\sqrt{|G|}\, G^{-1(\tilde{m}\tilde{n})} \partial_{\tilde{n}} \tilde{X}^{\underline{m}} \right) \delta \tilde{X}_{\underline{m}}\,, \end{aligned} \tag{28}$$

where

$$G^{-1\;mn} = G^{-1(mn)} + G^{-1[mn]}$$

is the matrix inverse to the open string metrics (25).

To describe the interacting system of the open string and the *dynamical* D-brane on the level of action functional one has to provide the counterpart of (13) with *coordinate functions of D-brane* instead of constants $\tilde{X}^i$. Namely we have to impose an identification

$$\hat{X}^{\underline{m}} \left(\hat{\xi}(\tau) \right) = \tilde{X}^{\underline{m}} \left(\hat{\zeta}(\tau) \right), \tag{29}$$

where $\zeta^{\tilde{m}} = \hat{\zeta}^{\tilde{m}}(\tau)$ defines an embedding of the worldline(s) of string endpoints (i.e. the boundary of the worldsheet) into the worldvolume of the Dp-brane

$$\zeta^{\tilde{m}} = \hat{\zeta}^{\tilde{m}}(\tau): \qquad \partial\mathcal{M}^2 \quad \Rightarrow \quad \mathcal{M}^{p+1}\,. \tag{30}$$

The problem is that, when (29) is taken into account, the variations $\delta\hat{X}^{\underline{m}}$ and $\delta\tilde{X}^{\underline{m}}|_{\zeta=\hat{\zeta}(\tau)}$ cannot be treated as independent on the intersection $\mathcal{M}^{1+p}\cap\mathcal{M}^2=\partial\mathcal{M}^2$. Indeed, varying (29) one finds

$$\delta\hat{X}^{\underline{m}}|_{\xi=\tilde{\xi}(\tau)}+\delta\tilde{\xi}^{m}(\tau)\partial_m\hat{X}^{\underline{m}}\left(\tilde{\xi}(\tau)\right)=\delta\tilde{X}^{\underline{m}}|_{\zeta=\hat{\zeta}(\tau)}+\delta\hat{\zeta}^{\tilde{m}}(\tau)\partial_{\tilde{m}}\tilde{X}^{\underline{m}}\left(\hat{\zeta}(\tau)\right). \tag{31}$$

A method to implement (29) into the action principle was proposed in Ref.[1] It consists in imposing the identification (29) with Lagrange multipliers one-form $P_1^{\underline{m}}=d\tau\mathcal{P}^{\underline{m}}(\tau)$ into the action (cf. with (22)). Thus the *action for a coupled system* of open string and *dynamical* Dp-brane can be written as

$$S=S_{str}+S_{Dp}+S_{int}\,. \tag{32}$$

Here S_{str} is determined by (1) but with *open* worldsheet whose boundary is a set of two (or more!) worldlines which are embedded into the worldvolume of the Dp–brane (30). S_{Dp} is determined by Eq. (24). The last term in (32)

$$S_{int}=\int\limits_{\partial\mathcal{M}^2}\left(\hat{A}+P_{1\underline{m}}\left(\hat{X}^{\underline{m}}(\tilde{\xi}(\tau))-\tilde{X}^{\underline{m}}(\hat{\zeta}(\tau))\right)\right), \tag{33}$$

describes, in particular, the interaction of the string endpoints with the worldvolume gauge field of the Dp–brane $A=d\zeta^{\tilde{m}}A_{\tilde{m}}(\zeta)$ whose pull-back on the intersection is $\hat{A}=d\hat{\zeta}^{\tilde{m}}(\tau)A_{\tilde{m}}(\hat{\zeta}(\tau))=d\tau\partial_\tau\hat{\zeta}^{\tilde{m}}A_{\tilde{m}}(\hat{\zeta})$.

The variation of the interaction term is

$$\begin{aligned}\delta S_{int}=&\int\limits_{\partial\mathcal{M}^2}\delta P_{1\underline{m}}\left[\hat{X}^{\underline{m}}(\tilde{\xi}(\tau))-\tilde{X}^{\underline{m}}(\hat{\zeta}(\tau))\right]\\&+\int\limits_{\partial\mathcal{M}^2}P_{1\underline{m}}\left[\delta\hat{X}^{\underline{m}}(\tilde{\xi}(\tau))-\delta\tilde{X}^{\underline{m}}(\hat{\zeta}(\tau))\right]\\&+\int\limits_{\partial\mathcal{M}^2}d\hat{\zeta}^{\tilde{n}}(\tau)\delta A_{\tilde{n}}\left(\hat{\zeta}(\tau)\right)\\&+\int\limits_{\partial\mathcal{M}^2}\delta\hat{\zeta}^{\tilde{m}}(\tau)\left[i_{\tilde{m}}\mathcal{F}-P_{1\underline{m}}\partial_{\tilde{m}}\tilde{X}^{\underline{m}}\right]\\&+\int\limits_{\partial\mathcal{M}^2}\delta\tilde{\xi}^{m}(\tau)\left[P_{1\underline{m}}\partial_m\hat{X}^{\underline{m}}-d\xi^p\varepsilon_{pn}\sqrt{-g}g^{nk}\partial_k\hat{X}^{\underline{m}}\partial_m\hat{X}_{\underline{m}}\right].\end{aligned} \tag{34}$$

The forth and the fifth lines include the variation with respect to coordinate functions $\hat{\zeta}^{\tilde{m}}(\tau)$ and $\tilde{\xi}^{m}(\tau)$ defining the embedding of the string boundary worldlines in the Dp–brane worldvolume (30) and in the worldsheet (10).

To vary the whole action it is convenient to introduce the following current density distribution form (see Bandos and Kummer[1] and refs. therein)[b]

$$j_p = d\zeta^{\wedge p}_{\tilde{m}} \int\limits_{\partial\mathcal{M}^{p+1}} d\hat{\zeta}^{\tilde{m}}(\tau)\delta^{p+1}\left(\zeta - \hat{\zeta}(\tau)\right) = (-)^p d\zeta^{\wedge p}_{\tilde{n}} j^{\tilde{n}} \,. \tag{35}$$

As $j_p \wedge d\zeta^{\tilde{m}} = d^{p+1}\zeta \; j^{\tilde{m}}$, it is easy to see that

$$\int\limits_{\mathcal{M}^{1+3}} j_3 \wedge \tilde{A}_1 = \int\limits_{\partial\mathcal{M}^{1+1}} \hat{A}_1 \,. \tag{36}$$

With the use of (35) and the worldsheet current density distribution (14) one can write the variation of the action (32) in the form $(G_{\tilde{m}\tilde{n}} \equiv \partial_{\tilde{m}}\tilde{X}^{\underline{m}}(\zeta)\partial_{\tilde{n}}\tilde{X}_{\underline{m}}(\zeta) - \mathcal{F}_{\tilde{m}\tilde{n}}$ (25), $G^{-1\tilde{m}\tilde{q}}G_{\tilde{q}\tilde{n}} = \delta^{\tilde{m}}_{\tilde{n}})$

$$\begin{aligned}
\delta S \;=\; & \frac{1}{2}\int\limits_{\mathcal{M}^2} d^2\xi\sqrt{-g}\delta g^{mn}(\xi)\left(\partial_m\hat{X}^{\underline{m}}\partial_n\hat{X}_{\underline{m}} - \frac{1}{2}g_{mn}g^{kl}\partial_k\hat{X}^{\underline{m}}\partial_l\hat{X}_{\underline{m}}\right) \\
& - \int\limits_{\mathcal{M}^2} \delta\hat{X}_{\underline{m}}\left[d^2\xi\partial_m\left(\sqrt{-g}g^{mn}\partial_n\hat{X}^{\underline{m}}\right)\right. \\
& \left. - j_1 \wedge \left(P^{\underline{m}}_1 + d\xi^m\varepsilon_{mn}\sqrt{-g}g^{nk}\partial_k\hat{X}^{\underline{m}}\right)\right] \\
& + \int\limits_{\partial\mathcal{M}^2} \delta P_{1\underline{m}}\left[\hat{X}^{\underline{m}}(\tilde{\xi}(\tau)) - \tilde{X}^{\underline{m}}(\hat{\zeta}(\tau))\right] \\
& - \int\limits_{\mathcal{M}^{p+1}} d^{p+1}\zeta\left[\partial_{\tilde{m}}\left(\sqrt{|G|}G^{-1[\tilde{m}\tilde{n}]}\right) - j^{\tilde{n}}\right]\delta A_{\tilde{n}}(\zeta) \\
& - \int\limits_{\mathcal{M}^{p+1}} \left[d^{p+1}\zeta\partial_{\tilde{m}}\left(\sqrt{|G|}G^{-1(\tilde{m}\tilde{n})}\partial_{\tilde{n}}\tilde{X}^{\underline{m}}\right) + j_p \wedge P^{\underline{m}}_1\right]\delta\tilde{X}_{\underline{m}} \,.
\end{aligned} \tag{37}$$

Then the equations of motion become evident. They are

- The Born–Infeld equation (with G_{mn} defined by (25))

$$\partial_m\left(\sqrt{|G|}G^{-1[\tilde{m}\tilde{n}]}\right) = j^{\tilde{n}} \equiv \int\limits_{\partial\mathcal{M}^2} d\hat{\zeta}^{\tilde{n}}(\tau)\delta^{p+1}\left(\zeta - \hat{\zeta}(\tau)\right) . \tag{38}$$

---

[b]In our conventions $d\zeta^{\tilde{m}_1} \wedge \ldots \wedge d\zeta^{\tilde{m}_{p+1}} = \varepsilon^{\tilde{m}_1\ldots\tilde{m}_{p+1}}(d\zeta)^{\wedge(p+1)} \equiv (-)^p\varepsilon^{\tilde{m}_1\ldots\tilde{m}_{p+1}}d^{p+1}\zeta$, $d\zeta^{\wedge p}_{\tilde{m}_1} \equiv \varepsilon_{\tilde{m}_1\tilde{m}_2\ldots\tilde{m}_{p+1}}d\zeta^{\tilde{m}_2} \wedge \ldots \wedge d\zeta^{\tilde{m}_{p+1}}, \ldots$

It includes the definite source localized on the intersection (and thus outside the intersection the equation is free).

- Equations for the Dp-brane coordinate functions (with G_{mn} defined by Eq. (25)) [c]

$$\partial_m \left(\sqrt{|G|} G^{-1(\tilde{m}\tilde{n})} \partial_{\tilde{n}} \tilde{X}^{\underline{m}} \right) = - \int_{\partial \mathcal{M}^2} d\tau \mathcal{P}^{\underline{m}}(\tau) \delta^{p+1} \left(\zeta - \hat{\zeta}(\tau) \right) . \quad (39)$$

- The equation for the string coordinate functions (with $g_{mn}(\xi)$ defined by (7) or, equivalently, by (4) and the Weyl symmetry (5))

$$d^2\xi \partial_m \left(\sqrt{-g} g^{mn}(\xi) \partial_n \hat{X}^{\underline{m}}(\xi) \right) = j_1 \wedge \left(P_1^{\underline{m}} + d\xi^m \varepsilon_{mn} \sqrt{-g} g^{nk} \partial_k \hat{X}^{\underline{m}} \right) . \quad (40)$$

5 Reparametrization Symmetry of the Coupled System

The variations of the action (32) with respect to the worldvolume and world-sheet embedding coordinate functions $\delta\hat{\zeta}(\tau)$ (30) and $\tilde{\xi}^m(\tau)$ (10) is

$$\begin{aligned} \delta_{\tilde{\xi},\hat{\zeta}} S &= \int_{\partial \mathcal{M}^2} \delta\hat{\zeta}^{\tilde{m}}(\tau) \left[i_{\tilde{m}} \mathcal{F} - P_{1\underline{m}} \partial_{\tilde{m}} \tilde{X}^{\underline{m}} \right] \\ &+ \int_{\partial \mathcal{M}^2} \delta\tilde{\xi}^m(\tau) \left[P_{1\underline{m}} \partial_m \hat{X}^{\underline{m}} - d\tilde{\xi}^p(\tau) \varepsilon_{pn} \sqrt{-g} g^{nk} \partial_k \hat{X}^{\underline{m}} \partial_m \hat{X}_{\underline{m}} \right] . \end{aligned} \quad (41)$$

The corresponding equations of motion

$$i_{\tilde{m}} \mathcal{F}|_{\partial \mathcal{M}^2} = P_{1\underline{m}} \partial_{\tilde{m}} \tilde{X}^{\underline{m}}(\zeta(\tau)) , \quad (42)$$

$$d\hat{\xi}^p(\tau) \varepsilon_{pn} \sqrt{-g} g^{nk} \partial_k \hat{X}^{\underline{m}} \partial_m \hat{X}_{\underline{m}} = P_{1\underline{m}} \partial_m \hat{X}^{\underline{m}} , \quad (43)$$

are dependent.

To prove the dependence of Eq. (43) one should contract Eq. (40) with $\partial_l \hat{X}_{\underline{m}}(\xi)$ on the target space indices. Then, writing the left hand part as

$$\partial_m \left(\sqrt{-g} g^{mn}(\xi) \partial_n \hat{X}^{\underline{m}}(\xi) \right) \partial_l \hat{X}^{\underline{m}}(\xi) \quad (44)$$

---

[c] To arrive at the expression for the r.h.s. one should perform some formal extension of the property (36), as the form $P_1 = d\tau \mathcal{P}(\tau)$ has not been defined as a pull–back of any 1-form on $\mathcal{M}^{p+1}$. Let us assume that such a form $\tilde{P}_{1\underline{m}} = d\zeta^{\tilde{m}} \tilde{\mathcal{P}}_{\underline{m}}$ is introduced. Then, by definition, $\tilde{P}_{1\underline{m}}(\hat{\zeta}(\tau)) \equiv d\tau \partial_\tau \hat{\zeta}^{\tilde{m}} \tilde{\mathcal{P}}_{\underline{m}}(\hat{\zeta}(\tau)) = d\tau \mathcal{P}_{\underline{m}}(\tau)$ and one finds $\int_{\mathcal{M}^{p+1}} j_p \wedge \tilde{P}_{1\underline{m}} = \int_{\partial \mathcal{M}^2} d\tau \mathcal{P}_{\underline{m}}(\tau) . \delta^{p+1}(\zeta - \hat{\zeta}(\tau))$.

$$= \partial_m \left(\sqrt{-g} g^{mn}(\xi) \partial_n \hat{X}^{\underline{m}}(\xi) \partial_l \hat{X}^{\underline{m}}(\xi)\right) - \left(\sqrt{-g} g^{mn}(\xi) \partial_n \hat{X}^{\underline{m}}(\xi)\right) \partial_l \partial_m \hat{X}^{\underline{m}}(\xi)$$

and using Eq. (4) (which remains the same for the coupled system), one obtains (in arbitrary space-time dimension D)

$$d^2\xi \sqrt{-g} \frac{D-2}{4} \partial_l \left(\sqrt{-g} g^{mn}(\xi) \partial_m \hat{X}^{\underline{m}}(\xi) \partial_n \hat{X}^{\underline{m}}(\xi)\right) \tag{45}$$

$$= - j_1 \wedge \left(P_1^{\underline{m}} + d\xi^m \varepsilon_{mn} \sqrt{-g} g^{nk} \partial_k \hat{X}^{\underline{m}}\right).$$

However, using the Weyl symmetry, one can fix the gauge (6), where the l.h.s. of Eq. (45) vanishes, and we arrive at (43).

To prove the dependence of Eq. (42) one should consider the longitudinal part of Eq. (39), i.e the result of the contraction of Eq. (39) with $\partial_{\tilde{m}} \tilde{X}^{\underline{m}}$ on the target space indices

$$\partial_m \left(\sqrt{|G|} G^{-1(\tilde{m}\tilde{n})} \partial_{\tilde{n}} \tilde{X}^{\underline{m}}\right) \partial_{\tilde{l}} \tilde{X}_{\underline{m}} = -p^{\underline{m}} \partial_{\tilde{l}} \tilde{X}_{\underline{m}} \, . \tag{46}$$

Here we, for shortness, introduced the notation $\int_{\partial \mathcal{M}^2} d\tau \mathcal{P}^{\underline{m}}(\tau) \delta^{p+1}\left(\zeta - \hat{\zeta}(\tau)\right)$ $= p^{\underline{m}}$. Dealing with the l.h.s. of Eq. (46) in the same way as in (44) and using the identities

$$\delta^{\tilde{m}}_{\tilde{n}} = G^{-1(\tilde{m}\tilde{k})} \partial_{\tilde{k}} \tilde{X}^{\underline{m}} \partial_{\tilde{n}} \tilde{X}_{\underline{m}} + G^{-1[\tilde{m}\tilde{k}]} F_{\tilde{k}\tilde{n}} \, ,$$

$$G^{-1[\tilde{m}\tilde{k}]} \partial_{\tilde{k}} \tilde{X}^{\underline{m}} \partial_{\tilde{n}} \tilde{X}_{\underline{m}} = -G^{-1(\tilde{m}\tilde{k})} F_{\tilde{k}\tilde{n}} \, ,$$

(which follow from $G^{-1\tilde{m}\tilde{k}} G_{\tilde{k}\tilde{n}} = \delta^{\tilde{m}}_{\tilde{n}}$) one arrives at

$$\partial_m \left(\sqrt{|G|} G^{-1(\tilde{m}\tilde{n})} \partial_{\tilde{n}} \tilde{X}^{\underline{m}}\right) \partial_{\tilde{l}} \tilde{X}_{\underline{m}} = -\partial_m \left(\sqrt{|G|} G^{-1[\tilde{m}\tilde{k}]}\right) F_{\tilde{k}\tilde{n}} \, . \tag{47}$$

Substituting Eqs. (46) and (38) we obtain from Eq. (47)

$$\int_{\partial \mathcal{M}^2} \left(-i_{\tilde{m}} F_2 + P_{1\underline{m}} \partial_{\tilde{m}} \tilde{X}^{\underline{m}}(\zeta(\tau))\right) \delta^{p+1} \left(\zeta - \hat{\zeta}(\tau)\right) = 0. \tag{48}$$

Eq. (48) is equivalent to (42).

The dependence of the equations (42), (43) should be regarded as Noether identities which reflect some gauge symmetries of the action.

Reversing the arguments, we can consider (42), (43) as independent equations, but conclude that the longitudinal parts of Eqs. (39), (40) are dependent as they are in the case of the free string and Dp-brane. From such point of view we can state that the dependence of Eq. (43) and Eq. (42) reflects the reparametrization symmetries of the string worldsheet and the Dp-brane worldvolume which, hence, survive when coupling is 'switched on'.

6 Concluding Remarks

Thus, applying the method of Ref. [1] we obtained the equations of motion (38) -(40), (29) for the coupled system of an open bosonic string and a dynamical bosonic Dp–brane where the string endpoints live. In such a way we illustrated the key issues of the approach [1,2] and we are able to comment on its relation to the standard description of Dp–branes as rigid planes. [3]

The investigation of Eqs. (38) -(40), (29) and their supersymmetric generalizations will be the subject of the forthcoming paper. Note that such a study simplifies essentially when the Lorentz–harmonic formulations of the (super–) string [10] and (super–)Dp–brane actions [11] are used instead of (1), (24).

Acknowledgments

The authors are grateful to D. Sorokin and M. Tonin for interest in this work and many useful conversations. One of the authors (I.B.) thanks Padova Section of INFN for hospitality in Padova, where a part of this work has been done. A partial support from the INTAS Grant **96-308** and from the Ukrainian GKNT Grant **2.5.1/52** is acknowledged.

References

1. I. Bandos and W. Kummer, *Phys. Lett.* **B462**, 254 (1999), (hep-th/9905144).
2. I. Bandos and W. Kummer, *Superstring 'ending' on super-D9-brane: a supersymmetric action functional for the coupled brane system*, *Preprint*, TUW/99-11, hep-th/9906041, *Nucl. Phys.* **B** (in press).
3. J. Polchinski, *Phys. Rev. Lett.* **75**, 4724 (1995); *Tasty lectures on D–branes*, hep-th/9611050 *and references therein.*
4. P.K. Townsend, *Brane democracy*, hep-th/9507048; *Phys. Lett.* **B373**, 68 (1998).
5. E. Witten, *Nucl. Phys.* **B443**, 85 (1995); **B460**, 335 (1996).
6. A. Tseytlin, *Nucl. Phys.* **B501**, 41-52 (1997); *Born-Infeld action, supersymmetry and string theory*, hep-th/9908105.
7. M. Green, J. Schwarz and E. Witten Superstring theory. V. 1,2. CUP, 1987.
8. M. Cederwall, A. von Gussich, B.E.W. Nilsson, A. Westerberg, *Phys. Lett.* **B390**, 148 (1997), hep-th/9606173, *Nucl. Phys.* **B490**, 163 (1997), hep-th/9611159;
   M. Aganagic, C. Popescu, J.H. Schwarz, *Phys. Lett.* **B393**, 311 (1997), hep-th/9610249;

M. Cederwall, A. von Gussich, B.E.W. Nilsson, P. Sundell and A. Westerberg, *Nucl. Phys.* **B490**, 179 (1997), hep-th/9610148;
E. Bergshoeff, P.K. Townsend, *Nucl. Phys.* **B490**, 145 (1997), hep-th/9611173;
M. Aganagic, C. Popescu and J.H. Schwarz, *Nucl. Phys.* **B490**, 202 (1997), hep-th/9612080.

9. N. Seiberg and E. Witten, *JHEP* **9909**, 032 (1999), hep-th/9908142.
10. I. A. Bandos and A. A. Zheltukhin, *JETP. Lett.* **54**, 421 (1991); *Phys. Lett.* **B288**, 77 (1992); *Phys. Part. Nucl.* **25**, 453 (1994); *Class. Quantum Grav.* **12**, 609 (1995).
11. I. Bandos, D. Sorokin and M. Tonin, *Nucl. Phys.* **B497**, 275 (1997).

THE $OSp(1|4)$ SUPERPARTICLE AND EXOTIC BPS STATES[a]

I. BANDOS
Institute for Theoretical Physics,
National Science Center
"Kharkov Institute of Physics and Technology", Kharkov, 61108, Ukraine

J. LUKIERSKI[b]
Institute for Theoretical Physics University of Wroclaw,
50-204 Wroclaw, Poland

D. SOROKIN [c]
INFN, Sezione Di Padova
Via F. Marzolo, 8, 35131 Padova, Italia

We discuss the dynamics of a superparticle in a superspace whose isometry is generated by the superalgebra $OSp(1|4)$ or its central–charge contraction. Extra coordinates of the superspace associated with tensorial central charges are shown to describe spin degrees of freedom of the superparticle, so quantum states form an infinite tower of (half)–integer helicities. A peculiar feature of the model is that it admits BPS states which preserve 3/4 of target–space supersymmetries.

In this contribution we present results of work done on studying superparticle models whose symmetry and physical properties are defined by a supersymmetric algebra extended by tensorial charges. [1,2,3]

A motivation for this study has been to understand the physical meaning of tensorial "central" charges when they are associated not with superbranes but with relativistic (point–like) particles.

As we shall see, in the case of superparticles tensorial central charges have different physical meaning than that of superbranes. They correspond to spin degrees of freedom of the superparticles, while it is well known that in the case of branes tensorial charges describe the coupling of branes to tensor gauge fields of target–space supergravity. That is, brane tensorial charges are similar to electric and magnetic charges of particles.

For instance, a membrane in $D = 4$ or $D = 11$ target space couples to a three–form gauge field $A_{mnp}(x)$. In the membrane worldvolume action [4] this

---

[a]Talk given by D. Sorokin

[b]Supported in part by KBN grant 2P03B130.12

[c]On leave from Kharkov Institute of Physics and Technology, Kharkov, 61108, Ukraine

coupling is described by the Wess–Zumino term

$$S_{coupl} = \int_{\mathcal{M}_3} dx^m dx^n dx^p A_{mnp}(x)\,, \tag{1}$$

where the integral is taken over the membrane worldvolume $\mathcal{M}_3$ of the pull back of the gauge field. As was shown in de Azcarraga et al.,[5] the membrane charge associated with this coupling is a two–rank tensor of the form

$$Z^{mn} = \int_{\mathcal{M}_2} dx^m \wedge dx^n\,, \tag{2}$$

where the integral is taken over the two–dimensional surface of the membrane.

Using the supermembrane action[4] de Azcarraga et al.[5] derived the form of the superalgebra generated by the Noether charges of the supermembrane. The supertranslation algebra thus obtained was shown to contain the contribution of the membrane charge (2) to the r.h.s. of the anticommutator of the fermionic supercharges

$$\{Q_\alpha, Q_\beta\} = -2(C\gamma_m)_{\alpha\beta} P^m + (C\gamma_{mn})_{\alpha\beta} Z^{mn}\,, \tag{3}$$

where P_m is the standard momentum (or bosonic translation) generator.

Thus, the worldvolume actions for the superbranes imply that the underlying supersymmetry of superbranes is described by a superalgebra extended by tensorial central charges.[5,6,7]

Recent analysis carried out in Craps et al., and Ferrara et al.[8,9] has demonstrated that when superbranes propagate in anti–de–Sitter superbackgrounds, the superalgebra of their Noether supercharges gets extended to a corresponding maximal OSp superalgebra.

For instance, for branes in $D = 10$ and $D = 11$ target superspaces with AdS geometry the Noether charges generate the $OSp(1|32)$ superalgebra. In Bars[10] it has been assumed that the underlying superalgebra of M–theory should be even larger, namely, $OSp(1|64)$ which is the minimal simple superalgebra which contains the supertranslation algebra of M–theory with a two–form and a five–form central charge as a *subsuperalgebra*.

If the $OSp(1|32)$ and $OSp(1|64)$ supergroup are related to M–theory, it seems instructive to find and study simple dynamical systems whose properties would be governed by $OSp(1|2n)$ supergroups. And this has been another motivation for our work.

To simplify consideration let us take the $OSp(1|4)$ supergroup which is the isometry of $N = 1, D = 4$ superspace having the four–dimensional AdS_4 space as a bosonic subspace (see Bandos et al.[3] for the generic $OSp(1|2n)$ case). The $OSp(1|4)$ superalgebra has the following form

$$[M_{ab}, M_{cd}] = i(\eta_{ad}M_{bc} + \eta_{bc}M_{ad} - \eta_{ac}M_{bd} - \eta_{bd}M_{ac}), \tag{4}$$

$$[M_{ab}, P_c] = i(\eta_{bc}P_a - \eta_{ac}P_b), \tag{5}$$

$$[P_a, P_b] = \frac{i}{R^2}M_{ab}, \tag{6}$$

$$\{Q_\alpha, Q_\beta\} = -2(C\gamma^a)_{\alpha\beta}P_a + \frac{1}{R}(C\gamma^{ab})_{\alpha\beta}M_{ab},$$

$$[M_{ab}, Q_\alpha] = -\frac{i}{2}Q_\beta\,(\gamma_{ab})^\beta{}_\alpha, \quad \gamma_{ab} = \frac{1}{2}(\gamma_a\gamma_b - \gamma_b\gamma_a), \tag{7}$$

$$[P_a, Q_\alpha] = -\frac{i}{2R}Q_\beta\,(\gamma_a)^\beta{}_\alpha. \tag{8}$$

The generators M_{ab} form the $SO(1,3)$ subalgebra (4), and M_{ab} and P_a (a=0,1,2,3) generate the $SO(2,3) \sim Sp(4)$ subalgebra of $OSp(1|4)$. Q_α are four Majorana spinor generators of $OSp(1|4)$. The parameter R is the AdS_4 radius, and $C_{\alpha\beta}$ is a charge conjugation $Sp(4)$–invariant matrix.

The $N = 1, D = 4$ supertranslation algebra with tensorial central charges (3) is a contraction of the $OSp(1|4)$ superalgebra carried out in the following way. One rescales the $SO(1,3)$ generators as

$$M_{ab} = RZ_{ab}, \tag{9}$$

and takes the limit $R \to \infty$, which corresponds to the limit where AdS_4 becomes asymptotically flat. Then Z_{ab} and P_a become Abelian and commute with Q_α, and the anticommutator of Q_α reduces to that in (3).

Consider now a simple "oscillator" realization of $OSp(1|4)$. For this let us combine P_a and M_{ab} into a symmetric spin–tensor generator

$$M_{\alpha\beta} = -2(C\gamma_a)_{\alpha\beta}P^a + (C\gamma_{ab})_{\alpha\beta}Z^{ab}, \tag{10}$$

so that

$$[M_{\alpha\beta}, M_{\gamma\delta}] = -\frac{4i}{R}[C_{\gamma(\alpha}M_{\beta)\delta} + C_{\delta(\alpha}M_{\beta)\gamma}],$$
$$[M_{\alpha\beta}, Q_\gamma] = -\frac{4i}{R}C_{\gamma(\alpha}Q_{\beta)}, \quad \{Q_\alpha, Q_\beta\} = M_{\alpha\beta}. \tag{11}$$

If we introduce a Grassmann–even spinor operator λ_α, which forms the Heisenberg algebra

$$[\lambda_\alpha, \lambda_\beta] = \frac{2i}{R} C_{\alpha\beta}\,, \tag{12}$$

and a Grassmann–odd scalar ψ such that

$$\psi^2 = \frac{1}{2}, \qquad \psi\lambda_\alpha - \lambda_\alpha\psi = 0\,, \tag{13}$$

we can realize $M_{\alpha\beta}$ and Q_α as follows

$$M_{\alpha\beta} = \frac{1}{2}(\lambda_\alpha\lambda_\beta + \lambda_\beta\lambda_\alpha)\,, \qquad Q_\alpha = \lambda_\alpha\psi\,. \tag{14}$$

Using the commutation relations for λ_α and ψ it is not hard to check that Q_α and $M_{\alpha\beta}$ represented in this way generate the $OSp(1|4)$ superalgebra, $\mathcal{Z}_\mathcal{A} \equiv (\lambda_\alpha, \psi)$ forming the fundamental representation of $OSp(1|4)$.

Note that in the limit $R \to \infty$, λ_α become commuting quantities (see (12)). Then we observe that

$$P_a = \lambda\gamma_a\lambda\,, \tag{15}$$

in which one recognizes the Cartan–Penrose relation implying that

$$P_a P^a \equiv 0\,, \tag{16}$$

due to $D = 4$ γ–matrix identities. We thus assume that at $R \to \infty$ the "oscillator" realization of $OSp(1|4)$ may correspond to a massless $D = 4$ superparticle with λ_α playing the role of a twistor–like variable.

Before contraction such a superparticle propagates in the supergroup manifold $OSp(1|4)$ parametrized by coordinates x^a, y^{ab} and θ^α associated, respectively, with the generators P_a, M_{ab} and Q_α.

An action for this superparticle can be constructed in a way similar to that used for the first time by Ferber [11] for developing the supertwistor formulation of supersymmetric field theories.

In our case the superparticle worldline $Z^M(\tau) = (x^a(\tau), y^{ab}(\tau), \theta^\alpha(\tau))$ on the $OSp(1|4)$ manifold is parametrized by the time variable τ. To construct the action we pick the worldline pull back of left–invariant $OSp(1|4)$ Cartan one–forms taking values in the $SO(2,3)$ subalgebra of $OSp(1|4)$

$$\Omega^{\alpha\beta}(Z) = dZ^M(\tau)\Omega_M^{\alpha\beta}(Z)\,, \tag{17}$$

and contract the spinor indices with *commuting* spinor variables $\lambda_\alpha(\tau)$ which are classical counterparts of the spinor operators (12) used to realize the generators (14) of $OSp(1|4)$. λ_α become non–commutative upon solving for second–class constraints of the model and passing from Poisson to Dirac brackets.

The action we thus obtain has the twistor–like form

$$S = \int d\tau \lambda_\alpha \lambda_\beta \partial_\tau Z^M \Omega_M^{\alpha\beta} \,. \tag{18}$$

By construction (18) is invariant under the $OSp(1|4)$ transformations of the coordinates $Z^M = (x^a, y^{ab}, \theta^\alpha)$ and under the $SO(2,3)$ subgroup of $OSp(1|4)$ acting in the tangent space of the supermanifold $OSp(1|4)$. To better understand the symmetry structure of the action (18) we note that the isometry of the supergroup manifold $OSp(1|4)$ is the direct product of two supergroups $OSp(1|4)_L \times OSp(1|4)_R$. By using in (18) the $OSp(1|4)$–left–invariant Cartan forms corresponding only to $SO(2,3)_R$ we break "right–acting" $OSp(1|4)_R$ down to $SO(2,3)$.

To be able to analyze the action (18) one should know an explicit expression for the Cartan forms $\Omega^{\alpha\beta}$. This can be obtained by substituting an appropriate parametrization of the $OSp(1|4)$ supergroup element $G(Z)$ into the definition of the Cartan forms

$$\Omega = -iG^{-1}dG \equiv \Omega^{\alpha\beta}(Z) M_{\alpha\beta} + E^\alpha(Z) Q_\alpha \,, \tag{19}$$

where E^α is a spinorial Cartan form associated with the supercharge generators.

In Bandos et al. [3] we have found a parametrization of $G(Z)$ which allowed us to obtain simpler expressions for the Cartan forms of $OSp(1|4)$, and generically of $OSp(1|2n)$, than those derived in earlier papers. [12,13,14] We have got

$$\begin{aligned} \Omega^{\alpha\beta} &= v^\alpha{}_\gamma(y) v^\beta{}_\delta(y) \Big[(\gamma_a)^{\gamma\delta} e^a(x) + R(\gamma_{ab})^{\gamma\delta} \omega^{ab}(x) \\ &\quad + \theta^{(\gamma} \mathcal{D} \theta^{\delta)} + (dvv^{-1})^{\gamma\delta} \Big] \,, \end{aligned} \tag{20}$$

where $v^\alpha{}_\gamma(y)$ are $SO(1,3)$ matrices in the spinor representation, $e^a(x)$ and $\omega^{ab}(x)$ are, respectively, the vierbein and the spin connection on the bosonic coset space $AdS_4 = \frac{SO(2,3)}{SO(1,3)}$, and $\mathcal{D} = d + \frac{1}{2}\omega^{ab}(x)\gamma_{ab} + \frac{1}{2R} e^a(x)\gamma_a$ is the covariant AdS_4 differential.

Substituting (20) into the superparticle action (18) and making the redefinition of $\lambda_\alpha(\tau)$

$$\Lambda_\alpha(\tau) = \lambda_\beta v^\beta{}_\alpha(y) \,, \tag{21}$$

we rewrite the action in the following form

$$S = \int_{\mathcal{M}_\tau} \Big[\Lambda \gamma_a \Lambda (e^a - i\theta\gamma^a \mathcal{D}\theta) + \Lambda \gamma_{ab} \Lambda$$

$$\times \left(R\omega^{ab} + \frac{i}{2}\theta\gamma^{ab}\mathcal{D}\theta + tr(dvv^{-1}\gamma^{ab}) \right) \Bigg] . \qquad (22)$$

Note that the $SO(1,3)$ coordinates y^{ab} enter this action only through the last term dvv^{-1}. All other terms depend on the bosonic x^a and fermionic θ^α coordinates of the coset superspace $\frac{OSp(1|4)}{SO(1,3)}$ whose bosonic subspace is AdS_4.

If in (22) we skip the terms with $\Lambda\gamma_{ab}\Lambda$ we will get the action describing a massless superparticle propagating in the AdS_4 superspace. [3]

On the other hand, if we take the limit $R \to \infty$ then, as we have discussed above, the $OSp(1|4)$ superalgebra gets contracted to the super–Poincaré algebra with the tensorial central charge, and the $OSp(1|4)$ superparticle action reduces to the action which describes a superparticle propagating in $N = 1, D = 4$ flat superspace (x^a, θ^α) extended by tensorial coordinates y^{ab} associated now with central charge momentum generators Z_{ab} (9). The superparticle action takes the form

$$S = \int \left[\Lambda\gamma_a\Lambda(dx^a - i\theta\gamma^a d\theta) + \Lambda\gamma_{ab}\Lambda(dy^{ab} + i\theta\gamma^{ab}d\theta) \right] . \qquad (23)$$

This action, which by construction obeys supersymmetry with tensorial charges, was proposed in Bandos and Lukierski. [1] The massless $N = 1, D = 4$ superparticle described by this action possesses quite unusual features.

One of them is that the action is invariant under fermionic (so called κ–symmetry) transformations with *three* independent parameters. The κ–symmetry transformations have the following form

$$\delta_\kappa\theta^\alpha = \kappa^I(\tau)\mu_I^\alpha , \qquad (I = 1, 2, 3) ,$$

$$\delta_\kappa x^a = i\theta\gamma^a\delta_\kappa\theta , \qquad \delta_\kappa y^{ab} = -i\theta\gamma^{ab}\delta_\kappa\theta , \qquad (24)$$

where μ_I^α are three linearly independent commuting spinors orthogonal to λ_α, i.e. $\lambda_\alpha\mu_I^\alpha = 0$.

Remember, that standard superparticle and, in general, superbrane actions are invariant under κ–symmetry transformations with the number of independent parameters being half the number of the spinor coordinates of target superspace. So in $N = 1, D = 4$ superspace the standard massless superparticle is invariant under *two* independent κ-symmetries. As was realized in Sorokin et al., [15] in the twistor–like formulation κ-symmetries can be made irreducible and traded for $n = 2$ extended worldline supersymmetry with transformation properties of θ^α and x^a being

$$\delta\theta^\alpha = \epsilon_1(\tau)\lambda^\alpha + \epsilon_2(\tau)(\gamma_5\lambda)^\alpha , \qquad \delta x^a = \theta\gamma^a\delta\theta . \qquad (25)$$

One can easily observe the difference between the transformations (24) and (25). The latter explicitly contain λ^α, while the former involve spinors orthogonal to λ^α.

The invariance of the superbrane action with respect to local fermionic transformations implies that there exist superbrane configurations which preserve the number of target space supersymmetries which is equal to or less than the number of independent κ–symmetries. Such supersymmetric states saturate the Bogomol'nyi–Prassad–Sommerfeld energy bound. Thus in the case of the standard superbranes the number of unbroken target–space supersymmetries is not higher than $\frac{1}{2}$ supersymmetry of the target–space vacuum.

For instance, in the case of the standard $N = 1, D = 4$ massless superparticle only two of four target–space supersymmetries are unbroken. While in the case of the superparticle with tensorial central charges there are BPS superparticle configurations with *three* target–space supersymmetries, i.e. $\frac{3}{4}$ of supersymmetry remain unbroken. [1]

Recently the possibility of the existence of exotic BPS brane configurations preserving more than $\frac{1}{2}$ supersymmetry has been discussed in Gauntlett and Hull. [16] The superparticle model based on the action (23) is an example of such configurations.

This unusual property is also characteristic of the superparticle propagating on the whole supergroup manifold $OSp(1|4)$ described by the action (22). [3] The algebraic reason for such an exotic situation is that, as we have discussed above, the $OSp(1|4)$ superalgebra and its central charge contraction is realized in such a way that the anticommuting Poisson brackets of the Noether supercharges Q_α derived from (22) is equal to the λ_α–spinor belinears

$$\{Q_\alpha, Q_\beta\} = M_{\alpha\beta} = \lambda_\alpha\lambda_\beta \,. \tag{26}$$

The matrix $\lambda_\alpha\lambda_\beta$ is degenerate and has the rank one. Hence only one of four supergenerators Q_α (14) has nonzero anticommutator. To single out this supercharge, let us introduce a basis $(\mu^\alpha, \mu_I^\alpha)$ $(I = 1, 2, 3)$ in the spinor space such that (compare with (24))

$$\mu^\alpha\lambda_\alpha = 1\,, \qquad \mu_I^\alpha\lambda_\alpha = 0\,.$$

Than, in view of (13) and (14),

$$Q = \mu^\alpha Q_\alpha = \psi \quad \Rightarrow \quad Q^2 = \frac{1}{2}$$

corresponds to one broken supersymmetry and three supercharges

$$Q_I = \mu_I^\alpha Q_\alpha \quad \Rightarrow \quad \{Q_I, Q_J\} = \{Q_I, Q\} = 0$$

anticommute with themselves and with Q. Hence, Q_I act trivially on BPS superparticle states and correspond to three unbroken supersymmetries.

Another feature of the superparticle model with the tensorial charge coordinates is the physical meaning of these extra variables. As the Hamiltonian analysis and the quantization of this superparticle model have shown, [2] only one of the six tensorial charge coordinates y^{ab} is independent due to a large number of constraints. This coordinate takes discrete integer values $n = 2s$ and labels half–integer and integer helicities ($s = 0, \pm\frac{1}{2}, \pm 1, ..., \infty$) of massless quantum states of the superparticle in $D = 4$ space–time.

Let us consider this in more detail. In the Weyl representation of spinors

$$\lambda_\alpha = (\lambda_A, \bar\lambda^{\dot A}), \qquad \theta_\alpha = (\theta_A, \bar\theta^{\dot A}), \qquad A = 1, 2; \qquad \dot A = 1, 2,$$

the action (23) takes the form

$$\begin{aligned} S &= \int \Big[\lambda_A \bar\lambda_{\dot A}(dx^{A\dot A} - i\theta^A d\bar\theta^{\dot A} + id\theta^A \bar\theta^{\dot A}) \\ &\quad + \lambda_A \lambda_B (dy^{AB} - i\theta^A d\theta^A) + c.c.\Big], \end{aligned} \tag{27}$$

where

$$x^{A\dot A} = x^a \sigma_a^{A\dot A}, \qquad y^{AB} = y^{ab}(\sigma_{ab})^{AB}, \qquad y^{\dot A\dot B} = y^{ab}(\bar\sigma_{ab})^{\dot A\dot B},$$

and $\sigma_a^{A\dot A}$ are the Pauli matrices $(\sigma_{ab})_A{}^B = 1/2i\left(\sigma_{aA\dot B}\tilde\sigma_b^{\dot B B} - (a \leftrightarrow b)\right)$.

From (27) we obtain that the canonical momenta associated with $D = 4$ coordinates x^a and tensorial charge coordinates y^{ab} are, respectively,

$$p_{A\dot A} = \lambda_A \bar\lambda_{\dot A}, \tag{28}$$

$$Z_{AB} = \lambda_A \lambda_B, \quad Z_{\dot A\dot B} = \bar\lambda_{\dot A}\bar\lambda_{\dot B}.$$

Notice, in particular, that $p^{A\dot A}p_{A\dot A} \equiv 0$, and, hence, the superparticle is massless.

The Cartan–Penrose relation (28) establishes the correspondence between three independent components of lightlike $p_a = \sigma_a^{A\dot A}p_{A\dot A}$ and three components of $\lambda_A, \bar\lambda_{\dot A}$. Only the phase of λ

$$\lambda_A = e^{i\varphi(\tau)}\lambda_A^0 \tag{29}$$

remains undetermined.

If $y^{ab}=0$, we deal with a twistor superparticle considered in Ferber and Shirafuji,[11,17] its action being

$$S=\int \lambda_A \bar{\lambda}_{\dot{A}}(dx^{A\dot{A}} - i\theta^A d\bar{\theta}^{\dot{A}} + id\theta^A \bar{\theta}^{\dot{A}}). \tag{30}$$

In addition to all symmetries of the action (27), the action (30) is invariant under local $U(1)$ transformations

$$\lambda_A \quad \rightarrow \quad e^{i\varphi(\tau)}\lambda_A\,, \qquad \bar{\lambda}_{\dot{A}} \quad \rightarrow \quad e^{-i\varphi(\tau)}\bar{\lambda}_{\dot{A}}\,. \tag{31}$$

This gauges away the phase component of λ_A and establishes the one–to–one correspondence between the independent components of the twistor superparticle momentum p_a and λ_A.

As the quantization of the action (30) has shown,[11,17] the quantum states of the $N=1, D=4$ twistor superparticle form chiral supermultiplets of physical states with helicity 0 and $\frac{1}{2}$. These supermultiplets are described by chiral superfields $\Phi(x^a - i\theta\sigma^a\bar{\theta}, \theta_A)$.

In the case of the superparticle model (27) with additional tensorial coordinates y^{ab} there is no local $U(1)$ symmetry (31). The compact phase component of λ_A becomes a ***physical*** momentum degree of freedom which corresponds to a single independent component of tensorial charge momenta Z_{ab}.

So the superparticle wave function in the momentum representation now becomes

$$\Phi(p_a,\varphi,\theta^\alpha)=\Sigma_{n=0}^{\infty}\left[e^{in\varphi}\Phi_n(p_a,\theta^\alpha)+e^{-in\varphi}\bar{\Phi}_n(p_a,\theta^\alpha)\right]\,. \tag{32}$$

Integer n is associated with an independent discrete (or quantized) component of the central charge coordinates y^{ab}, which is the Fourier image of the compact phase momentum component φ. This resembles a "dual" Kaluza–Klein effect when instead of a spatial direction, compactified is the corresponding momentum coordinate of the phase space.

In the Lorentz–covariant form the first–quantized wave function of the superparticle in the momentum representation looks as follows[2]

$$\Phi=\Sigma_{n=0}^{\infty}\Phi^{\alpha_1\ldots\alpha_n}(p_m)\lambda_{\alpha_1}\ldots\lambda_{\alpha_n}+\chi\Sigma_{n=0}^{\infty}\Psi^{\alpha_1\ldots\alpha_n}(p_m)\lambda_{\alpha_1}\ldots\lambda_{\alpha_n}\,, \tag{33}$$

where $\chi=\theta^\alpha\lambda_\alpha, \chi^2\equiv 0$. Each component of this series forms an irreducible representation of the Lorentz group $SO(1,3)$ and describes a massless state with integer or half–integer superhelicity depending whether n is even or odd.

Thus, the superparticle model with tensorial charges produces, upon quantization, an infinite tower of massless fields of higher spin. The structure of the wave function which describes higher–spin states is similar to that used in the formalism developed by Vasiliev (for a recent review see Vasiliev [18]) to construct the theory of higher–spin fields. Hence, the model which we have briefly described can be assumed to be a classical counterpart of the field theory of higher spins.

To conclude, we have demonstrated that tensorial charges appearing in extensions of supertranslation algebras may have different meaning than that one got accustomed to in superbrane models. They may describe spinning degrees of freedom of a dynamical system. Superparticle models on supergroup manifolds $OSp(1|2n)$ and their contractions to flat superspaces with tensorial coordinates may give rise to exotic BPS configurations which preserve more than $\frac{1}{2}$ supersymmetry.

We have seen that the AdS_4 space is an intrinsic part of the construction of the superparticle action (22) on $OSp(1|4)$, which upon the contraction describes free higher–spin states. It is well known that to switch on interactions of higher–spin fields one needs the space–time to be of AdS geometry. [18] So an interesting problem to study is whether the superparticle model on $OSp(1|4)$ may help to make a progress in constructing the theory of interacting fields of higher spin.

References

1. I. Bandos and J. Lukierski, Mod. Phys. Lett. **14**, 1257 (1999).
2. I. Bandos, J. Lukierski and D. Sorokin, *Superparticle Models with Tensorial Central Charges*, hep-th/9904109, *Phys. Rev.* **D**, in press.
3. I. Bandos, J. Lukierski, C. Preitschopf and D. Sorokin, *OSp supergroup manifolds, superparticles and supertwistors*, hep-th/9907113, *Phys. Rev.* **D**, in press.
4. E. Bergshoeff, E. Sezgin and P. K. Townsend, *Phys. Lett.* **189B**, 75 (1987); *Ann. Phys.* **185**, 330 (1988).
5. J. A. de Azcarraga et. al., *Phys. Rev. Lett.* **63**, 2443 (1989).
6. D. Sorokin and P.K. Townsend, *Phys. Lett.* **412B**, 265 (1997).
7. H. Hammer, *Nucl. Phys.* **B521**, 503 (1998).
8. B. Craps, J. Gomis, D. Mateos and A. Van Proeyen, JHEP **9904**, 004 (1999).
9. S. Ferrara and M. Porrati, *Phys. Lett.* **B458**, 43 (1999).
10. I. Bars, C. Deliduman and D. Minic, *Phys. Lett.* **B457**, 275 (1999).
11. A. Ferber, *Nucl. Phys.* **B132**, 55 (1977).
12. B. Zumino, *Nucl. Phys.* **B127**, 189 (1977).
13. F. Gürsey and L. Marchildon, *Phys. Rev.* **D17**, 2038 (1978).

14. E. A. Ivanov and A. S. Sorin, *J. Phys. A.: Math. Gen.* **13**, 1159 (1980).
15. D. Sorokin, V. Tkach and D. Volkov, *Mod. Phys. Lett.* **A4**, 901 (1989).
16. J. Gauntlett and C.M. Hull, *BPS States with Extra Supersymmetry*, hep-th/9909098.
17. T. Shirafuji, *Progr. Theor. Phys.* **70**, 18 (1983).
18. M. Vasiliev, *Higher Spin Gauge Theories: Star–Product and AdS Space*, hep-th9910096.

SOLITONS AND EXTENDED OBJECTS

E.A. BERGSHOEFF
Institute for Theoretical Physics,
University of Groningen, Nijenborgh 4,
9747 AG Groningen, The Netherlands

Nonperturbative extended objects in string and/or M-theory manifest themselves as soliton-like solutions to the low-energy supergravity equations of motion. The dynamics of these objects is described by a corresponding worldvolume action. Similarly, intersections of extended objects can be described by soliton-like solutions of the worldbrane equations of motion. In this talk I first review the above sketched relation between solitons and extended objects. Next, as an example, I discuss solitons on the eleven-dimensional supermembrane.

1 Extended Objects

It is a natural question to ask oneselves whether the idea of strings can be extended to higher-dimensional extended objects. In the mid-eighties it was proposed that the eleven-dimensional supergravity, ignored by string theory, was related to the existence of a supermembrane in eleven dimensions.[1] However, the quantization of such a supermembrane turned out to be problematic. In the mid-nineties the supermembrane, and other extended objects, made a comeback via the study of nonperturbative string theory. First of all, it turned out that nonperturbative string theory contains special nonperturbative states, the so-called BPS states, which are p-branes, i.e. extended objects with p spatial extended dimensions. Secondly, the strong coupling limit of IIA supergravity, the low-energy limit of the Type IIA superstring, is given by eleven-dimensional supergravity, which nowadays is considered to be the low-energy limit of 'M-theory' which is probably related to a supermembrane theory in eleven dimensions.

It is well-known that nonperturbative extended objects in string and/or M-theory manifest themselves as soliton-like solutions to the low-energy supergravity equations of motion. It is therefore important to study these soliton-like solutions. An important class of solutions is given by the so-called "two-block" solutions. The relevant part of the Lagrangian carrying the solution is given by (we use the mostly plus signature):

$$\mathcal{L}(D,a,p) = \frac{\sqrt{-g}}{2\kappa_D^2}\left[-R - \frac{4}{D-2}(\partial\phi)^2 - \frac{1}{2(p+2)!}e^{a\phi}F^2_{(p+2)}\right], \qquad (1)$$

where κ_D is the gravitational coupling parameter in D dimensions, $g_{\mu\nu}$ the metric, ϕ the dilaton scalar and $F_{(p+2)}$ the curvature of a (p+1)-form gauge potential.

The metric of a two-block solution is given by

$$\begin{aligned} ds^2 &= H^{\alpha} dx^2_{(p+1)} + H^{\beta} dx^2_{(D-p-1)}\,, \\ e^{\phi} &= H^{\frac{(D-2)a}{4\Delta}}\,, \\ F_{0\cdots pm} &= \pm\sqrt{\frac{4}{\Delta}}\partial_m H^{-1}\,, \end{aligned} \tag{2}$$

with

$$\begin{aligned} \alpha &= \frac{-4(D-p-3)}{(D-2)\Delta}\,, \qquad \beta = \frac{4(p+1)}{(D-2)\Delta}\,, \\ \Delta &= \frac{(D-2)a^2}{8} + 2\frac{(p+1)(D-p-3)}{(D-2)}\,. \end{aligned} \tag{3}$$

The function H is a harmonic function of the transverse directions (with integration constant r_0):

$$H = 1 + \left(\frac{r_0}{r}\right)^{D-p-3}\,, \qquad D-p-3 \neq 0, -2\,. \tag{4}$$

Note that we have taken $g_s = e^{\phi(\infty)} = 1$. The first p+1 directions in the two-block metric (2) are the worldvolume (WV) directions, the remaining D-p-1 ones refer to the transverse directions (see Figure 1).

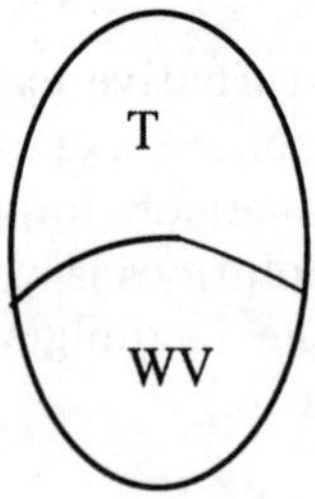

Figure 1: Any two-block solution with metric as given in (2) divides the target spacetime directions into p+1 worldvolume (WV) and D-p-1 transverse (T) directions.

As an example, relevant for the following discussion we give the supermembrane (M2) solution in eleven dimensions:

$$\begin{aligned} ds^2 &= H^{-2/3}dx^2_{(3)} + H^{1/3}dx^2_{(8)}\,, \\ F^{(4)}_{012m} &= \pm\partial_m H^{-1}\,, \qquad\qquad m = 3,4,\cdots,10\,, \end{aligned} \tag{5}$$

with the harmonic function H given by

$$H = 1 + \left(\frac{r_0}{r}\right)^6 . \tag{6}$$

This solution preserves half the supersymmetry with Killing spinor ϵ given by

$$(1 \pm \gamma_{012})\epsilon_\infty = 0\,, \qquad\qquad \epsilon = H^{-1/6}\epsilon_\infty\,. \tag{7}$$

The position of the membrane is summarized by the following figure

$$\text{M2} \;:\; \times|\times\times- - - - - - - - \tag{8}$$

Another solution, not of the two-block kind, that will play an important role in this talk is the M-theory Kaluza-Klein (KK) monopole. The solution has an off-diagonal metric:

$$\begin{aligned} ds^2 &= dx^2_{(7)} + H^{-1}\left(dz + A_i dx^i\right)^2 + H dx_i^2\,, \\ F_{ij} &= \epsilon_{ijk}\partial_k H\,, \qquad\qquad i = 1,2,3, \end{aligned} \tag{9}$$

with H harmonic over the $x_i (i = 1,2,3)$ directions:

$$H = 1 + \frac{1}{r}\,. \tag{10}$$

The solution is summarized by the following figure:

$$\text{KK} \;:\; \times|\times\times\times\times\times\times\; z\; - - - \tag{11}$$

Here z refers to a special U(1) isometry direction in the transverse space. This implies the existence of a spatial (tri-holomorphic) Killing vector $k = \partial/\partial z$ with

$$k^2 = k^\mu k^\nu g_{\mu\nu} \equiv (R/\ell_p)^2 \,. \tag{12}$$

Note that for $r \to 0$ the monopole solution is not singular. Instead the metric is given by (ignoring the first constant term in the harmonic function)

$$ds^2 = dx_{(7)}^2 + \frac{r}{r_0}(dz + \mathbf{A})^2 + \frac{r_0}{r}(dr^2 + r^2 d\Omega_2^2) \,, \tag{13}$$

which, after making the coordinate transformation

$$\rho = 2(r_0 r)^{1/2} \tag{14}$$

can be written in the form

$$ds^2 = dx_{(7)}^2 + d\rho^2 + \tfrac{1}{4}\rho^2 \left(d\Omega_2^2 + \frac{1}{r_0^2}(dz + \mathbf{A})^2 \right) . \tag{15}$$

This is the metric of flat Minkowski space $\mathbb{E}^{1,9}$, and not a cone, provided that

$$\frac{z}{2r_0} \sim \frac{z}{2r_0} + 2\pi \quad \text{or} \quad z \sim z + 4\pi r_0 \,. \tag{16}$$

Assuming that z parametrizes a circle direction with radius R, i.e.

$$z \sim z + 2\pi R \,, \tag{17}$$

we find that the integration constant r_0 in the harmonic H is related to the radius R via

$$r_0 = \tfrac{1}{2} R \,. \tag{18}$$

The KK monopole metric (9) can be written as

$$ds^2 = ds^2(\mathbb{E}^{(1,6)}) + ds_4^2 \,, \tag{19}$$

where ds_4^2 is the Taub-NUT metric

$$ds_4^2 = H^{-1}(dz + \mathbf{A})^2 + H\, ds^2(\mathbb{E}^3) \,. \tag{20}$$

The 1-form $\mathbf{A}$ on $\mathbb{E}^3$ satisfies $\nabla H = \nabla \times \mathbf{A}$, which implies that H is harmonic on $\mathbb{E}^3$. The vector field $k = \partial/\partial z$ is Killing and triholomorphic. The orbits of k are Kaluza-Klein (KK) circles which shrink to points at singularities of H. Let $\mathbf{X}$ be Cartesian coordinates on $\mathbb{E}^3$ and $\mathbf{X}_0$ a constant 3-vector from the origin. The simplest choice of H that will play a role later in this talk is

$$H = 1 + \frac{1}{|\mathbf{X} + \mathbf{X}_0|} + \frac{1}{|\mathbf{X} - \mathbf{X}_0|}, \tag{21}$$

which describes a two-centered KK-monopole of M-theory (see Figure 2).

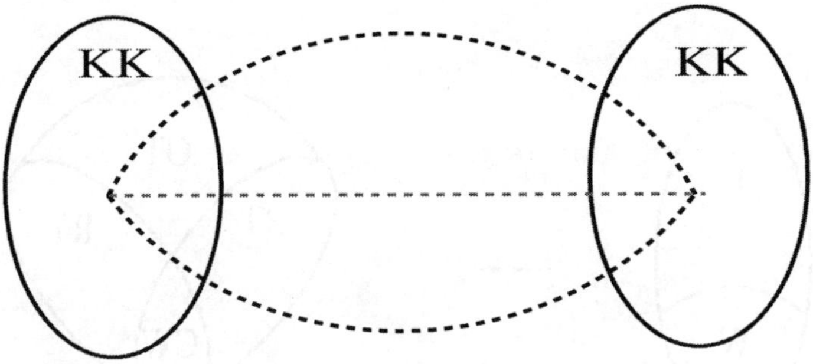

Figure 2: The two-centred Kaluza-Klein monopole solution. The two monopoles, with positions $\mathbf{X} = \pm\mathbf{X}_0$, are indicated by the two ellipses at the left and the right. The distance, in $\mathbb{E}^3$, between the two monopoles is indicated by the horizontal line. The other two curved lines represent the KK circle with shrinking radius at the position of the two monopoles.

2 Worldvolume Solitons

So far, we have discussed branes as solitons of the target space action. The dynamics of these branes is described by a so-called worldvolume action. In a covariant formulation the transverse position of the brane is given by embedding coordinates that are nonzero in both transverse and worldvolume directions. The ones in the worldvolume directions can be gauged away due to the reparametrization invariance on the brane worldvolume. The generic form of the worldvolume action is given by

$$\begin{aligned} S_{\mathrm{WV}} &= \tau_p \int d^{p+1}\xi e^{\alpha\phi} \sqrt{-\mathrm{det}\partial_i X^\mu \partial_j X^\nu g_{\mu\nu}} \\ &\quad + \mu_p \int d^{p+1}\xi \, C^{(p+1)} + \cdots \end{aligned} \tag{22}$$

where τ_p is the brane tension, i.e. the mass per unit p-volume, μ_p the charge per unit p-volume and $C^{(p+1)}$ the gauge field that couples to the p-brane. The parameter α is a dilaton-coupling parameter. In a supersymmetric formulation there are additional fermionic fields.

Given a worldvolume action we can now search for soliton-like solutions corresponding to this action. In the same way as solitons of the target space action can be interpreted as branes moving in the target spacetime, the worldvolume solitons can be interpreted as describing intersections of two branes, see Figure 3.

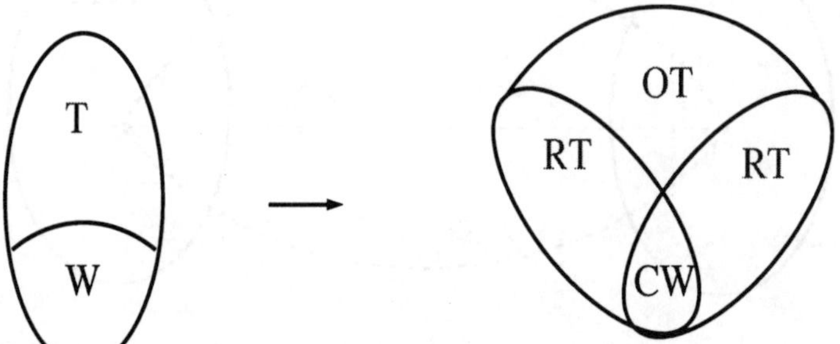

Figure 3: Intersecting branes. Whereas a single brane divides the target space directions into transverse (T) and worldvolume (WV) ones, two intersecting branes lead to four inequivalent directions. The directions that lie in the worldvolume of both branes are called common worldvolume (CW) directions. The directions transverse to both branes are called overall transverse (OT) directions. The directions that are in the worldvolume of one brane but transverse to the other brane are called relative transverse (RT) directions.

The first example of such a worldvolume soliton, is the Dirac Born-Infeld (DBI) soliton constructed in Callan and Maldacena [3] and Gibbons. [4] This solution describes the intersection of a fundamental string F1 on a D3–brane, see Figure 4.

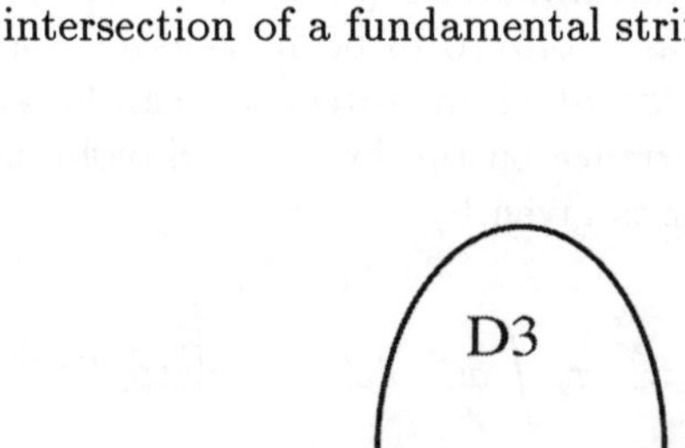

Figure 4: The DBI soliton. This solution describes the intersection of a fundamental string (F1) with a D3–brane.

The solution is carried by a nontrivial BI worldvolume vector, describing the charged endpoint of the fundamental string moving in the D3-brane worldvolume, and one active scalar describing the direction of the fundamental string. The harmonic of the solution only depends on the 3 spacelike directions in the D3-brane worldvolume that are transverse to the particle-like endpoint of the fundamental string. Other worldvolume solitons can be described in a similar way by a worldvolume gauge field and/or a number of active scalars. For instance, the M5-brane has a selfdual string soliton which is carried by the 2-form selfdual worldvolume potential and one active scalar. [5] It has also a 3-brane soliton with two active scalars corresponding to the following intersection: [5]

$$
(3|\mathrm{M5},\mathrm{M5}) \; = \; \left\{ \begin{array}{c|l} \times & \times\times\times\times\times-----\\ \times & \times\times\times--\times\times--- \end{array} \right.
$$

Our goal is to extend the solitons $\leftrightarrow$ branes correspondence to sigma-model lumps in 2+1 dimensions. We first summarize the sigma-model lumps in 2+1 dimensions. They are solutions of the sigma-model action:

$$
\mathcal{L} \;=\; \tfrac{1}{2}\int d^3x\;\eta^{ij}\partial_i X^I \partial_j X^J g_{IJ}(X)\,, \qquad\qquad i=0,1,2 \tag{23}
$$

where g_{IJ} is Kähker. The energy density of a static, i.e. $P_I = 0$, solution is given by

$$
\begin{aligned}
\epsilon \;&=\; \nabla X^I \cdot \nabla X^J g_{IJ}\\
&=\; (\nabla X^I \pm *\nabla X^K I_K{}^I)\,(\nabla X^J \pm *\nabla X^L I_L{}^J)\, g_{IJ} + X^{IJ}\omega_{IJ}
\end{aligned}
$$

where $X^{IJ} \equiv \nabla X^I \times \nabla X^J$. This leads to the BPS bound

$$
\epsilon \geq |X^{IJ}\omega_{IJ}|\,, \tag{24}
$$

which is satisfied by

$$
\nabla X^I = \mp * \nabla X^J I_J{}^I\,. \tag{25}
$$

The hyper-Kähler case is similar. There is now a triplet of Kähler 2-forms and the BPS bound is given by

$$\epsilon \geq |X^{IJ}\omega_{IJ}| . \tag{26}$$

The relevant case to consider is a hyper Kähler background given by the two-centred KK monopole, see Figure 2. In the next section we show that the sigma-model lump can be extended to a worldvolume lump of the full supermembrane action.

3 Solitons on the Supermembrane

We consider the supermembrane in a background for which the 4-form field-strength vanishes and the 11-metric takes the form (19). Our starting point for finding soliton solutions as minimum energy configurations of the supermembrane will be its Hamiltonian formulation.[2] Let $\xi^i = (t, \sigma^a)$ be the worldvolume coordinates, with σ^a the worldspace coordinates, and let X^m be the D=11 spacetime coordinates. The supermembrane Lagrangian, omitting fermions, can then be written as

$$\mathcal{L} = P_m \dot{X}^m - s^a P_m \partial_a X^m - \tfrac{1}{2} v \left[P^2 + \det(g_{ab})\right] , \tag{27}$$

where

$$g_{ab} = \partial_a X^m \partial_b X^n g_{mn} \tag{28}$$

is the induced worldspace metric, P_m is the 11-momentum conjugate to X^m, and s^a and v are Lagrange multipliers. Let $X^m = (Y^i, X^I)$ $(i = 1, 2, 3)$ so that

$$ds^2_{11} = dY^i dY^j \eta_{ij} + dX^I dX^J g_{IJ} , \tag{29}$$

where η is the 3-dimensional Minkowski metric. We make the gauge choice $Y^i(\xi) = \xi^i$. This implies that

$$g_{ab} = \eta_{ab} + \partial_a X^I \partial_b X^J g_{IJ} . \tag{30}$$

It also implies that

$$P_m = \left(-\varepsilon - 1, -P_I \partial_a X^I, P_I\right) \tag{31}$$

where ε is the energy density relative to the brane vacuum (which is taken to have unit tension). The Hamiltonian constraint imposed by v can be solved for ε

$$
\begin{aligned}
(\varepsilon+1)^2 &= 1+\nabla X^I\cdot\nabla X^J g_{IJ}+(g^{IJ}+\nabla X^I\cdot\nabla X^J)P_IP_J \\
&\quad +\tfrac{1}{2}(\nabla X^I\times\nabla X^J)(\nabla X^K\times\nabla X^L)g_{IK}g_{JL}\,,
\end{aligned}
\tag{32}
$$

where we have used standard 2D vector calculus notation for worldspace derivatives. This expression differs in several respects from the corresponding expression for the sigma-model energy density. Firstly, the supermembrane expression is quadratic in ε; this is because the sigma-model approximation is a kind of non-relativistic approximation to the supermembrane (they differ in the same way that the energies of a relativistic and non-relativistic particle differ). Secondly, the supermembrane expression involves terms quartic in derivatives that are absent in the sigma-model case.

We now aim to rewrite the above expression for the energy density in the form [6]

$$
\begin{aligned}
(\varepsilon+1)^2 &= \left(1\pm\tfrac{1}{2}\nabla X^I\times\nabla X^J\omega_{IJ}\right)^2 \\
&\quad +\tfrac{1}{2}\left(\nabla X^I\pm *\nabla X^K I_K{}^I\right)\left(\nabla X^J\pm *\nabla X^L I_L{}^J\right)g_{IJ} \\
&\quad +\tfrac{1}{4}\sum_{r=1}^{R}\left(\nabla X^I\times\nabla X^J\Omega^{(r)}_{IJ}\right)^2\,,
\end{aligned}
\tag{33}
$$

where $*\nabla=(\partial_2,-\partial_1)$ if $\nabla=(\partial_1,\partial_2)$. We assume that $I_I{}^J$ is a complex structure, that the 8-metric g_{IJ} is Hermitian with respect to it and that $\omega_{IJ}=I_I{}^K g_{KJ}$ is the corresponding closed Kähler 2-form. For the moment we leave unspecified the R 2-forms $\Omega^{(r)}$. These conditions are already sufficient to ensure that all but the quartic terms in ∇X of (32) are reproduced. To reproduce the quartic terms too we require that

$$
X^{IJ}X^{KL}\left[\omega_{IJ}\omega_{KL}+\sum_{r=1}^{R}\Omega^{(r)}_{IJ}\Omega^{(r)}_{KL}-2g_{KI}g_{JL}\right]=0\,,
\tag{34}
$$

where

$$
X^{IJ}\equiv\nabla X^I\times\nabla X^J\,.
\tag{35}
$$

Note that X^{IJ} is an antisymmetric 8×8 matrix. If none of its 4 skew-eigenvalues vanish, then (34) implies that

$$
\omega_{I(J}\omega_{K)L}+\sum_{r=1}^{R}\Omega^{(r)}_{I(J}\Omega^{(r)}_{K)L}=g_{I(K}g_{J)L}-g_{KJ}g_{IL}\,.
\tag{36}
$$

For a membrane in flat space this condition is satisfied by taking the matrices

$$I_I{}^J \equiv \omega_{IK} g^{KJ}\,, \qquad (J^{(r)})_I{}^J \equiv \Omega^{(r)}_{IK} g^{KJ} \tag{37}$$

to be the seven complex structures of $\mathbb{E}^8$.

For every vanishing skew eigenvalue of X^{IJ} the dimension of the transverse space is effectively reduced by two. In this reduced space, we must again have (36) but it may now be possible to choose some of the six J matrices to vanish. For example, if X^{IJ} has two vanishing skew-eigenvalues then the transverse space is effectively 4-dimensional; in other words, there are four 'active scalars'. We may now set all but two of the J matrices to zero. The other two, together with I can be taken to be the three almost complex structures of the transverse 4-manifold (these will be covariantly constant if this transverse 4-space is hyper-Kähler, but we need not assume any special properties at this point). If X^{IJ} has three vanishing skew-eigenvalues, corresponding to two active scalars, then the transverse space is effectively two-dimensional, and we may take all the J matrices to vanish.

Given (34) we deduce that

$$\varepsilon \geq \tfrac{1}{2}|X^{IJ}\omega_{IJ}| \tag{38}$$

with equality when

$$\nabla X^I = \mp * \nabla X^J I_J{}^I \tag{39}$$

and

$$X^{IJ}\Omega^{(r)}_{IJ} = 0\,, \qquad r = 1, \cdots, R\,. \tag{40}$$

The condition (39) is the statement that in complex coordinates Z^α, adapted to the complex structure I, the functions $Z^\alpha(z)$ are holomorphic on worldspace, with $z = \sigma^1 + i\sigma^2$. The conditions (40) are implied by (39) if the matrices $J^{(r)}$ are such that

$$IJ^{(r)} + J^{(r)}I = 0\,, \qquad r = 1, \cdots, R\,. \tag{41}$$

This is true when $I, J^{(r)}$ are the 7 complex structures of $\mathbb{E}^8$. It is also satisfied if $I, J^{(1)}, J^{(2)}$ are the three almost complex structures of a 4-dimensional space, with the other J matrices vanishing. This is the case of most interest here because we may obviously reduce the transverse 8-space to an effective transverse 4-space by requiring all scalars to vanish except those associated with the ds_4^2 metric in (19). This restriction still allows configurations with either two or four active scalars. The situation is summarized in Table 1.

n=4	R=6	$\mathbb{E}^8$
n=2	R=2	hyper-Kähler
n=1	R=0	Kähler

Table 1: The first column indicates the number 2n of active scalars. The second column gives the number R of nonzero J-matrices. The third column indicates the geometry of the active 2n-dimensional transverse space.

The brane interpretation of the worldvolume soliton is as follows. There are 2 M2-branes involved. One M2-brane represents the vacuum configuration. The other intersecting M2-brane describes the lump worldvolume soliton. In order to give the worldvolume soliton a finite energy the intersecting M2-brane must wrap around a non-singular 2-cycle. The 2 cycle is provided by adding a two-centred KK monopole as a background configuration. This background configuration contains the required two-cycle as indicated in Figure 2. The relative positions of the two M2-branes, and the precise way in which the KK monopole is added to the two intersecting M2-branes is summarized by the diagrams below.

$$
n=1 \quad : \quad \left\{ \begin{array}{c|cccccccccc} \times & \times & \times & - & - & - & - & - & - & - & - \\ \times & - & - & \times & \times & - & - & - & - & - & - \end{array} \right.
$$

$$
\Downarrow
$$

$$
n=1 \quad : \quad \left\{ \begin{array}{c|cccccccccc} \times & \times & \times & - & - & - & - & - & - & - & - \\ \times & - & - & - & - & - & - & \times & \times & - & - \\ \times & \times & \times & \times & \times & \times & \times & z & - & - & - \end{array} \right.
$$

4 Conclusions

The brane interpretation of the WV lump is consistent with the WV supersymmetry algebra. Consider the N=8, D=3 WV supersymmetry algebra:

$$
\{Q^{\tilde{I}}_\alpha, Q^{\tilde{J}}_\beta\} = \delta^{\tilde{I}\tilde{J}} P_{\alpha\beta} + Z^{(\tilde{I}\tilde{J})}_{(\alpha\beta)} + \varepsilon_{\alpha\beta} Z^{[\tilde{I}\tilde{J}]} \,, \tag{42}
$$

with $\alpha = 1, \cdots, 8$ and $\tilde{I} = 1, \cdots 8$. The central charge $Z^{[\tilde{I}\tilde{J}]}$ is a WV 0-form and a TS 2-form. As a 8×8 matrix it has at most 4 skew-eigenvalues. These eigenvalues are in one to one correspondence with the skew-eigenvalues of the matrix X^{IJ} defined in (35).

In our paper we also constructed time-dependent solutions, the Q-kinks. More work in this direction has been done recently in Lambert and Tong.[7]

Acknowledgments

The work presented in this talk is based on the paper **hep-th/9904020** published in JHEP05 (1999) 021. I thank Paul Townsend for the stimulating collaboration on this project. This work is supported by the European Commission TMR program ERBFMRX-CT96-0045, in which I am associated to the University of Utrecht.

References

1. E. Bergshoeff, E. Sezgin and P.K. Townsend, *Supermembranes and eleven-dimensional supergravity*, *Phys. Lett.* **189B**, 75 (1987); *Properties of the eleven-dimensional supermembrane theory*, *Ann. Phys.* (N.Y.) **185**, 330 (1998).
2. E. Bergshoeff, E. Sezgin and Y. Tanii, *Hamiltonian formulation of the supermembrane*, *Nucl. Phys.* **B298**, 187 (1988).
3. C. Callan and J. Maldacena, *Brane dynamics from the Born-Infeld action*, *Nucl. Phys.* **B513**, 198 (1998).
4. G.W. Gibbons, *Born-Infeld particles and Dirichlet p-branes*, *Nucl. Phys.* **B514**, 603 (1998).
5. P.S. Howe, N.D. Lambert and P.C. West, *The Self-Dual String Soliton*, *Nucl. Phys.* **B515**, 203 (1998); *ibidem*, *The Threebrane Soliton of the M-Fivebrane*, *Phys. Lett.* **B419**, 79 (1998).
6. E. Bergshoeff and P.K. Townsend, *Solitons on the Supermembrane*, JHEP05 (1999) 021, hep-th/9904020.
7. N.D. Lambert and D. Tong, *Kinky D-strings*, hep-th/9907098.

1/4 PBGS AND SUPERPARTICLE ACTIONS[a]

F. DELDUC
Laboratoire de Physique,
Groupe de Physique Théorique ENS Lyon
46, allée d'Italie, F - 69364 - Lyon CEDEX 07

E. IVANOV AND S. KRIVONOS
Bogoliubov Laboratory of Theoretical Physics, JINR
141 980, Dubna, Moscow Region, Russian Federation

We construct the worldline superfield massive superparticle actions which preserve 1/4 portion of the underlying higher-dimensional supersymmetry. We consider the cases of $N = 4 \to N = 1$ and $N = 8 \to N = 2$ partial breaking. In the first case we present the corresponding Green-Schwarz type target superspace action with one κ-supersymmetry. In the second case we find out two possibilities, one of which is a direct generalization of the $N = 4 \to N = 1$ case, while another is essentially different.

1 Introduction

The most attractive feature of the description of superbranes based on idea of the partial spontaneous breaking of global supersymmetry (PBGS) [1–14] is the manifest off-shell realization of the worldvolume supersymmetry. In this approach, the physical worldvolume multiplets are interpreted as Goldstone superfields realizing spontaneous breaking of the full brane supersymmetry group down to its unbroken worldvolume subgroup. The invariant Goldstone superfields actions, after passing to the component fields, coincide with the static gauge forms of the relevant Green-Schwarz (GS) type actions.

Until present, only the examples of 1/2 breaking of supersymmetry corresponding to the standard BPS p-branes and D-branes were treated in the literature on the PBGS. It is interesting to extend the PBGS framework to the 1/4 breaking and other fractional patterns (see, e.g., References [15–20] for the discussion of such options at the algebraic level).

In the present work we describe several examples of the 1/4 PBGS superfield actions in the simplest case of massive superparticles, namely, the actions corresponding to the PBGS patterns $N = 4 \to N = 1$ and $N = 8 \to N = 2$. In the first case we construct a worldline superfield action and show its equivalence to the GS-type target superspace action with one fermionic κ symmetry. In the second case we find out two different models. For one of them we find the

---

[a]Talk given by S. Krivonos

bosonic part of the worldline superfield action and the corresponding GS-type action with two κ symmetries. A common feature of all the cases considered is that the algebras of their underlying spontaneously broken $d=1$ supersymmetries are dimensionally-reduced forms of $N=1$ and $N=2$ $D=4$ Poincaré superalgebras extended by tensorial central charges. [21,22,19,16]

2 N=4 → N=1 PBGS

Our goal here will be to construct a $N=1, d=1$ superfield action which respects three extra spontaneously broken supersymmetries. Thus, the minimal multiplet should include at least three $N=1$ fermionic Goldstone superfields $\psi^i(t,\theta)(i=1,2,3)$ given on $N=1, d=1$ superspace with the coordinates $\{t,\theta\}, (\bar{t}=-t, \bar{\theta}=\theta)$. They should inhomogeneously transform under the broken supersymmetries. One can check that this requirement is met in a minimal way and the algebra of transformations gets closed at cost of adding one additional fermionic $N=1$ superfield $\Upsilon(t,\theta)$. The broken supersymmetry transformations read

$$\delta\psi^i = \epsilon^i\left(1-\mathcal{D}\Upsilon\right) - \varepsilon^{ijk}\epsilon^j\mathcal{D}\psi^k , \qquad \delta\Upsilon = \epsilon^i\mathcal{D}\psi^i , \tag{1}$$

where

$$\mathcal{D} = \frac{\partial}{\partial\theta} + \theta\partial_t , \qquad \{\mathcal{D},\mathcal{D}\} = 2\partial_t . \tag{2}$$

They form the following algebra[b]

$$\{Q,Q\} = 2P , \qquad \{S^i,S^j\} = 2\delta^{ij}P , \qquad \{Q,S^i\} = 0 . \tag{3}$$

We wish to have a superparticle model with the worldline scalar $N=1$ multiplets containing physical bosonic fields. Thus we are led to introduce bosonic superfields v^i

$$\psi^i = \frac{1}{2}\mathcal{D}v^i , \qquad \left(\bar{\psi}^i = \psi^i , \, \bar{v}^i = -v^i\right) , \tag{4}$$

$$\delta v^i = -2\epsilon^i\left(\theta - \Upsilon\right) + \varepsilon^{ijk}\epsilon^j\mathcal{D}v^k , \qquad \delta\Upsilon = \frac{1}{2}\epsilon^i\partial_t v^i . \tag{5}$$

Due to the explicit presence of θ in the transformations (5), the anticommutators of Q and the spontaneously broken supersymmetry generators S^i acquire

---

[b]Hereafter, we deal with the algebras of the superfield variations. The superalgebras of the supercharges constructed by the PBGS actions following the Nöther procedure are different: they inevitably include some constant central charges which are crucial for evading[1,2] the famous Witten's no-go theorem.[23]

active central charges Z^i in the right-hand side:

$$\{Q, S^i\} = 2Z^i \,. \tag{6}$$

The central charge generators act as pure shifts of v^i, suggesting the interpretation of v^i as Goldstone superfields parametrizing transverse directions in a four-dimensional space where Z^i, P act as the translation operators.

Surprisingly, the superalgebra (6) cannot be interpreted as a dimensional reduction of the standard $N = 1$ Poincaré superalgebra in $d = 4$, with Z^i, P being the components of full 4- momentum. One should proceed not from the standard $N = 1$, $D = 4$ super Poincaré algebra, but from its extension by tensorial central charges. [21,22,19,16] The generators Z^i turn out to partly come from these central charges and partly from the extra components of 4-momentum. Namely,

$$P = \frac{1}{2} P_0 \,, \qquad Z^1 = P_2 \,, \qquad Z^2 = -P_1 \,, \qquad Z^3 = \frac{i}{4}(\overline{T}_{22} - T_{22}) \,, \tag{7}$$

where the original $N = 1$, $D = 4$ superalgebra is defined by the relations

$$\begin{aligned} &\{Q_\alpha, \bar{Q}_{\dot\beta}\} = 2\,(\sigma^m)_{\alpha\dot\beta} P_m \,, \\ &\{Q_\alpha, Q_\beta\} = 2\,T_{(\alpha\beta)} \,, \qquad \{\bar{Q}_{\dot\alpha}, \bar{Q}_{\dot\beta}\} = 2\,\overline{T}_{(\dot\alpha\dot\beta)} \,. \end{aligned} \tag{8}$$

Let us construct invariant action for the system under consideration. The field $\Upsilon(v)$ is a good candidate for the Lagrangian density

$$S_v = \int dt d\theta \Upsilon(v) \,, \tag{9}$$

in view of its transformation property (1). Then the question is how to covariantly express Υ in terms of ψ^i and, further, v^i. This can be done rather easily.

The most general ansatz for Υ is as follows

$$\begin{aligned} \Upsilon \;=\; & \psi^i \mathcal{D}\psi^i A + \psi^{2i}\psi^i_t B + \psi^{2i}\mathcal{D}\psi^i \psi^j_t \mathcal{D}\psi^j C \\ & + \psi^i \psi^i_t \psi^j \mathcal{D}\psi^j E + \psi^3 \mathcal{D}\psi^i_t \mathcal{D}\psi^i F + \psi^3 \psi^{2i}_t \mathcal{D}\psi^i G \,, \end{aligned} \tag{10}$$

where $A, B, \ldots, G$ are as yet undetermined functions of X, and we use the following notations

$$\psi^i_t = \partial_t \psi^i \,, \qquad \psi^{2i} = \varepsilon^{ijk}\psi^j \psi^k \,, \qquad \psi^3 = \varepsilon^{ijk}\psi^i \psi^j \psi^k \,. \tag{11}$$

Now, using (1), (10) we can write $\delta\psi^i$ in terms of ψ^i. Then we explicitly evaluate $\delta\Upsilon$ and require it to be equal to $\epsilon^i \mathcal{D}\psi^i$ in accordance with the transformation law (1). After rather lengthy calculations we get the system of *algebraic* equations for the unknowns $A, \ldots, G$

$$A = \frac{2}{1+\sqrt{1-4\mathcal{D}\psi^i\mathcal{D}\psi^i}}\,, \qquad B = \frac{A^2}{2(A-2)}\,, \qquad C = -\frac{A^4}{2(A-2)}\,,$$
$$E = \frac{A^3}{A-2}\,, \qquad F = \frac{A^3(A-4)}{6(A-2)^2}\,, \qquad G = -\frac{A^5(A-4)}{6(A-2)^2}\,. \tag{12}$$

The integral (9) with Υ defined by (10), (12) provides us with the action for the system under consideration.

We can greatly simplify this action. First, the B and C terms in (10) can be absorbed into the F term. All the remaining E, F, G terms can be reduced to the single A term, redefining the superfields v^i as follows

$$v^i \to \phi^i = v^i + \psi^3 \varepsilon^{ijk}\psi^j_t \mathcal{D}\psi^k H_1 + \psi^3\psi^i_t H_2 + \varepsilon^{ijk}\psi^{2j}\mathcal{D}\psi^k H_3\,, \tag{13}$$

where H_1, H_2, H_3 are some functions of X. These functions can be given explicitly, but to know their precise structure is of no need for our purposes. The action in terms of the redefined bosonic superfield ϕ^i takes the very simple form

$$S_\phi = \int dt d\theta \frac{2\xi^i \mathcal{D}\xi^i}{1+\sqrt{1-4\mathcal{D}\xi^j\mathcal{D}\xi^j}}\,, \qquad \xi^i \equiv \frac{1}{2}\mathcal{D}\phi^i\,. \tag{14}$$

By construction, it is guaranteed to be invariant.

Thus we have found the correct Goldstone superfields action describing the PBGS pattern $N=4 \to N=1$.

Let us end this section by noting that the bosonic core of the action (14)

$$S^{bos}_\phi = \frac{1}{2}\int dt\left(1-\sqrt{1-\partial_t\phi^i\partial_t\phi^i}\right) \tag{15}$$

is the standard massive $D=4$ particle action in the static gauge.

3 Target Space Action with One κ-Supersymmetry

To clarify the situation with $N=4 \to N=1$ PBGS, we construct the target space action which possesses only one κ supersymmetry and reduces to the action (14) in a fixed gauge.

We shall deal with the $N=4$ superalgebra (6). In accord with the standard strategy of constructing GS-type actions for massive superparticles (see References [26,3,4,27,6]) we introduce bosonic $X^0(t), Y^i(t)$ and fermionic $\Theta(t), \Psi^i(t)$

$d = 1$ fields, the coordinates of a target $N = 4$ superspace, with the standard transformation properties under $N = 4$ supersymmetry (6)

$$\delta X^0 = -\epsilon\Theta - \epsilon^i\Psi^i\,, \qquad \delta Y^i = -\epsilon^i\Theta - \epsilon\Psi^i, \qquad \delta\Theta = \epsilon\,, \qquad \delta\Psi^i = \epsilon^i\,, \quad (16)$$

and construct the invariants Π^0, Π^i

$$\Pi^0 = \partial_t X^0 + \Theta\partial_t\Theta + \Psi^i\partial_t\Psi^i\,, \qquad \Pi^i = \partial_t Y^i - \partial_t\Theta\Psi^i + \Theta\partial_t\Psi^i\,. \quad (17)$$

After some guess-work, the target sigma-model action invariant under the global target space supersymmetry (16), local t reparametrizations and one local fermionic κ symmetry was found to have the following form

$$S_{gs} = -\int dt\sqrt{\Pi^0\Pi^0 - \Pi^i\Pi^i} - \int dt\left(\Theta\partial_t\Theta - \Psi^i\partial_t\Psi^i\right)\,. \quad (18)$$

The κ symmetry transformations are given by

$$\delta\Theta = \kappa\,, \qquad \delta\Psi^i = \kappa\,\frac{\Pi^i}{\Pi^0 - \sqrt{\Pi^0\Pi^0 - \Pi^i\Pi^i}}$$
$$\delta X^0 = -\Theta\delta\Theta - \Psi^i\delta\Psi^i, \quad \delta Y^i = -\Psi^i\delta\Theta - \Theta\delta\Psi^i\,, \quad (19)$$

where $\kappa(t)$ is an arbitrary fermionic gauge parameter.

The action (18) possesses only one κ supersymmetry and therefore provides a "space-time" realization of the $N = 4 \to N = 1$ PBGS phenomenon.

To prove that there are no any other local fermionic symmetry in (18) apart from κ-symmetry (19), we need to study the algebra of the constraints in the Hamiltonian formalism. We first introduce the einbein $e(t)$ and rewrite the action (18) as

$$S_{gs} = \int dtL = -\int dt\left[\frac{1}{2e}\left(\Pi^0\Pi^0 - \Pi^i\Pi^i\right) + \frac{e}{2}\right]$$
$$- \int dt\left(\Theta\partial_t\Theta - \Psi^i\partial_t\Psi^i\right)\,. \quad (20)$$

Then we compute canonically conjugated variables

$$P^0 = -\frac{\Pi^0}{e}\,, \quad P^i = \frac{\Pi^i}{e}\,, \quad P_e = 0\,, \quad \Omega = \left(\frac{\Pi^0}{e} + 1\right)\Theta - \frac{\Pi^i}{e}\Psi^i\,,$$
$$\Omega^i = \left(\frac{\Pi^0}{e} - 1\right)\Psi^i - \frac{\Pi^i}{e}\Theta\,. \quad (21)$$

The canonical hamiltonian reads

$$H = -\frac{e}{2}\left(P^0P^0 - P^iP^i - 1\right)\,. \quad (22)$$

There is one primary bosonic constraint, P_e, and four fermionic constraints

$$\tau^0 = \Omega + (P^0 - 1)\Theta + P^i\Psi^i \ , \quad \tau^i = \Omega^i + (P^0+1)\Psi^i + P^i\Theta \ . \tag{23}$$

When taking the Poisson bracket of the primary bosonic constraint with the canonical hamiltonian, we obtain the secondary bosonic constraint

$$P^0P^0 - P^iP^i = 1 \ . \tag{24}$$

We now have to determine which of the fermionic constraints $\tau^\mu = (\tau^0, \tau^i)$ are first class, and thus generate gauge symmetries, and which are second class. We compute the matrix of the Poisson brackets of the fermionic constraints

$$\{\tau^\mu, \tau^\nu\} = C^{\mu\nu} \ , \quad C = 2\begin{pmatrix} P^0 - 1 & \vec{P}^t \\ \vec{P} & (P^0+1)\mathbf{1} \end{pmatrix} \ , \tag{25}$$

where $\mathbf{1}$ is the 3×3 unit matrix. The eigenvalues of C are easily computed to be P^0+1, P^0+1, $P^0 + \sqrt{\vec{P}^2+1}$, $P^0 - \sqrt{\vec{P}^2+1}$. On the constraint surface, the last of these eigenvalues vanishes, and the other three remain non zero. Thus, there is only one first class constraint which may be chosen to be

$$\kappa = \tau^0 - \frac{1}{P^0+1}\vec{P}\vec{\tau} . \tag{26}$$

Its Poisson brackets with the constraints read

$$\{\kappa, \tau^0\} = \frac{2(P^0P^0 - P^iP^i - 1)}{P^0+1} \ , \qquad \{\kappa, \tau^i\} = 0 \ . \tag{27}$$

This constraint generates the unique local fermionic symmetry (19) through the Poisson bracket.

In the static gauge the action (18) reads

$$S_{gs} = -\int dt \left[\sqrt{(1+\Psi^i\partial_t\Psi^i)^2 - \partial_tY^i\partial_tY^i} - \Psi^i\partial_t\Psi^i\right] \ . \tag{28}$$

It is straightforward to show that it is related to the component form of the action (14) by a field redefinition.

4 N=8 → N=2 PBGS

To construct a superparticle model which would exhibit $N=8 \to N=2$ PBGS we should, before all, examine how 6 broken supersymmetries could be realized on a set of $N=2,\ d=1$ superfields. We succeeded in finding two such realizations.

Case I. In the first realization the basic set of $N=2, d=1$ superfields consists of seven bosonic superfields: a general real superfield Φ and two conjugated triplets of chiral-anti-chiral superfields $\bar{v}^i, v^i$:

$$Dv^i = \bar{D}\bar{v}^i = 0 \,, \qquad i = 1,2,3 \,,$$

where

$$D = \frac{\partial}{\partial\theta} + \frac{1}{2}\bar{\theta}\partial_t \,, \qquad \bar{D} = \frac{\partial}{\partial\bar{\theta}} + \frac{1}{2}\theta\partial_t \,. \tag{29}$$

The broken supersymmetry transformations read

$$\delta v^i = -2\left(\bar{\theta} - D\Phi\right)\epsilon^i + \varepsilon^{ijk}\bar{\epsilon}^j D\bar{v}^k \,, \qquad \delta\Phi = \frac{1}{2}\left(\epsilon^i D\bar{v}^i + \bar{\epsilon}^i \bar{D} v^i\right) \,. \tag{30}$$

Together with the manifest supersymmetry, they form the algebra with six central charges $Z^i, \bar{Z}^i$

$$\begin{aligned} \{Q,\bar{Q}\} &= P \,, \qquad \{S^i,\bar{S}^j\} = \delta^{ij}P \,, \\ \{Q,S^i\} &= 2Z^i \,, \qquad \{\bar{Q},\bar{S}^i\} = 2\bar{Z}^i \,. \end{aligned} \tag{31}$$

The fermionic chiral superfields defined by

$$\psi^i = -\frac{1}{2}\bar{D}v^i \,, \qquad \bar{\psi}^i = \frac{1}{2}D\bar{v}^i$$

are transformed under (30) as

$$\delta\psi^i = \left(1 - \bar{D}D\Phi\right)\epsilon^i + \varepsilon^{ijk}\bar{\epsilon}^j\bar{D}\bar{\psi}^k, \ \delta\Phi = \epsilon^i\bar{\psi}^i - \bar{\epsilon}^i\psi^i \,. \tag{32}$$

So they are Goldstone superfields corresponding to the linear realization of six spontaneously broken supersymmetries with the parameters $\epsilon^i, \bar{\epsilon}^i$. The bosonic superfields $\bar{v}^i, v^i$ are the Goldstone ones associated with the spontaneously broken central charges transformations.

Once again, the superfield Φ, in accord with its transformation properties, can be chosen as the Lagrangian density for this PBGS pattern.

To express Φ in terms of the Goldstone superfields $\psi^i, \bar{\psi}^i$, one can apply the method of Refs.[24,25] It basically consists in passing to another superfield basis by performing a finite spontaneously broken supersymmetry transformation with the Goldstone fermionic superfields as the parameters. The redefined superfield $\tilde{\Phi} = \Phi + O(\psi,\bar{\psi})$ transforms homogeneously and so it can be put equal to zero with preserving covariance under all supersymmetries. This produces equations allowing one to express Φ in terms of $\psi^i, \bar{\psi}^i$.

The straightforward application of this method to the present case yields a rather complicated system of equations. It can be easily solved in the limit of vanishing fermions, yielding the static gauge action for a massive particle in a 7-dimensional space-time as the bosonic part of the full superfield action

$$S_v^{bos} = \frac{1}{2}\int dt \left(1 - \sqrt{1 + \partial_t v^i \partial_t \bar{v}^i}\right) . \tag{33}$$

The GS formulation for this case is very similar to the case of $N = 4 \to N = 1$ PBGS. We define the standard realization of $N = 8$ superalgebra (31) in the superspace with seven bosonic $X^0, Y^i, \bar{Y}^i$ and eight fermionic $\Theta, \bar{\Theta}, \Psi^i, \bar{\Psi}^i$ coordinates:

$$\begin{aligned} \delta\Theta &= \epsilon\ ,\ \delta\Psi^i = \epsilon^i\ ,\ \delta Y^i = -2\epsilon^i\Theta\ , \\ \delta X^0 &= -\frac{1}{2}\left(\epsilon\bar{\Theta} + \bar{\epsilon}\Theta + \epsilon^i\bar{\Psi}^i + \bar{\epsilon}^i\Psi^i\right) . \end{aligned} \tag{34}$$

Defining the invariants

$$\begin{aligned} \Pi^0 &= \partial_t X^0 + \frac{1}{2}\left(\Theta\partial_t\bar{\Theta} + \bar{\Theta}\partial_t\Theta + \Psi^i\partial_t\bar{\Psi}^i + \bar{\Psi}^i\partial_t\Psi^i\right) , \\ \Pi^i &= \partial_t Y^i + 2\Psi^i\partial_t\Theta\ ,\quad \bar{\Pi}^i \equiv (\overline{\Pi^i})\ , \end{aligned} \tag{35}$$

we can construct the unique action

$$S_{gs} = -\int dt\sqrt{\Pi^0\Pi^0 - \Pi^i\bar{\Pi}^i} + \int dt\left(\Theta\partial_t\bar{\Theta} - \Psi^i\partial_t\bar{\Psi}^i\right) , \tag{36}$$

with two κ supersymmetries:

$$\begin{aligned} \delta X^0 &= -\frac{1}{2}\left(\bar{\Theta}\delta\Theta + \Theta\delta\bar{\Theta} + \bar{\Psi}^i\delta\Psi^i + \Psi^i\delta\bar{\Psi}^i\right) , \\ \delta Y^i &= -2\Psi^i\delta\Theta\ , \\ \delta\Psi^i &= \frac{\Pi^i\delta\bar{\Theta}}{\Pi^0 + \sqrt{\Pi^0\Pi^0 - \Pi^i\bar{\Pi}^i}} . \end{aligned} \tag{37}$$

The Hamiltonian analysis, which repeats the basic steps of the analysis in the $N = 4 \to N = 1$ case, shows that there are no further gauge fermionic symmetries in the action (36).

In the static gauge, $X^0 = t, \Theta = 0$, the action (36) takes the very simple form

$$S_{gs} = -\int dt\left[\sqrt{\left(1 + \frac{1}{2}\Psi^i\partial_t\bar{\Psi}^i + \frac{1}{2}\partial_t\Psi^i\bar{\Psi}^i\right)^2 + \partial_t Y^i\partial_t\bar{Y}^i} + \Psi^i\partial_t\bar{\Psi}^i\right] . \tag{38}$$

Case II. The second realization of $N = 8, d = 1$ supersymmetry with six spontaneously broken supersymmetries can be constructed in terms of general bosonic $N = 2$ superfield Φ and six chiral and anti-chiral Goldstone fermions $\{\psi_\alpha, \bar{\psi}_\alpha, \xi, \bar{\xi}\}$, $\alpha = 1,2$

$$\bar{D}\psi_\alpha = \bar{D}\xi = 0 \,, \qquad D\bar{\psi}_\alpha = D\bar{\xi} = 0 \,, \tag{39}$$

which form two doublets and two singlets with respect to $SO(2)$ automorphism group. The appropriate closed set of the broken supersymmetry transformations reads

$$\begin{aligned}
\delta\xi &= \left(1 + \bar{D}D\Phi\right)\nu + \varepsilon_{\alpha\beta}\bar{\mu}_\alpha \bar{D}\bar{\psi}_\beta \,, \\
\delta\Phi &= \bar{\nu}\xi - \nu\bar{\xi} - \bar{\mu}_\alpha\psi_\alpha + \mu_\alpha\bar{\psi}_\alpha \,, \\
\delta\psi_\alpha &= \varepsilon_{\alpha\beta}\left(\bar{\nu}\bar{D}\bar{\psi}_\beta + \bar{\mu}_\beta\bar{D}\bar{\xi}\right) + \left(1 - \bar{D}D\Phi\right)\mu_\alpha \,.
\end{aligned} \tag{40}$$

To reveal the underlying central-charges extended supersymmetry algebra and to gain physical bosonic fields, we need to pass as before to bosonic superfields. The minimal realization amounts to introducing two real scalar superfields u_α:

$$\psi_\alpha = -\frac{1}{2}\bar{D}u_\alpha \,, \quad \bar{\psi}_\alpha = \frac{1}{2}Du_\alpha \,. \tag{41}$$

To learn what kind of "prepotential" one should introduce for the remaining Goldstone superfield ξ, let us examine the relation between $U(1)$ charges of spinor superfields which follows from (40)

$$q_\xi = -2q_\psi - q_D \,. \tag{42}$$

Here q_D is the $U(1)$ charge of the covariant derivative D ($q_D = -1$ if one ascribes the charge $+1$ to θ). From this relation and Eq. (41) it follows that the only way to introduce the bosonic superfield v for ξ is to choose it complex and having the $U(1)$ charge $-2q_D$

$$\xi = -\frac{1}{2}\bar{D}v \,, \quad \bar{\xi} = \frac{1}{2}D\bar{v} \,, \quad Dv = \bar{D}\bar{v} = 0 \,. \tag{43}$$

In terms of $v, \bar{v}$ the supersymmetry transformations become:

$$\begin{aligned}
\delta v &= -2\left(\bar{\theta} + D\Phi\right)\nu + \varepsilon_{\alpha\beta}\bar{\mu}_\alpha Du_\beta \,, \\
\delta u_\alpha &= \varepsilon_{\alpha\beta}\left(\bar{\nu}Du_\beta + \nu\bar{D}u_\beta + \bar{\mu}_\beta D\bar{v} + \mu_\beta\bar{D}v\right) \\
&\quad + 2\left(\theta + \bar{D}\Phi\right)\bar{\mu}_\alpha - 2\left(\bar{\theta} - D\Phi\right)\mu_\alpha \,, \\
\delta\Phi &= -\frac{1}{2}\left(\nu D\bar{v} + \bar{\nu}\bar{D}v - \bar{\mu}_\alpha\bar{D}u_\alpha - \mu_\alpha Du_\alpha\right) \,.
\end{aligned} \tag{44}$$

Denoting the generators of the broken supersymmetry by $S_\alpha, \bar{S}_\alpha$ and $S, \bar{S}$, and the generators of the manifest $N = 2$ supersymmetry by $Q, \bar{Q}$, one can write the full supersymmetry algebra pertinent to this case as

$$\begin{gathered}\{Q,\bar{Q}\} = \{S,\bar{S}\} = P\,, \qquad \{S_\alpha,\bar{S}_\beta\} = \delta_{\alpha,\beta}P\,, \\ \{Q,\bar{S}\} = 2\bar{Z}\,, \qquad \{\bar{Q},S\} = 2Z\,, \qquad \{Q,\bar{S}_\alpha\} = 2Z_\alpha\,, \\ \{\bar{Q},S_\alpha\} = 2Z_\alpha\,, \qquad \{S,\bar{S}_\alpha\} = 2\varepsilon_{\alpha\beta}Z_\beta\,, \qquad \{\bar{S},S_\alpha\} = 2\varepsilon_{\alpha\beta}Z_\beta\,.\end{gathered} \tag{45}$$

Once again, we can take the superfield Φ as the Lagrangian density. To covariantly express Φ in terms of the Goldstone fermions or Goldstone bosons, we may again stick to the general method of Refs. [24,25] However it gives rather complicated equations which we for the time being were unable to solve. We could try to find at least the bosonic part of the action. We get the following quartic equation for the bosonic part of the action which we denote by X:

$$(X^2 - X + a)(X^2 + a - 1) + 2D\xi\bar{D}\bar{\xi} = 0\,, \tag{46}$$

where

$$a = D\xi\bar{D}\bar{\xi} + \bar{D}\psi_\alpha D\bar{\psi}_\alpha\,.$$

The general solution of this equation exists (we require it to vanish in the limit when all fields are put equal to zero), but it looks not too illuminating to present it here. In the two limits, $\psi_\alpha = 0$ or $\xi = 0$, it takes the familiar form of the static gauge actions of massive particles moving on some 3-dimensional target manifolds

$$\begin{aligned} S_v^{bos} &= \frac{1}{2}\int dt\left(1 - \sqrt{1 + \partial_t v \partial_t \bar{v}}\right)\,, \\ S_u^{bos} &= \frac{1}{2}\int dt\left(1 - \sqrt{1 + \partial_t u_\alpha \partial_t u_\alpha}\right)\,. \end{aligned} \tag{47}$$

In the generic case there is a non-trivial cross-interaction between the bosonic fields appearing in (47). It can hopefully be interpreted in terms of intersection of the trajectories of two different superparticles, with the physical worldline scalar multiplets represented by the superfields u_α and $v, \bar{v}$, respectively.

The fact that the bosonic part of the action cannot be written in the standard static gauge Nambu-Goto form seriously obscures the construction of the GS formulation for this case. We believe that the better understanding of this case would be helpful for studying the 1/4 PBGS systems with higher-dimensional worldvolumes.

5 Conclusions

In this work we have presented the manifestly worldline supersymmetric superparticle actions exhibiting hidden spontaneously broken supersymmetries the number of which is four times the number of the linearly realized manifest ones. We treated in detail the case of $N = 4 \to N = 1$ partial breaking and discussed some basic features of the more complicated $N = 8 \to N = 2$ case. The common unusual feature of the superparticle systems considered is that their space-time interpretation is possible only within the superspaces corresponding to higher-dimensional supersymmetries with tensorial central charge generators. It would be of interest to understand whether this is the general property of systems with fractional PBGS.

Acknowledgments

We thank A. Kapustnikov, J. Lukierski, D. Lüst, A. Pashnev, P. Pasti, C. Preitschopf, D. Sorokin, M. Tonin and B. Zupnik for many useful discussions. E.I. and S.K. are grateful to Organizers of XIV-th Max Born Symposium for inviting them to present this talk. This work was supported in part by the PICS Project No. 593, RFBR-CNRS Grant No. 98-02-22034, RFBR Grant No. 99-02-18417, NATO Grant No. PST.CLG 974874 and INTAS Grants INTAS-96-0538, INTAS-96-0308.

References

1. J. Hughes, J. Polchinski, *Nucl. Phys.* **B278**, 147 (1986).
2. J. Hughes, J. Liu, J. Polchinski, *Phys. Lett.* **B180**, 370 (1986).
3. A. Achucarro, J. Gauntlett, K. Itoh, P.K. Townsend, *Nucl. Phys.* **B314**, 129 (1989).
4. J.P. Gauntlett, K. Itoh, P.K. Townsend, *Phys. Lett.* **B238**, 65 (1990).
5. E. Ivanov, A. Kapustnikov, *Phys. Lett.* **B252**, 439 (1990), **B267**, 541E; (1991); *ibid* **B267**, 175 (1991); *Int. J. Mod. Phys.* **A7**, 2153 (1992).
6. J.P. Gauntlett, C.F. Yastremiz, *Class. Quantum Grav.* **7**, 2089 (1990).
7. J. Bagger, A. Galperin, *Phys. Lett.* **B336**, 25 (1994).
8. J. Bagger, A. Galperin, *Phys. Rev.* **D55**, 1091 (1997).
9. J. Bagger, A. Galperin, *Phys. Lett.* **B412**, 296 (1997).
10. T. Adawi, M. Cederwall, U. Gran, M. Holm, B.E.W. Nilsson, *Int. J. Mod. Phys.* **A13**, 4691 (1998).
11. M. Roček, A. Tseytlin, *Phys. Rev.* **D59**, 106001 (1999).
12. F. Gonzalez-Rey, I.Y. Park, M. Roček, *Nucl. Phys.* **B544**, 243 (1999).
13. S. Bellucci, E. Ivanov, S. Krivonos, *Phys. Lett.* **B460**, 348 (1999).

14. E. Ivanov, S. Krivonos, *Phys. Lett.* **B453**, 237 (1999).
15. M. Berkooz, M.R. Douglas, R.G. Leigh, *Nucl. Phys.* **B480**, 265 (1996).
16. J.P. Gauntlett, G.W. Gibbons, G. Papadopoulos, P.K. Townsend, *Nucl. Phys.* **B500**, 133 (1997).
17. A.A. Tseytlin, *Class. Quantum Grav.* **14**, 2085 (1997).
18. N. Ohta, P.K. Townsend, *Phys. Lett.* **B418**, 77 (1998).
19. I. Bandos, J. Lukierski, *Mod. Phys. Lett.* **A14**, 1257 (1999); I. Bandos, J. Lukierski, D. Sorokin, "Superparticle models with tensorial central charges", FTUV/99-07, IFIC/99-07, TUW-99-06, HUB-EP-99/15 (hep-th/9904109).
20. J.P. Gauntlett, C.M. Hull, "BPS States with Extra Supersymmetry", QMW-PH-99-13, hep-th/9909098.
21. J.A. de Azcárraga, J.P. Gauntlett, J.M. Izquierdo, P.K. Townsend, *Phys. Rev. Lett.* **63**, 2443 (1989).
22. S. Ferrara, M. Porrati, *Phys. Lett.* **B423**, 255 (1998).
23. E. Witten, *Nucl. Phys.* **B188**, 513 (1981).
24. E.A. Ivanov, A.A. Kapustnikov, *J. Phys.* **A11**, 2375 (1978); *J. Phys.* **G8**, 167 (1982).
25. E.A. Ivanov, A.A. Kapustnikov, *Phys. Lett.* **B143**, 155 (1984); *Nucl. Phys.* **B333**, 439 (1990).
26. J.A. de Azcárraga, J. Lukierski, *Phys. Lett.* **B113**, 170 (1982); *Phys. Rev.* **D28**, 1337 (1983).
27. P.K. Townsend, *Phys. Lett.* **B202**, 53 (1988).

INFINITEDIMENSIONAL SYMMETRIES AND PARTICLE MODELS

A. PASHNEV
JINR–Bogoliubov Laboratory of Theoretical Physics
Dubna, Head Post Office, P.O.Box 79, 101 000 Moscow, Russia
e-mail: pashnev@thsun1.jinr.ru

It is shown that vielbeins and connections of any spacetime are naturally described in terms of nonlinear realizations of the corresponding infinite - dimensional diffeomorphism groups. It is shown that dynamics of $D+2$ elements of the diffeomorphism group in one dimension describes the D - dimensional massless relativistic particles. The coordinates of this elements ($D+2$ einbeins, $D+2$ connections and 1 additional common coordinate of higher dimensionality) play the role of coordinates, momenta and Lagrange multiplier, needed for the manifestly conformal and reparametrization invariant description of the D - dimensional particle in terms of the $D+2$ - dimensional spacetime.

Introduction

As was shown in Borisov and Ogievetsky, [1] gravity can be described as a nonlinear realization of the four dimensional diffeomorphism group. The consideration was based on the fact that infinite dimensional diffeomorphism group in four dimensional space can be represented as the closure of two finite dimensional groups - conformal and affine. [2] As a consequence of such representation of the diffeomorphism group, the basic field in this consideration was the symmetric tensor field of the second rank - the metric field $g_{m,n}$, which corresponds to symmetric generators of the affine group. The generalization of this approach to the case of superspace was given in Ivanov. [3]

In the present work we consider nonlinear realization of the whole infinite dimensional diffeomorphism group of the arbitrary space. Among the coordinates parametrizing the group element (coset space) in such realization there are present usual coordinates of the space. The vielbein and connection are naturally represented as the functions of other coordinates of the coset space.

As is well known there exist several equivalent formulations of the massless relativistic particles. The second order and first order formalisms are examples of them. One more example is the conformally invariant description, [4] which starts from $D+2$ dimensional spacetime. The existence of alternative approaches always sheds some new light on the nature of the physical system. In particular, the conformally invariant description from the very beginning considers the particle coordinates and einbein on the equal footings.

In the present work we consider the natural description of the massless relativistic particle in terms of nonlinear realization of the infinite dimensional diffeomorphism group of the one dimensional space. We show that the spacetime coordinates and einbein are realized as dilatons of one dimensional diffeomorphism group. We consider simultaneously the dynamics of $D+2$ different points in the group space, hence they contain the same number of dilatons. The corresponding $(D+2)$ components of momentum are connected with the Cristoffel symbols. One more parameter of the group, having higher dimension, is the same for all $(D+2)$ points. It plays the role of Lagrange multiplier and effectively reduces the number of spacetime coordinates from $D+2$ to D ones.

Firstly we consider the case of the usual D - dimensional bosonic space with the coordinates s^m, $m = 0, 1, ..., D-1$. The generators of the corresponding diffeomorphism group regular at the origin can be written in the coordinate representation as

$$P^{m_1,m_2,\ldots,m_n}{}_m = is^{m_1}s^{m_2}\ldots s^{m_n}\frac{\partial}{\partial s^m}\,. \tag{1}$$

With the help of the representation (1) one can calculate the commutation relations between the generators and forget about the auxiliary coordinates s^m.

Let us consider the following parametrization of the group element:

$$G = e^{ix^m P_m}e^{i\phi^l{}_{l_1,l_2}P^{l_1,l_2}{}_l}e^{i\phi^k{}_{k_1,k_2,k_3}P^{k_1,k_2,k_3}{}_k}\ldots e^{i\phi^n{}_{n1}P^{n_1}_n}\,. \tag{2}$$

It is convenient to take the element of the finite dimensional group $GL(D)$, generated by $P^{n_1}_n$, as the last multiplier in the expression (2). Such structure of the group element simplifies the evaluation of the variations $\delta\phi^l{}_{l_1,\ldots,l_n}$ under the infinitesimal left action

$$G' = (1+i\epsilon)G, \epsilon = \epsilon^m P_m + \epsilon^m{}_{m_1}P^{m_1}{}_m + \epsilon^m{}_{m_1,m_2}P^{m_1,m_2}{}_m + \ldots\,. \tag{3}$$

The coordinates in (2) transform through the infinitesimal transformation parameters $\epsilon^m{}_{m_1,\ldots m_k}$ and coordinates, which are placed to the left from given ones in (2). The only exception is the transformation law for $\phi^n_{n_1}$, which includes only ϵ, x^m and $\phi^n_{n_1}$ itself. At this stage it is natural to consider all parameters as the fields in D - dimensional space parametrized by coordinates x^m.

The simplest transformation law have the dimension-one coordinates x^m. They transform as the coordinates of the D -dimensional space under the reparametrization:

$$\delta x^m = \varepsilon^m(x) \equiv \epsilon^m + \epsilon^m_{m_1}x^{m_1} + \epsilon^m_{m_1m_2}x^{m_1}x^{m_2} + \ldots\,. \tag{4}$$

The next parameters in the coset transform as a Cristoffel symbol:

$$\delta\phi^m{}_{m_1m_2} = \frac{\partial\varepsilon^m}{\partial x^n}\phi^n{}_{m_1m_2} - \frac{\partial\varepsilon^n}{\partial x^{m_2}}\phi^m{}_{m_1n} - \frac{\partial\varepsilon^n}{\partial x^{m_1}}\phi^m{}_{nm_2} + \frac{1}{2}\frac{\partial^2\varepsilon^m}{\partial x^{m_1}\partial x^{m_2}} . \quad (5)$$

In general the transformation law for parameter with n lower indices will contain the term with n-th derivative of the parameter $\epsilon^m(x)$. Only variations of the last parameters $\phi^n{}_{n_1}$ need the separate consideration. One can show[5] that the exponent of the matrix ϕ^m_n:

$$(e^{i\phi^n{}_{n_1}P^{n_1}_n})^l_k = (e^{-\phi})^l_k \equiv E^l_k , \quad (6)$$

transforms as the covariant vector with respect to its lower index

$$\delta E^l_k = -\frac{\partial\varepsilon^m(x)}{\partial x^k}E^l_m . \quad (7)$$

Simultaneously, its upper index is inert. This is the transformation law of the vielbein.

The fact that E^l_k is indeed the vielbein becomes evident if we consider the Cartan's differential form

$$\Omega = G^{-1}dG = i\Omega^a P_a + i\Omega^b_a P^a_b + i\Omega^b_{a_1a_2}P^{a_1a_2}_b + \dots , \quad (8)$$

which simultaneously with its components $(\Omega^a, \Omega^a_b, \dots)$ is invariant with respect to the left transformation (3). We emphasize the fact of invariance by using the letters from the beginning of the alphabet for indices. The explicit expressions for the components of the Ω -form are:

$$\begin{aligned}
\Omega^a &= E^a_m dx^m , & (9)\\
\Omega^a_b &= -E^m_b dE^a_m - 2dx^k\phi^m_{kn}E^n_b E^a_m , & (10)\\
\Omega^a_{bc} &= (d\phi^m_{kn} - dx^l\phi^m_{il}\phi^i_{kn} \\
&\quad + dx^l\phi^m_{in}\phi^i_{kl} + dx^l\phi^m_{ik}\phi^i_{ln} - 3dx^l\phi^m_{knl})E^a_m E^k_b E^n_c , \dots & (11)
\end{aligned}$$

The first of this forms is one-form vielbein. Under the right gauge transformation belonging to $GL(D)$

$$G' = G\{1 - ih(x)\} = G\{1 - ih^a_b(x)P^b_a\} \quad (12)$$

vielbein one -form $E^a \equiv \Omega^a$ transforms as vector

$$\delta E^a = h^a_b E^b . \quad (13)$$

So, this $GL(D)$ can be considered as the gauge group in the tangent space. All Ω -forms with higher number of indices transform homogeneously as corresponding tensors. The only exception is the differential one-form (10) which transforms inhomogeneously:

$$\delta\Omega^a_b = h^a_c\Omega^c_b - \Omega^a_c h^c_b - dh^a_b \, . \tag{14}$$

This is exactly the transformation law of the connection one-form and Ω^a_b is the natural candidate for the connection in the absence of any other tensors of second rank, which could be, in principle, added to the connection. So, the "minimal" one-form connection is given by (10) in terms of vielbein E^a_m and Cristoffel symbol ϕ^m_{kn}. The corresponding curvature two -form

$$R^a_b = d\Omega^a_b + \Omega^a_c\Omega^c_b \tag{15}$$

transforms as a tensor of second rank.

Due to its definition, the Ω -form satisfy the Maurer-Cartan equation

$$d\Omega + \Omega\wedge\Omega = 0 \, , \tag{16}$$

or, in components

$$d\Omega^a + \Omega^a_b\Omega^b = 0 \, , \tag{17}$$

$$d\Omega^a_b + \Omega^a_c\Omega^c_b + 2\Omega^a_{bc}\Omega^c = 0 \, , \tag{18}$$

$$d\Omega^a_{bc} + \Omega^a_d\Omega^d_{bc} + \Omega^a_{bd}\Omega^d_c + \Omega^a_{cd}\Omega^d_b + 3\Omega^a_{bcd}\Omega^d = 0, \ldots \, . \tag{19}$$

The lefthand side of first of these equations (17) is the covariant differential of the vielbein with the connection Ω^a_b. The fact of its equality to zero means vanishing of the corresponding torsion. Second equation represents the curvature two-form (15) in terms of vielbein and Ω-form with three indices

$$R^a_b = -2\Omega^a_{bc}\Omega^c \, . \tag{20}$$

The rest of equations express covariant differentials of Ω-forms in terms of other Ω-forms.

The following expression for the action

$$S = \int R^{a_1}_b \eta^{ba_2}\Omega^{a_3} \ldots \Omega^{a_D}\varepsilon_{a_1 a_2 \ldots a_D} \tag{21}$$

leads to Einstein - Hilbert action

$$S = \int d^D x\sqrt{-g}R, \qquad g^{mn} = \eta^{ab}E^m_a E^n_b \, , \tag{22}$$

after elimination of ϕ^m_{kn}. Using its equation of motion we can express then in terms of g^{mn}. Some comments are needed here. Up to now the gauge group acting in the tangent space (considered as right transformations (12)), was the group $GL(D)$ of general linear transformations in D - dimensions. In principle, one can construct action invariant under the whole $GL(D)$ gauge group, for example $\int R^b_a R^a_b$ in four dimensions. The presence in the action (21) of two constant tensors - absolutely antisymmetric tensor $\varepsilon_{a_1 a_2 \ldots a_D}$ and tangent space flat metric $\eta^{ab} = diag(1,1,...,1,-1)$ means that the invariance group of the action (21) is the subgroup of $GL(D)$, namely, the group $SO(D-1,1)$. So, the choice of the gauge group in the tangent space depends on the structure constants in the action, which can break $GL(D)$ down to its subgroup.

Now for convenience of reader we remind the conformally invariant description of the relativistic particle. [4] The action for massless relativistic particle in D - dimensional spacetime can be written in terms of $D+2$ coordinates $x_{\mathcal{A}}$, $\mathcal{A} = 0, 1, \ldots D+1$, of the spacetime with the signature

$$\Sigma_{\mathcal{A}} = (-\underbrace{++\ldots++}_{D}-): \tag{23}$$

$$S = \int d\tau (\frac{1}{2}\dot{x}^2 - \frac{1}{2}\lambda x^2)\,. \tag{24}$$

Besides of the $SO(D,2)$ invariance, it is gauge-invariant under the transformations

$$\delta x \;=\; \epsilon\dot{x} - \frac{1}{2}\dot{\epsilon}x\,, \tag{25}$$

$$\delta\lambda \;=\; \epsilon\dot{\lambda} + 2\dot{\epsilon}\lambda + \frac{1}{2}\dddot{\epsilon}\;. \tag{26}$$

The relation of the action 24 with the usual D - dimensional action is established by solving the equation of motion for the Lagrange multiplier λ

$$x^{\mathcal{A}} x_{\mathcal{A}} \equiv x^a x_a + 2x_+ x_- = 0;\;\; a = 0, \ldots, D-1;\;\; x_\pm = \frac{1}{\sqrt{2}}(x_D \pm x_{D+1})\,. \tag{27}$$

In terms of new variables

$$\tilde{x} = \frac{x}{x_+}\,,\quad e = \frac{1}{x_+^2}\,,\quad (x_- = -\frac{x^a x_a}{2x_+})\,, \tag{28}$$

the Lagrangian becomes

$$L = \frac{1}{2}\frac{\dot{\tilde{x}}^2}{e}\,. \tag{29}$$

Its reparametrization invariance

$$\delta\tilde{x} = \epsilon\dot{\tilde{x}}, \quad \delta e = \dot{\epsilon}e + \epsilon\dot{e} \tag{30}$$

is the consequence of (25-26).

So, the Lagrange multiplier λ plays the crucial role in the conversion of the $D+2$ - dimensional action into D - dimensional one. Nevertheless, their geometrical meaning as well as the nature of initial $D+2$ coordinates $x_{\mathcal{A}}$ is unclear. In the following we will show that all this functions of τ have an interpretation in terms of parameters of diffeomorphism groups.

Consider the auxiliary 1 - dimensional bosonic space with the coordinate s. The generators of the corresponding diffeomorphism group form the Virasoro algebra without central charge $[L_n, L_m] = -i(n-m)L_{n+m}$. The group element is parametrized as

$$G = e^{i\tau L_{-1}} \cdot e^{iU^1 L_1} \cdot e^{iU^2 L_2} \cdot e^{iU^3 L_3} \ldots e^{iU L_0}\,. \tag{31}$$

The transformation laws of the coordinates in (31) under the infinitesimal left action (3) are

$$\delta\tau \;=\; \varepsilon(\tau) \equiv \epsilon^0 + \epsilon^1\tau + \epsilon^2\tau^2 + \ldots\,, \tag{32}$$

$$\delta U \;=\; \dot{\varepsilon}(\tau)\,, \tag{33}$$

$$\delta U^1 \;=\; -\dot{\varepsilon}(\tau)U^1 + \frac{1}{2}\ddot{\varepsilon}(\tau)\,, \tag{34}$$

$$\delta U^2 \;=\; -2\dot{\varepsilon}(\tau)U^2 + \frac{1}{6}\dddot{\varepsilon}(\tau)\,. \tag{35}$$

At this stage it is natural to consider all parameters as the fields in one dimensional space parametrized by coordinate τ. It means the following active form of the transformations of the parameters $U(\tau), U^m(\tau)$

$$\delta U(\tau) \;=\; -\varepsilon(\tau)\dot{U}(\tau) + \dot{\varepsilon}(\tau)\,, \tag{36}$$

$$\delta U^1(\tau) \;=\; -\varepsilon(\tau)\dot{U}^1(\tau) - \dot{\varepsilon}(\tau)U^1(\tau) + \frac{1}{2}\ddot{\varepsilon}(\tau)\,, \tag{37}$$

$$\delta U^2(\tau) \;=\; -\varepsilon(\tau)\dot{U}^2(\tau) - 2\dot{\varepsilon}(\tau)U^2(\tau) + \frac{1}{6}\dddot{\varepsilon}(\tau)\,. \tag{38}$$

One can easily verify that the functions $x = e^{U(\tau)/2}$ and $\lambda = -3U^2(\tau)$ have exactly the transformation laws (25-26) with $\varepsilon(\tau) = -\epsilon$. Simultaneously $U^1(\tau)$ transforms as one dimensional Cristoffel symbol.

The independence of δU^2 from U and U^1 means that one can consider more than one group elements

$$G_{\mathcal{A}} = e^{i\tau L_{-1}} \cdot e^{iU^1_{\mathcal{A}} L_1} \cdot e^{iU^2 L_2} \cdot e^{iU^3_{\mathcal{A}} L_3} \ldots e^{iU_{\mathcal{A}} L_0}, \quad \mathcal{A} = 0, 1 \ldots, D+1\,, \tag{39}$$

which have identical values of parameters τ and $U^2(\tau)$ and differ in the values of all other parameters $U^m_{\mathcal{A}}$. This property is valid when all of them are transformed with the same infinitesimal transformation parameter $\varepsilon(\tau)$.

Consider the Cartan's differential forms for each value of $\mathcal{A}$ $\Omega_{\mathcal{A}} = G^{-1}_{\mathcal{A}} dG_{\mathcal{A}} = i\Omega^{-1}_{\mathcal{A}} L_{-1} + i\Omega^{0}_{\mathcal{A}} L_0 + i\Omega^{1}_{\mathcal{A}} L_1 + \ldots$. All their components $(\Omega^{-1}_{\mathcal{A}}, \Omega^{0}_{\mathcal{A}}, \Omega^{1}_{\mathcal{A}}, \ldots)$ are invariant with respect to the left transformation (3). The explicit expressions for the components of the Ω -form are:

$$\Omega^{-1}_{\mathcal{A}} = e^{-U_{\mathcal{A}}} d\tau\,, \tag{40}$$

$$\Omega^{0}_{\mathcal{A}} = DU_{\mathcal{A}} - 2d\tau U^1_{\mathcal{A}}\,, \tag{41}$$

$$\Omega^{1}_{\mathcal{A}} = (dU^1_{\mathcal{A}} + d\tau (U^1_{\mathcal{A}})^2 - 3d\tau U^2) e^{U_{\mathcal{A}}}, \ldots . \tag{42}$$

The most interesting is the form $\Omega^1_{\mathcal{A}}$. The action

$$\begin{aligned} S &= -\frac{1}{2}\int \sum_{\mathcal{A}} \Sigma_{\mathcal{A}} \Omega^1_{\mathcal{A}} \\ &= -\frac{1}{2}\int d\tau \sum_{\mathcal{A}} \Sigma_{\mathcal{A}} e^{U_{\mathcal{A}}} (\dot{U}^1_{\mathcal{A}} + (U^1_{\mathcal{A}})^2 - 3U^2)\,, \end{aligned} \tag{43}$$

where $\Sigma_{\mathcal{A}}$ is the signature (23), is invariant under the transformation (3) and corresponds to the first order formalism for the action (24). Indeed, up to the total derivative, after the change of variables

$$x_{\mathcal{A}} = e^{U_{\mathcal{A}}(\tau)/2}\,, \quad p_{\mathcal{A}} = e^{U_{\mathcal{A}}(\tau)/2} U^1_{\mathcal{A}}\,, \quad \lambda = -3U^2(\tau)\,, \tag{44}$$

it becomes

$$S_f = \int d\tau (\dot{x}p - \frac{1}{2}p^2 - \frac{1}{2}\lambda x^2)\,. \tag{45}$$

This action is invariant under the gauge transformations (25-26) supplied by

$$\delta p = \epsilon \dot{p} + \frac{1}{2}\dot{\epsilon} p - \frac{1}{2}\ddot{\epsilon} x\,. \tag{46}$$

After the elimination of $p_{\mathcal{A}}$ with the help of its equation of motion $p_{\mathcal{A}} = \dot{x}_{\mathcal{A}}$ the action (45) coincides with the action (24).

Conclusions

We have considered the nonlinear realizations of infinite - dimensional diffeomorphism groups of any space. The parameters of coset space in a very natural manner include the coordinates, vielbeins and connections of the corresponding space. The geometrical and physical meaning of higher parameters $u^M{}_{M_1 \ldots M_n}$ with $n \geq 3$ is still unclear. Construction of the differential Ω -forms invariant under the action of diffeomorphism group is straightforward. At the same time the $GL(D)$ gauge group, considered as the right action on the group element, plays the role of gauge group in the tangent space. The most of the Ω-forms transform as tensors of this $GL(D)$. The only Ω-form with two tangent indices Ω_b^a plays the role of connection which automatically is torsionless.

Such an invariant differential Ω - forms can be considered as building blocks for construction of integral invariants like action. In D-dimensional space (21) gives the expression for the gravity action. The same Ω - forms in 1-dimensional space are used to construct the conformally invariant action (43), which describes the massless particle in D dimensions. The only difference is, that in this last case one needs $D+2$ copies of the group elements.

In Pashnev [6] the conformally and reparametrization invariant actions for $N=1$ spinning particle in arbitrary dimension D were constructed. It would be interesting to apply the method developed here and in Pashnev [5,6] to other infinite dimensional symmetries, such as diffeomorphism groups of extended superspaces and higher dimensional spaces, W-algebras and so on.

Acknowledgments

I would like to thank Ch. Preitshopf for useful discussions. This investigation has been supported in part by the Russian Foundation of Fundamental Research, grant 99-02-18417, joint grant RFFR-DFG 99-02-04022, and INTAS, grant 96-0538.

References

1. A.B. Borisov, V.I. Ogievetsky, *Theor. Math. Phys.*, **21**, 329 (1974).
2. V.I. Ogievetsky, *Lett. Nuovo Cim.*, **8**, 988 (1973).
3. E.A. Ivanov, I. Niederle, *Phys. Rev.*, bf D45, 4545 (1992).
4. R. Marnelius, *Phys. Rev.*, **D20**, 2091 (1979).
5. A. Pashnev, *Nonlinear realizations of the (super)diffeomorphism groups, geometrical objects and integral invariants in the superspace*, Preprint JINR E2-97-122. e-Print archive hep-th/9704203.
6. A. Pashnev, *Massless and spinning particles as dynamics in one dimensional (super)diffeomorphism groups*, Preprint JINR E2-99-42, e-Print archive hep-th/9902143.

Integrable Models and Related Topics

STOCHASTIC MATRIX ALGEBRAS APPLIED TO MODELS OF AGGLOMERATION AND GROWTH[a]

R. ALDROVANDI[b] AND R. KERNER
Laboratoire de Gravitation et Cosmologie Rélativistes
Université Pierre et Marie Curie CNRS/ESA 7065
Tour 22/12 4ème étage BP 142
4 Place Jussieu Cedex 05 75252 Paris France

1 Introduction

A light ray following its curved trajectory through interstellar space to be caught by the awaiting astronomer's telescope and help to verify General Relativity predictions is a token of modernity. It makes use, however, of a very special property of light, its responsiveness to the variations in the structure of the space it travels in. In that way, it reminds of ancient observations of light bent by unknown vagaries in the composition of the oldest non-trivial media known to mankind, glasses. We believe to have, today, a good knowledge of how the geometry of spacetime is modified by the presence of momentum-energy sources. We know much less about the origin and peculiarities of glass internal geometry.

Maybe one day we shall understand how the geometry of glasses, as that simplest curvature generated by refraction indexes, comes up from its atomic structure. For the time being, we shall commit ourselves to understanding the very formation of amorphous solids.

Vitrification is a non-equilibrium process. But it is a slow one, and can be described by a step–by–step approach. It is a rare opportunity to look at a non-equilibrium process as at a slow-motion picture. As if we could take the underlying dynamical process by surprise. The approach to equilibrium is a delicate matter, to be considered with extreme care. Recent advances in the understanding of the vitrification process follow that rule of prudence, which is further justified by their very success. They consider [1] a system whose initial equilibrium is disturbed by the insertion of nucleation centers, of destabilizing seeds. Each seed is a primaeval agglomerate, which presents to the exterior medium the valences of its rim constituents. The medium remains in equilibrium, and provides a thermal bath with a well-defined temperature. And equilibrium also prevails locally, at the level of interatomic encounters. This

[a]Talk given by R. Kerner

[b]On leave of absence from Instituto de Física Teórica, UNESP, São Paulo, Brazil.

means that binding probabilities are given by Boltzmann factors. What is out of equilibrium are the agglomerates themselves, which grow continuously, then join each other to occupy more and more of the medium space and cover it all at the end of the transition. Most of the experimentally confirmed results come from an additional stability assumption: That the concentrations of the atom species involved are, at the end of the process, the same as those in the medium. Imposition of this hypothesis requires the knowledge of the asymptotic behavior of the evolution process. We shall illustrate these ideas through a particularly simple example, that of a Selenium glass modified by the presence of Arsenic. If the initial medium is a Se_3As_2 liquid, the final glass will recover the same proportion of Se and As. Actually, old glasses are mostly SiO_2–based, but Se_3As_2 is simpler, though sophisticated enough to exhibit all the main points.

Evolution takes place by the binding of free atoms of the medium to the available bonds at the surface of a seed agglomerate. In the stochastic approach we adopt here, the state of a typical agglomerate's rim is represented by a probability distribution vector, which gives the relative frequencies of all possible sites on the rim. Each step in the evolution process is described by the transition matrix M, which must be chosen once and for all.[2] The state of the medium is fixed, and given by the concentrations of the different atomic species and the temperature. The progressive growth of the initial agglomerate is modeled by the action of successive powers of M. The final configuration, actually an intricate lace-like network of atoms and bonds, is provided by the asymptotic matrix $M^{\infty} = \lim_{m\to\infty} M^m$. It is necessary to fix beforehand what will be understood by a state and by a step. We can take as a new state that which is obtained by the addition of one single atom (bond saturation), or that resulting from the complete saturation of all available bonds (rim saturation). We shall examine both cases (Sections 2.1 and 2.2), and discover that they lead to the same results. The transition matrices used have very special properties: they are actually stochastic matrices.[6] Thus, the basic tenets once acquired, we formalize the whole procedure in a short introduction to stochastic matrices (Section 3). Special attention is called to the many consequences of a powerful result from the theory of positive operators, the Perron-Frobenius theorem. Some of the findings of the Se–As examples appear then as very general characteristics of Markov chains. Considerations on the asymptotic behavior of matrices and states follow then, and the notion of detailed balancing is shown to have been implicitly used. The system will be shown to be microscopically out of equilibrium, but in equilibrium at a large enough scale. That scale is given by the size of an entirely saturated rim.

2 Of Arsenic, and New Lace

Consider then a Selenium glass "poisoned" with Arsenic, which acts as modifier. The basic cells are selenium and arsenic atoms, indicated respectively by (—o—) and (—•⟨). The concentration of free *As* atoms in the exterior medium will be called c and that of *Se*, $(1-c)$. Three kinds of possible bonds can be exposed at the surface, as shown in the left columns of Figures 1 and 2 below.

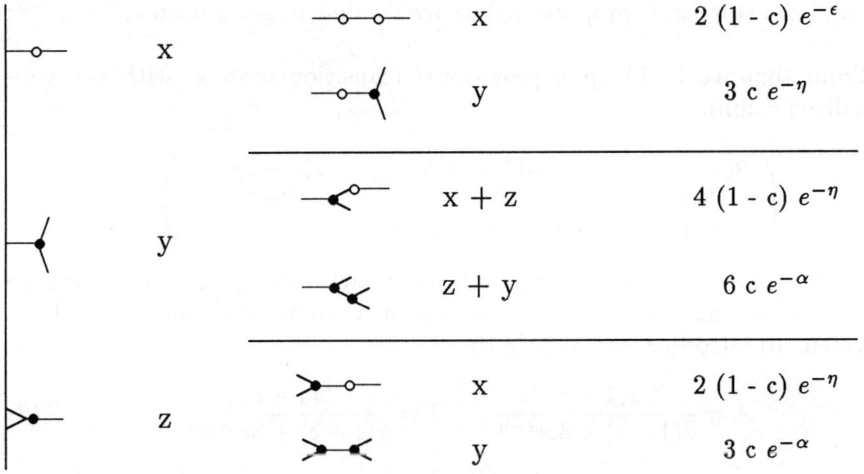

Figure 1: *States, steps and matrix entries in the one-bond saturation case.*

Two choices are possible, in what concerns states and transition matrix. The first is illustrated in the columns of Figure 1. Three possible kinds of sites, with one or two free bonds offered at the rim will be indicated by $x =$ o—, $y =$ •⟨ and $z =$ •—. To each site one of the two basic cells can attach itself, reproducing one of the initial configurations, in the specific combinations shown in the next column of the Figure. The attachment of *one single basic cell*, or the saturation of one single bond, is a step in the evolution. In the second choice, each step is obtained by the complete saturation of all the bonds at the rim. Let us start with the first possibility.

2.1 *Bond Saturation*

The probabilities constituting the entries of the stochastic matrix will depend on (i) the number of valences of the incoming particle and at the rim site (ii) the concentration of the corresponding species (iii) the binding energies in each case, through Boltzmann factors. These factors will be of three kinds:

$$\circ\!\!-\!\!\circ \Longleftrightarrow e^{E_{SeSe}/kT} = e^{-\epsilon} \, ,$$
$$\circ\!\!-\!\!\bullet \Longleftrightarrow e^{E_{SeAs}/kT} = e^{-\eta} \, ,$$
$$\bullet\!\!-\!\!\bullet \Longleftrightarrow e^{E_{AsAs}/kT} = e^{-\alpha} \, .$$

For the one-atom-at-a-time approach, the unnormalized probability factors are shown in the last column. The probability of obtaining state y after one step, starting from state x, is proportional to $3ce^{-\eta}$; that to get y from z, to $3ce^{-\alpha}$; and so on.

From that we build up a provisional transition matrix, with not–yet–normalized columns,

$$\begin{pmatrix} 2(1-c)e^{-\epsilon} & 4(1-c)e^{-\eta} & 2(1-c)e^{-\eta} \\ 3ce^{-\eta} & 6ce^{-\alpha} & 3ce^{-\alpha} \\ 0 & 4(1-c)e^{-\eta}+6ce^{-\alpha} & 0 \end{pmatrix} , \tag{1}$$

and then proceed to normalize the entries in each column to have their sum equal to 1. To get the stochastic matrix in a simple final form, it will be convenient to introduce the combinations

$$A = \frac{2(1-c)}{2(1-c)+3ce^{\epsilon-\eta}} \, ; \quad B = \frac{2(1-c)}{2(1-c)+3ce^{\eta-\alpha}} \, . \tag{2}$$

The transition matrix governing the process takes on the following form:

$$M = \begin{pmatrix} A & \frac{B}{2} & B \\ 1-A & \frac{1-B}{2} & 1-B \\ 0 & \frac{1}{2} & 0 \end{pmatrix} . \tag{3}$$

A general state on which M acts is described by a probability distribution $P = (p_x, p_y, 1-p_x-p_y)$. Matrix M has the eigenvalues $\lambda_1 = 1$; $\lambda_{2,3} = \frac{2A-B-1\mp\sqrt{(2A-B-1)^2-8(B-A)}}{4}$. We can check that, for $A \leq 1$ and $B \leq 1$, the absolute values of λ_2 and λ_3 are smaller then 1. The eigenvector corresponding to the unit eigenvalue, which is

$$v = \frac{1}{3-3A+2B} \begin{pmatrix} 2B \\ 2(1-A) \\ (1-A) \end{pmatrix} , \tag{4}$$

will be preserved by higher powers of M, while the other will be progressively damped. Thus, the system evolution will move preferentially along that stable eigenvector, which represents the asymptotic state.

What we need, therefore, is the asymptotic expression $M^\infty = \lim_{m\to\infty} M^m$. It is a troublesome task to get the successive powers of M and try to recognize a closed expression from them. There is fortunately an alternative procedure leading to the same result. In effect, consider the characteristic equation of M:

$$P(\lambda) = (\lambda - \lambda_1)(\lambda - \lambda_2)(\lambda - \lambda_3) = 0 .$$

The Cayley–Hamilton theorem [8] says that M will satisfy the equation:

$$(M - \lambda_1 I)(M - \lambda_2 I)(M - \lambda_3 I) = 0 , \tag{5}$$

where I is the identity matrix. Multiple eigenvalues would lead to repeated factors, but they do not appear in the cases considered here. Le t us isolate one of the factors, *e.g.* that corresponding to the eigenvalue λ_1, with its eigenvector v_1, and look at the properties of the remaning expression, $\hat{Z}_1 = (M - \lambda_2 I)(M - \lambda_3 I)$. The equation says then that $\hat{Z}_1$ is an eigenmatrix of M with eigenvalue λ_1: $M\hat{Z}_1 = \lambda_1 \hat{Z}_1$. Consequently,

$$\hat{Z}_1 v_1 = (\lambda_1 - \lambda_2)(\lambda_1 - \lambda_3) v_1 \quad \text{and} \quad \hat{Z}_1 \hat{Z}_1 = (\lambda_1 - \lambda_2)(\lambda_1 - \lambda_3) \hat{Z}_1 .$$

The trace of $\hat{Z}_1$ is also $(\lambda_1 - \lambda_2)(\lambda_1 - \lambda_3)$. When the eigenvalues are distinct, we can divide $\hat{Z}_1$ by this value, so as to have

$$Z_1 = \frac{(M - \lambda_2 I)(M - \lambda_3 I)}{(\lambda_1 - \lambda_2)(\lambda_1 - \lambda_3)} . \tag{6}$$

This matrix is an eigenprojector: it is now an idempotent, $Z_1 Z_1 = Z_1$, its trace is equal to 1 and $Z_1 v_1 = v_1$. If v_2 and v_3 are the eigenvectors of M with eigenvalues λ_2 and λ_3, then $Z_1 v_2 = 0$, $Z_1 v_3 = 0$. The procedure can be repeated for each factor in (5), defining analogous matrices Z_2 and Z_3. The product of any two distinct Z_k is proportional to $P(M)$ and, consequently, vanish. In consequence,

$$Z_k \, Z_{k'} = \delta_{kk'} \, Z_k \quad \text{and} \quad tr\,(Z_k Z_{k'}) = \delta_{kk'} .$$

Direct computation shows also that $I = \sum_k Z_k$ and $M = \sum_k \lambda_k \, Z_k$. A decomposition of matrix space, orthogonal in the sense of the trace metric $g(M_1, M_2) = tr(M_1 M_2)$, is thereby obtained. We can say that Z_1 projects M

"along" the direction of eigenvector v_1. There is an extra bonus. From the above properties it follows that

$$M^m = \sum_k {\lambda_k}^m \; Z_k \; . \tag{7}$$

Consequently, given any power series function $F(z) = \sum_k c_k z^k$, a matrix function $F(M)$ is formally defined by

$$F(M) = \sum_k F(\lambda_k) \; Z_k \; . \tag{8}$$

We say "formally" because there are, of course, convergence conditions to make this expression meaningful. These conditions (see Gantmacher [9]) are that all the eigenvalues stay inside the convergence radius of the series.

In our case, $\lambda_1 = 1$ and $|\lambda_k| < 1$ for $k = 2,3$. From (7), we see that, while Z_1 remains always the same, Z_2 and Z_3 are progressively damped. The asymptotic limit of M is thus Z_1 itself, which can be calculated from (6):

$$M^\infty = Z_1 = \frac{2}{3(1-A)+2B} \begin{pmatrix} B & B & B \\ 1-A & 1-A & 1-A \\ \frac{1-A}{2} & \frac{1-A}{2} & \frac{1-A}{2} \end{pmatrix} . \tag{9}$$

Notice that each column is exactly the stationary eigenvector (4), and the matrix, being a projector, is singular of rank 1.

Some extreme cases are illustrative. The pure Selenium case $c = 0$ gives $A = B = 1$, eigenvalues $\lambda_1 = 1$, $\lambda_2 = 0$ and $\lambda_3 = 0$. The stable eigenvector corresponding to the eigenvalue $\lambda_1 = 1$ is the pure Selenium state $(1,0,0)$, consistent with what is seen in Figure 1. It is immediate to find that M^∞, applied to an arbitrary distribution $P = (p_x, p_y, 1 - p_x - p_y)$, gives precisely that eigenvector.

Though the realistic cases involve only small values of c, it is instructive to consider also the pure Arsenic case $c = 1$. It leads to $A = B = 0$, with eigenvalues $\lambda_{1,2,3} = 1, -1/2, 0$. The stable eigenvector, which is then $(0, 2/3, 1/3)$, is also the result of applying M^∞ to an arbitrary initial distribution.

Another extreme case is that of high temperatures. When kT is much larger than the energy differences appearing in (2), the Boltzmann factors tend to 1, and we have $A = B = 2(1-c)/(2+c)$, $\lambda_1 = 1$, $\lambda_2 = (a-1)/2$, $\lambda_3 = 0$ and

$$M^\infty = \frac{1}{4+5c} \begin{pmatrix} 4(1-c) & 4(1-c) & 4(1-c) \\ 6c & 6c & 6c \\ 3c & 3c & 3c \end{pmatrix} .$$

The stable eigenvector, corresponding to $\lambda_1 = 1$, is

$$v = \frac{1}{4+5c}\begin{pmatrix} 4(1-c) \\ 6c \\ 3c \end{pmatrix}.$$

In all the cases above, we find that the asymptotic transition matrix is an idempotent, that is, it satisfies $(M^\infty)^2 = M^\infty$.

Notice from Figure 1 that, on the agglomerate's surface, p_x is the Se concentration and $p_y + p_z$ is the As concentration. Now, the high homogeneity exhibited by known glass structures suggests that, even in relatively small clusters, deviations from the average modifier concentration c must be negligible. Thus, *in the bulk*, the As concentration should be equal to c. To compute the bulk concentration, however, we should count only the y–sites created on the surface during the agglomeration process. Indeed, counting the z–sites would amount to counting the same atoms twice, because all the z–sites are created out of previously counted y–sites. Therefore, the condition of minimal fluctuations in the bulk concentration would read

$$\frac{p_y}{p_x} = \frac{2(1-A)}{2B} = \frac{c}{1-c},$$

which leads to the explicit dependence $c(T_g)$:

$$c = \frac{2(3\xi - 2)}{12\xi - 9\mu - 4}. \tag{10}$$

where we have introduced the abbreviated notation $\xi = e^{\epsilon-\eta}$ and $\mu = e^{\epsilon-\alpha}$.

This equation can be checked against experiment. For example, we can evaluate the derivative $\frac{\partial T}{\partial c} = \left(\frac{\partial c}{\partial T}\right)^{-1}$ for a given value of c. In particular, as $c \to 0$, when we can neglect the As–As bond creation (equivalent to putting $\mu = 0$ in (10)), we get

$$\left[\frac{\partial T}{\partial c}\right]_{c=0} = \frac{T_{g_0}}{\ln(3/2)},$$

(where T_{g_0} is the glass transition temperature of pure Se). This is the present–case expression of the general formula given by the stochastic approach,

$$\left[\frac{\partial T}{\partial c}\right]_{c=0} = \frac{T_{g_0}}{\ln(m'/m)}, \tag{11}$$

(where m and m' are the valences of the basic glass former and of the modifier), which is in very good agreement with experimental data (see Kerner et al. and Gerardo et al.[4,5]).

2.2 Rim Saturation

We have so far considered the accretion of a single basic cell as the unit step in the system evolution. We take now an alternative approach, in which the unit step in the progression, described by the transition matrix, would be the complete saturation of all the available free bonds on the cluster rim. This is represented in Figure 2.

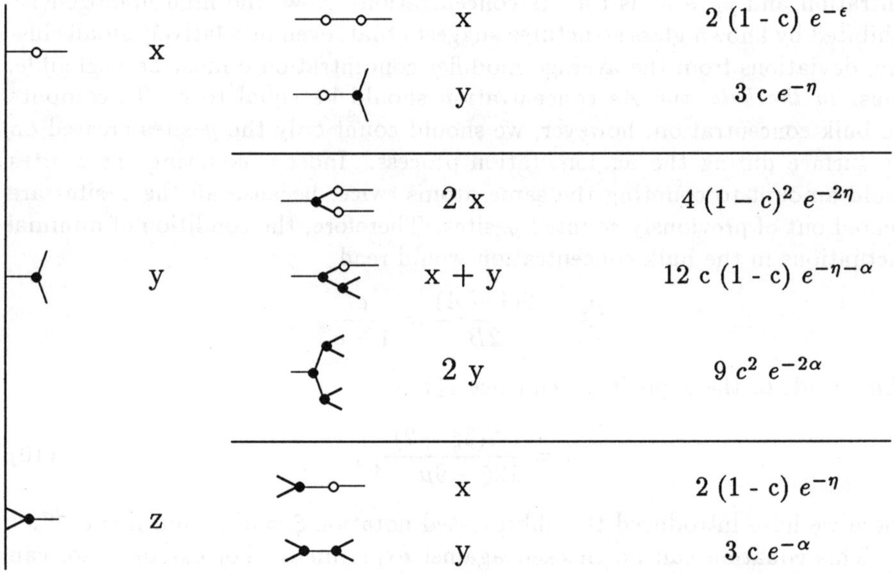

Figure 2: ***States, steps and matrix entries in the rim saturation case.***

Given the same initial state as before, the new state takes into account all the possible ways of saturating its bonds by the available external particles. The probabilities change, some of them becoming quadratic in the concentration. Notice that there is no final state z, so that the problem reduces to 2 dimensions.

The procedure is the same as before. The normalized transition matrix is written as

$$M = \begin{pmatrix} A & B \\ 1-A & 1-B \end{pmatrix}, \tag{12}$$

where A and B turn out to have exactly the same expressions (2) of the former

3×3 case. The eigenvalues are 1 and $A - B$. The asymptotic form is

$$M^{\infty} = \frac{1}{1 + B - A} \begin{pmatrix} B & B \\ 1 - A & 1 - A \end{pmatrix}, \tag{13}$$

and the stationary eigenvector is

$$v = \frac{1}{1 + B - A} \begin{pmatrix} B \\ 1 - A \end{pmatrix}, \tag{14}$$

which appears as the columns in M^{∞}. We find also here that $(M^{\infty})^2 = M^{\infty}$. We again impose that the asymptotic state is fixed by the external concentration, which means that the above eigenvector must equal the medium distribution vector $(1 - c, c)$. The solutions are $c = 0$, $c = 1$ and the nontrivial one

$$c = \frac{1 - A}{1 + B - A}, \tag{15}$$

which reduces to (10) when A and B are replaced by their values (2).

Thus, at least in what concerns this condition and its consequences, the two approaches — bond and rim saturation — lead to the same results. Notice that, here and in the previous section, it is not actually necessary to get the asymptotic eigenvector to arrive at this condition. It is enough to apply M directly to any average state of the cluster's rims. This coincidence can be interpreted as an example of ergodicity in the system under consideration: randomness in time is equivalent to randomness in the space of states.

Let us remark a few points. The transition matrices have been built up with positive entries equal or lesser than 1, and such that their sum along a column is always = 1. An eigenvalue = 1 has been found in both cases, and other eigenvalues with absolute values ≤ 1. The asymptotic transition matrices have been found to be idempotents. And, in each case, the columns of these asymptotic idempotents are just the stationary eigenvector.

Matrices with such properties are *stochastic matrices*. We shall now proceed to formalize and generalize what we have found in the above examples. This will allow us to insert the above evolution scheme cases into a very general framework, that of the Markov chains. Lest it be thought that we are extrapolating too far from only two cases, it should be said that the examples given are typical of a method which has been successfully applied to many other cases and always exhibit the features we have stressed. It has been shown to give good agreement with experiment not only for binary systems like the above ones, [4] but also to systems with 3 components. [5] And not only for the dendritic growth we have considered: it also works when rings are a dominant structure. [6]

3 Stochastic Matrices and Markov Chains

3.1 General Properties

An $N \times N$ matrix M is a stochastic matrix if its entries are real numbers satisfying the two conditions

$$0 \leq M_{\alpha\beta} \leq 1 \,, \quad \forall\, \alpha, \beta = 1, 2, \ldots N \,, \tag{16}$$

and

$$\sum_{\alpha=1}^{N} M_{\alpha\beta} = 1 \,. \tag{17}$$

As an immediate consequence, we have

$$\sum_{\alpha=1}^{N} {M^2}_{\alpha\beta} = \sum_{\alpha=1}^{N} \sum_{\gamma=1}^{N} M_{\alpha\gamma} M_{\gamma\beta} = 1 \,. \tag{18}$$

It follows that that M^2, as well as all the successive powers M^m, are also stochastic matrices. We have then a Markov chain (that is, a discrete Markov process) with ***transition matrix*** M. Suppose there is some physical system whose states are indicated by the discrete labels $\alpha, \beta, \gamma \ldots$ The entry $M_{\alpha\beta}$ is then the probability of transition from state β to state α and the power m is a discrete "time" parameter. We shall call a column-vector v a probability distribution when each entry satisfies $0 \leq v_\alpha \leq 1$ and further $\sum_\alpha v_\alpha = 1$. Thus, each column in a stochastic matrix is a probability distribution. In the formalism, the complete characterization of the state of the system is given by a similar distribution, whose meaning is that the system has probability v_α of being in state α, probability v_β of being in state β, etc.

If $P_n(\alpha)$ is the probability of the system being in state α at time n, its evolution is governed by the equation

$$P_{n+1}(\alpha) = \sum_{\beta=1}^{N} M_{\alpha\beta} P_n(\beta) \,. \tag{19}$$

Condition (17) implies that probability is conserved:

$$\sum_{\alpha=1}^{N} P_{n+1}(\alpha) = \sum_{\beta=1}^{N} P_n(\beta) \,. \tag{20}$$

In our examples we assumed a finite number N of states, but all that was said extends to a discrete countable set of states. A Markov process [11] is actually characterized by a transition matrix M and an initial distribution P_0, and by the Markov condition. This condition states that P_{n+1} depends directly on its immediate predecessor P_n, but not on any other previous distribution $P_{r\leq n}$. This "lack of memory" is the basic property of such processes, and is clearly respected in the given examples, in which only the rim — and not the previously formed interior of the agglomerate — is of any account in each step. The chain is indicated by (M, P_0), or simply $\{P_n\}$. A finite Markov chain $\{P_n, n = 1, 2, \ldots, N\}$ is *reversible* if $\{P_{N-n}\}$ is also a Markov chain.

The probability distribution $P(a)$ is an equilibrium distribution if it is an eigenvector of M with eigenvalue 1:

$$P(\alpha) = \sum_{\beta=1}^{N} M_{\alpha\beta} P(\beta) . \tag{21}$$

This describes a stationary solution of (19).

The spectrum of M is, of course, of primary interest. A first result comes trivially from the Cayley-Hamilton theorem. If the equation is written as

$$\Delta(\lambda) = \prod_{j=0}^{N} (\lambda - \lambda_j) = \sum_{j=0}^{N} c_j \lambda^j = 0 ,$$

then $\Delta(M) = 0$ and $\sum_{m=0}^{N} c_m M^m{}_{\alpha\beta} = 0$. Taking $\sum_\alpha$ and using (18), we find $\sum_{m=0}^{N} c_m = 0$, which says that 1 is an eigenvalue of M. Thus, a stochastic matrix has always at least one eigenvector of eigenvalue 1, that is, at least one equilibrium distribution.

The spectrum of positive, or non-negative matrices like M is the subject of a family of theorems going after the names of Perron and Frobenius.[7] The version which best fits our necessities runs as follows:

If M is a non-negative matrix, then

1. there exists a non-negative real number r which is an eigenvalue and is such that the absolute values of the other eigenvalues are at most $= r$;

2. there exists an eigenvector v corresponding to that maximal eigenvalue r whose components are all real non-negative;

3. if M is irreducible and of order > 1, the eigenspace corresponding to r is an 1-dimensional space spanned by a strictly positive vector;

4. if M is irreducible and of order > 1, the other eigenvalues with modulus $= r$ are *simple* eigenvalues, and have the form $re^{i2\pi/k}$ for some k.

Recall that a matrix M is irreducible when no permutation matrix P exists such that $P^{-1}MP$ is of the form $\begin{pmatrix} M_1 & M_2 \\ 0 & M_3 \end{pmatrix}$, with M_1 and M_3 matrices of lower orders.

We can learn a lot from this theorem after adapting it to our case, which is very special. For a non-negative matrix respecting conditions (16) and (17), the real number r is actually 1, so that the eventual additional eigenvalues with modulus 1 are k-th roots of unity (if M is irreducible). Let us see why the largest possible eigenvalue of M has modulus 1. Take the eigenvalue equation

$$\sum_{\beta=1}^{N} M_{\alpha\beta}\ v_\beta = \lambda\ v_\alpha\ . \tag{22}$$

A first consequence comes from summing over α. An alternative follows: either $\lambda = 1$ or $\sum_{\alpha=1}^{N} v_\alpha = 0$. This means that, of all the eigenvectors of the transition matrix, only those corresponding to $\lambda = 1$ can be probability distributions. A second consequence comes from taking absolute values:

$$\sum_{\beta=1}^{N} M_{\alpha\beta}\ |v_\beta| \geq |\lambda|\ |v_\alpha|\ . \tag{23}$$

Again, summation over α leads to

$$|\lambda| \leq 1\ . \tag{24}$$

Equality only holds if

$$\sum_{\beta=1}^{N} M_{\alpha\beta}\ |v_\beta| = \ |v_\alpha|\ , \qquad \forall\alpha\ , \tag{25}$$

that is, when $|v_\alpha|$ is an equilibrium distribution. In conjunction with the Perron-Frobenius theorem, which ensures the existence of a *real* maximal eigenvalue, the above result says that a stochastic matrix has always at least one eigenvalue $= 1$, as found in our examples. The theorem will then imply the existence of at least one entirely real eigenvector, with $\lambda = 1$ and $v_\alpha = |v_\alpha|$ for all α. The statements of items (4) and (3), by which the eigenvalues are simple and the space corresponding to the eigenvalue 1 is one-dimensional, ensure that an irreducible M will have a single eigenvalue $= 1$.

A bridge towards continuum evolution can be thrown by taking $M_{\alpha\beta} = \delta_{\alpha\beta} + \epsilon Q_{\alpha\beta}$, with $\epsilon \ll 1$. The evolution equation (19) becomes a difference equation,

$$\frac{1}{\epsilon}[P_{n+1}(\alpha) - P_n(\alpha)] = \sum_{\beta=1}^{N} Q_{\alpha\beta} P_n(\beta) \tag{26}$$

leading, in the limit of continuum time, to

$$\frac{\partial}{\partial t} P(t,\alpha) = \sum_{\beta=1}^{N} Q_{\alpha\beta} P(t,\beta)\,. \tag{27}$$

3.2 Projectors and Asymptotics

The discussion involving projectors in Section 2.1 can be put into more general terms. Suppose a function $F(\lambda)$ is given which can be expanded as a power series $F(\lambda) = \sum_{k=0}^{\infty} c_k (\lambda - \lambda_0)^k$ inside a convergence circle $|\lambda - \lambda_0| < r$. Then the function $F(M)$, whose argument is now some given $N \times N$ matrix M, is defined by $F(M) = \sum_{k=0}^{\infty} c_k (M - \lambda_0)^k$, and has a sense whenever the eigenvalues of M lie within the convergence circle. In our case, any series with convergence radius > 1 will lead to a well-defined matrix function. Given the eigenvalues $\lambda_1, \lambda_2, \cdots, \lambda_N$ of M, there exists a set of eigenprojectors $\{Z_j[M]\}$ which constitutes a matrix basis, in which M is written as

$$M = \sum_{j=1}^{N} \lambda_j \; Z_j[M]\,. \tag{28}$$

Provided the eigenvalues of M lie inside the convergence radius,[9] the function $F(M)$ will be given by the matrix

$$F(M) = \sum_{j=1}^{N} F(\lambda_j) \; Z_j[M]\,. \tag{29}$$

Thus, for example,

$$M^m = \sum_{j=1}^{N} \lambda_j{}^m Z_j[M]\,. \tag{30}$$

These projectors have many interesting properties.[13,14] For the case of a non-degenerate matrix M, (i) they are idempotents, $Z_j^2 = Z_j$; (ii) they can be normalized to $tr(Z_j) = 1$; (iii) they are orthonormal by the trace, $tr(Z_i Z_j) = \delta_{ij}$; (iv) they are eigenmatrices of M,

$$\sum_{\beta} M_{\alpha\beta} (Z_j)_{\beta\gamma} = \lambda_j (Z_j)_{\alpha\gamma}\,. \tag{31}$$

The closed form for each Z_j in terms of powers of M is

$$Z_j[M] = \frac{(M-\lambda_1 I)(M-\lambda_2 I)...(M-\lambda_{j-1}I)(M-\lambda_{j+1}I)...(M-\lambda_{N-1}I)(M-\lambda_N I)}{(\lambda_j-\lambda_1)(\lambda_j-\lambda_2)...(\lambda_j-\lambda_{j-1})(\lambda_j-\lambda_{j+1})...(\lambda_j-\lambda_{N-1})(\lambda_j-\lambda_N)} . \quad (32)$$

In the case of a stochastic matrix, (30) implies that all components related to eigenvalues with absolute values < 1 will be progressively damped down. If M is irreducible and has a single eigenvalue with unit absolute value, then its powers will tend to the corresponding projector,

$$M^\infty = Z_{(1)}[M] . \quad (33)$$

We had found idempotents for M^∞ in our examples. We see that this holds for any M with a single unit–modulus eigenvalue. In that case, the system will tend to the sole eigenstate v^∞ of $Z_{(1)}[M]$, which will be the final equilibrium state. We see in (31) that each column of $Z_{(1)}[M]$ is an eigenvector. If that state is single, all the columns must be equal to each other and to v^∞, which ties with the examples we have given. The Perron-Frobenius theorem also indicates which cases have indefinite asymptotics: those with more than one unit–modulus eigenvalue. The simplest example is the matrix $\begin{pmatrix} 0 & 1 \\ 1 & 0 \end{pmatrix}$, whose powers oscillate indefinitely.

3.3 Hamiltonian Language

We can rephrase much of what has been said above in a more "physical" language. Take the evolution equation (19) in the form

$$P_n(\alpha) = \sum_{\beta=1}^{N} (M^n)_{\alpha\beta}\, P_0(\beta) . \quad (34)$$

It gives the state at a time n, given the initial state at time 0. This can be put into a "Hamiltonian" form if we use (29) to define the Hamiltonian matrix $H = \ln M$,

$$H = \sum_{k=1}^{N} (\ln \lambda_k)\, Z_{(k)}[M] , \quad (35)$$

with λ_k's the eigenvalues of M. We have then

$$M = e^H = \sum_{k=1}^{N} e^{\ln \lambda_k}\, Z_{(k)}[M] , \quad (36)$$

and

$$P_n(\alpha) = \sum_{\beta=1}^{N} \left(e^{nH}\right)_{\alpha\beta} P_0(\beta) \, . \tag{37}$$

This form of the evolution equation bears a strong similarity to that of general dynamical systems. It has the aspect of the formal solutions of the Liouville equation,[15] and the unfolding of dynamic mappings in the development of chaotic systems.[14] Now, an eigenvalue 1 for M means an eigenvalue 0 for H. The corresponding eigenvector is a zero–energy state, a "vacuum", and does not suffer any evolution. The existence of a vacuum with real components is stated by the Perron-Frobenius theorem.

If M is reducible and has several eigenvalues = 1, the set of the corresponding eigenvectors will constitute a multiple vacuum. Only those components will remain which are along the vacuum projectors when $n \to \infty$. The other are damped down. The physical view becomes clearer now. The vacuum will correspond to the asymptotic limit of the system, the equilibrium state. If multiple, the system will tend to constitute several unconnected domains. In other words, once equilibrium is attained, each vacuum will determine a *phase* of the system. Notice, however, that such phases can be intermingled so as to elude macroscopic observation.

This Hamiltonian language can be extended to the continuum–time case. The continuum version of (30), for example, would be given by

$$M^t = e^{tH} = \sum_{k=1}^{N} e^{t \ln \lambda_k} Z_{(k)}[M] \, . \tag{38}$$

3.4 Detailed Balancing

Which stochastic processes would converge to a given equilibrium distribution $P(\alpha)$? One property enforcing convergence towards $P(\alpha)$ is the detailed balancing condition

$$M_{\alpha\beta} P(\beta) = M_{\beta\alpha} P(\alpha) \, . \tag{39}$$

Indeed, in this case $P(\alpha)$ is an equilibrium distribution:

$$P_{n+1}(\alpha) = \sum_{\beta=1}^{N} M_{\alpha\beta} P_n(\beta) = \sum_{\beta=1}^{N} M_{\beta\alpha} P_n(\alpha) = P_n(\alpha) \, . \tag{40}$$

We say then that M and P "are in detailed balance", or that the Markov chain (M, P) is in detailed balance. Things can then be redefined so as to have a transition matrix with real spectrum. In order to see it, define $\hat{P}_n(\alpha)$ by

$$P_n(\alpha) = \sqrt{P(\alpha)}\hat{P}_n(\alpha) .$$

Then (19) takes the form

$$\hat{P}_{n+1}(\alpha) = \sum_{\beta=1}^{N} \hat{M}_{\alpha\beta}\ \hat{P}_n(\alpha) , \tag{41}$$

with

$$\hat{M}_{\alpha\beta} = P(\alpha)^{-1/2}\ M_{\alpha\beta}\ P(\beta)^{1/2} . \tag{42}$$

Using (39),

$$\hat{M}_{\alpha\beta} = P(\alpha)^{-1/2}\ M_{\alpha\beta}\ P(\beta)\ P(\beta)^{-1/2} = P(\alpha)^{1/2}\ M_{\beta\alpha}\ P(\beta)^{-1/2} = \hat{M}_{\beta\alpha} .$$

$\hat{M}_{\alpha\beta}$ is a real symmetric matrix, therefore with real eigenvalues.

A theorem states that, if the matrix M is irreducible, then (M, P) is in detailed balance *if and only if* (M, P) is reversible.

4 Equilibrium and Non–Equilibrium

We are now in condition to discuss the question of equilibrium and non-equilibrium in the system. Condition (10) may be seen to come also from applying detailed balancing to M^∞, with the medium distribution vector $v = (1 - c, c - p_z, p_z)$. The same holds of condition (15), with the corresponding asymptotic transition matrix and medium distribution vector. But, in the second case, there is more. Not only the asymptotic limit, but matrix (12) itself satisfies detailed balancing with $v = (1 - c, c)$. This means two things. First, it is not necessary to *impose* condition (15). If it is the final condition, then it will be automatically respected step-by-step. Seen from the shell-to-shell evolution point of view, the system cannot escape stabilizing at the medium concentration. This is not true in the bond-to-bond case: matrix (3) does not satisfy detailed balancing with $v = (1 - c, c - p_z, p_z)$, even if its asymptotic limit does. The rim approach is more "macroscopic", and as much so that equilibrium is maintained at each step.

It looks as if the bulk were always in equilibrium, but not the rim. As long as isolated atoms are looking for a place to join the rim, non-equilibrium dominates. As soon as a shell is saturated, equilibrium of the agglomerate with the medium is attained. But new atoms turn up near the rim, and the system evolves until a new shell is complete. The system evolves in equilibrium from shell to shell, out of equilibrium during the completion of an unsaturated shell. By the last theorem quoted above, the process is reversible when seen from shell to shell, though not from bond saturation to bond saturation.

Acknowledgment

One of the authors (R.A.) is grateful to the Fundação de Amparo à Pesquisa do Estado de São Paulo, Brazil, for financial support.

References

1. R. Kerner, *J. Non-Cryst. Solids* **135**, 155 (1991).
2. R. Kerner, *Physica* **B215**, 267 (1995).
3. R. Kerner and M. Micoulaut, *J. Non-Cryst. Solids* **176**, 271 (1994).
4. R. Kerner and M. Micoulaut, *J. Non-Cryst. Solids* **210**, 298 (1997).
5. Gerardo G. Naumis and R. Kerner, *J. Non-Cryst. Solids* **231**, 111 (1998).
6. R.A. Barrio, R. Kerner, M. Micoulaut and G.G. Naumis, *Condens. Matter* **9**, 9219 (1997).
7. *Encyclopedic Dictionary of Mathematics*, ed. Kyiosi Itô, MIT Press, 1996. Sections 269.N and 310.H.
8. D.K. Faddeev and V.N. Faddeeva, *Computational Methods of Linear Algebra*, W.H. Freeman and Co., San Francisco, 1963.
9. F.R. Gantmacher, *The Theory of Matrices* , 2 vols., Chelsea Pub.Co., New York, 1990.
10. Jean Zinn-Justin, *Quantum field theory and critical phenomena*, Clarendon Press, Oxford , 1997.
11. J.R. Norris, *Markov Chains*, Cambridge University Press, Cambridge, 1997.
12. R. Kerner in *Current Problems in Condensed Matter*, Morán-Lopez (ed.), Plenum Press, 1998, p. 323.
13. R. Aldrovandi, A.L. Barbosa ansd L.P. Freitas *Int.J.Theor.Phys.* **36**, 3021 (1997).
14. R. Aldrovandi and L.P. Freitas *J. Math. Phys.* **39**, 5324 (1998).
15. J.L. McCauley, *Classical Mechanics*, Cambridge University Press, 1997.

TOROIDAL SOLITONS IN THE 3 + 1 DIMENSIONAL INTEGRABLE FIELD THEORY[a]

H. ARATYN[b]
Department of Physics
University of Illinois at Chicago
845 W. Taylor St., Chicago, IL 60607-7059

L.A. FERREIRA[c] AND A.H. ZIMERMAN[b]
Instituto de Física Teórica - IFT/UNESP
Rua Pamplona 145
01405-900, São Paulo - SP, Brazil

We review some recent results on finite energy static solitons with non-trivial Hopf indices in the scalar 3 + 1-dimensional field theory. The model we propose is a solvable, relativistic invariant field theory possessing an infinite number of local conservation laws. We construct for it a family of exact static toroidal soliton solutions labeled by two integers m and n and carrying Hopf topological charge equal to mn. The main feature of such theory, besides its integrability, is that the energy of static configuration is invariant under re-scaling of space variables, circumventing therefore Derrick's theorem. We also point out existence of a submodel of the Skyrme-Faddeev theory satisfying notion of higher-dimensional integrability.

1 Introduction

This paper discusses a new type of solitons appearing as classical solutions of scalar field theories in $3 + 1$ dimensions, defined on S^2, or equivalently on the symmetric space $SU(2)/U(1)$. The novel feature of such solitons is that they are not smooth lumps like the monopoles and instantons, but string like objects having knotted configurations. Requirement of finite energy for the static solutions imposes, in general, that fields should be constant at a spatial infinity. Consequently, space can be taken to be S^3, and the solutions define a mapping $S^3 \rightarrow S^2$, i.e. a Hopf map. Accordingly, the topological charges carried by solitons are determined by the Hopf index which is also expressed as the linking number of the pre images of a given pair of points of S^2. The soliton solutions classified by the Hopf numbers tend to be associated with extended, string like configurations and have, as is the case with the model considered here, a toroidal shape.

---

[a]Talk given by H. Aratyn
[b]Partially supported by NSF grant PHY-9820663
[c]Partially supported by CNPq-Brazil.

2 Integrable Models on $SU(2)/U(1)$

The models we shall be considering involve scalar fields living on the two dimensional sphere S^2. We denote them as $\vec{n} = (n_1, n_2, n_3)$, with $\vec{n}^2 = 1$. Alternatively, one can use the stereographic projection of S^2 and work with a complex scalar field u related to $\vec{n}$ through:

$$\vec{n} = \frac{1}{1+\mid u\mid^2}\left(u+u^*, -i\left(u-u^*\right), \mid u\mid^2 -1\right). \tag{1}$$

We will exploit a one-to-one mapping of the sphere S^2 into the symmetric space $SU(2)/U(1)$ to construct below the local zero curvature conditions. Let $T_3, T_\pm$ be the generators of $SU(2)$ ($U(1)$ is being generated by T_3):

$$[T_3\,, T_\pm] \;=\; \pm T_\pm\,, \qquad [T_+\,, T_-] = 2\,T_3\,. \tag{2}$$

Strictly speaking, these are the generators of $SL(2)$ while the generators of $SU(2)$ are T_i, $i = 1,2,3$ with $T_\pm = \frac{1}{2}\left(T_1 \pm iT_2\right)$.

We will also need commutation relations for arbitrary spin-j representation ($m = -j, -j+1, \ldots, j-1, j$) of $SU(2)$:

$$[T_3\,, P^{(j)}_m] \;=\; m\,P^{(j)}_m\,, \tag{3}$$

$$[T_\pm\,, P^{(j)}_m] \;=\; \sqrt{j(j+1)-m(m\pm 1)}\;\,P^{(j)}_{m\pm1}\,, \tag{4}$$

$$[P^{(j)}_m\,, P^{(j')}_{m'}] \;=\; 0\,. \tag{5}$$

Let us choose the group element W to be of the form:

$$W \equiv e^{iuT_+}\,e^{\varphi T_3}\,e^{iu^*T_-}\;;\quad R^{(1/2)}(W) = \frac{1}{\sqrt{1+\mid u\mid^2}}\begin{pmatrix} 1 & iu \\ iu^* & 1 \end{pmatrix}, \tag{6}$$

where $\varphi = \ln(1+\mid u\mid^2)$, and where $R^{(1/2)}(W)$ gives W in the defining (spinor) representation of $SU(2)$.

We then introduce the pure-gauge potential

$$\begin{aligned} A_\mu &= -\partial_\mu W\,W^{-1} \\ &= \frac{1}{1+\mid u\mid^2}\left(-i\partial_\mu u\,T_+ - i\partial_\mu u^*\,T_- + \left(u\partial_\mu u^* - u^*\partial_\mu u\right)T_3\right), \end{aligned} \tag{7}$$

and the tensor

$$\tilde{B}_\mu = \frac{1}{1+\mid u\mid^2}\left(\mathcal{K}_\mu\,P^{(1)}_1 - \mathcal{K}^*_\mu\,P^{(1)}_{-1}\right), \tag{8}$$

where $\mathcal{K}_\mu$ is a functional of the fields u and u^* and their derivatives and $P_{\pm 1}^{(1)}$ are in the triplet representation of $SU(2)$ ($j = 1$).

We require that the tensor $\tilde{B}^\mu$ satisfies:

$$D_\mu \tilde{B}^\mu = 0\,, \tag{9}$$

with the covariant derivative $D_\mu \cdot \equiv \partial_\mu \cdot + [A_\mu\,,\,\cdot]$. The relation (9) can be understood as *local* generalization to higher dimensions of the zero curvature condition in two dimensions. It leads to local conservation laws. Indeed, since the connection A_μ is chosen in (7) to be pure-gauge the zero curvature condition (9) implies for the currents:

$$J_\mu \equiv W^{-1}\,\tilde{B}^\mu\,W\,, \tag{10}$$

that they are conserved:

$$\partial_\mu\,J^\mu = 0\,. \tag{11}$$

Requiring furthermore

$$\mathrm{Im}\,(\mathcal{K}^\mu \partial_\mu u^*) = 0\,, \tag{12}$$

one obtains that the zero curvature condition (9) implies:

$$(1+\mid u\mid^2)\,\partial^\mu \mathcal{K}_\mu - 2u^* \mathcal{K}_\mu \partial^\mu u = 0\,, \tag{13}$$

as well as its complex conjugate equation.

According to (10) and (11) one gets three conserved currents corresponding to three states of the triplet representation. They are given by

$$J_\mu^{(1)} = \sum_{m=-1}^{1} J_\mu^{(1,m)} P_m^{(1)}\,, \tag{14}$$

with

$$J_\mu^{(1,1)} = \frac{\mathcal{K}_\mu + \mathcal{K}_\mu^*\,u^2}{(1+\mid u\mid^2)^2} = -J_\mu^{(1,-1)*}\,, \qquad J_\mu^{(1,0)} = \frac{i\sqrt{2}\,(\mathcal{K}_\mu^*\,u - \mathcal{K}_\mu\,u^*)}{(1+\mid u\mid^2)^2}\,. \tag{15}$$

Notice that the condition (12) implies that the term in the direction of $P_0^{(1)}$ vanishes, i.e. $\left[\partial_\mu u\,T_+ + \partial_\mu u^*\,T_-\,,\,\mathcal{K}_\mu\,P_1^{(1)} - \mathcal{K}_\mu^*\,P_{-1}^{(1)}\right] = 0$. Therefore, the equations of motion (13) are determined only by the way that $\tilde{B}_\mu$ transforms under the $U(1)$ subgroup generated by T_3. In fact, $\tilde{B}_\mu$ contains two irreducible representations of $U(1)$ which are the singlets $P_{\pm 1}^{(1)}$ of charges ± 1.

That means that if we change the representation of $SU(2)$ where $\tilde{B}_\mu$ lives, we do not change the equations of motion if $\tilde{B}_\mu$ still transforms under the same two singlets of the $U(1)$ subgroup. What can happen is that the commutator of the $T_\pm$ part of A_μ does not commute with $\tilde{B}_\mu$ anymore, and then we get some additional equations which should be considered as constraints on the model. Consequently, we would be dealing with submodels of the original theory. [1]

Taking into account Eq.(12), one observes that if one imposes additional constraint

$$\mathcal{K}_\mu \, \partial^\mu u = 0 \,, \tag{16}$$

the currents in (15) can be generalized to an arbitrary spin. In fact any current of the form

$$J_\mu \equiv \mathcal{K}_\mu \, \frac{\delta G}{\delta u} - \mathcal{K}^*_\mu \, \frac{\delta G}{\delta u^*} \,, \tag{17}$$

with G being any functional of u and u^* only (no derivatives), are conserved as a consequence of (12) and (16) and equation of motion (13) which simplifies greatly in view of constraint (16). Hence one can talk in that case about submodel defined by equations

$$\partial^\mu \mathcal{K}_\mu = 0 \,, \qquad \mathcal{K}_\mu \, \partial^\mu u = 0 \,, \tag{18}$$

and possessing infinitely many conserved currents (17).

3 Examples

The methods discussed above can be used to construct *integrable* models with an infinite number of conserved currents in a space-time of any dimension. The example of CP^1 in $(2+1)$ was discussed in Alvarez et al., [2] and corresponds to the choice $\mathcal{K}_\mu \to \partial_\mu u$. Examples involving other symmetric spaces (or homogeneous spaces) were considered. [1,3,4]

A particular class of models can be constructed using the quantity

$$K_\mu = (\partial^\nu u^* \partial_\nu u) \, \partial_\mu u - (\partial_\nu u)^2 \, \partial_\mu u^* \,, \tag{19}$$

since it automatically satisfies

$$K_\mu \partial^\mu u = 0 \,; \qquad \mathrm{Im}(K_\mu \partial^\mu u^*) = 0 \,. \tag{20}$$

Therefore, if $\mathcal{F}$ is any real functional of u, u^* and their derivatives, it follows that the choice

$$\mathcal{K}_\mu \to \mathcal{F}(u) \, K_\mu \,, \tag{21}$$

satisfies (12) and (16), and consequently leads to a class of models defined by the equations of motion (see (18))

$$\partial^{\mu}\left(\mathcal{F}\left(u\right)K_{\mu}\right)=0\,, \tag{22}$$

and possessing an infinite number of local conserved currents given by (17).

3.1 The Skyrme-Faddeev Model

The Skyrme-Faddeev (SF) model is defined by the Lagrangian

$$\mathcal{L}=m^{2}\left(\partial\vec{n}\right)^{2}-\eta_{0}\frac{1}{e^{2}}H_{\mu\nu}^{2}+\lambda\left(\vec{n}^{2}-1\right)\,, \tag{23}$$

with antisymmetric tensor:

$$H_{\mu\nu}\equiv\vec{n}\cdot\left(\partial_{\mu}\vec{n}\times\partial_{\nu}\vec{n}\right)\,, \tag{24}$$

where $\vec{n}$ are scalar fields living on S^2. In addition, $\eta_0=\pm1$ determines the signature of the Minkowski metric, $g_{\mu\nu}=\eta_0\,\mathrm{diag}\,(1,-1,-1,-1)$, and λ is a Lagrange multiplier. In terms of the complex field u introduced in (1), the equations of motion of (23) are given by

$$\left(1+\mid u\mid^{2}\right)\partial^{\mu}L_{\mu}-2u^{*}\left(L^{\mu}\partial_{\mu}u\right)=0 \tag{25}$$

and its complex conjugate. Here

$$L_{\mu}\equiv m^{2}\partial_{\mu}u-\eta_{0}\frac{4}{e^{2}}\frac{K_{\mu}}{\left(1+\mid u\mid^{2}\right)^{2}}\,, \tag{26}$$

and K_μ is defined in (19).

The energy for static configurations is given by

$$E\equiv\int d^{3}x\,\Theta_{00}=E_{1}+E_{2}\,, \tag{27}$$

with

$$E_{1}\equiv4m^{2}\int d^{3}x\frac{\mid\nabla u\mid^{2}}{\left(1+\mid u\mid^{2}\right)^{2}}\,;\;E_{2}\equiv\frac{32}{e^{2}}\int d^{3}x\frac{\left(\nabla u_{1}\right)^{2}\left(\nabla u_{2}\right)^{2}\left(1-\cos^{2}\gamma\right)}{\left(1+\mid u\mid^{2}\right)^{4}}\,,$$

where γ is the angle between the vectors ∇u_1 and ∇u_2, and so E_1 and E_2 are positive definite. One of the main properties of such theory is that these two terms compensate the scaling instabilities in the energy that each term would present if considered separate. It turns out that it is necessary to have $E_1 = E_2$ to get a stable configuration.

The importance of the Skyrme-Faddeev model for gauge theories was pointed out in Faddeev and Niemi [5] where (23) was suggested as the effective action at low energies for the pure $SU(2)$ Yang-Mills theory. If these claims prove to be correct, one should expect that the string like solitons must correspond to particle excitations of the quantized Yang-Mills theory. Consequently, such solitons resemble the skyrmions in the sense that they are not classical solutions of the original Lagrangean, like monopoles and instantons, but of an effective theory.

However, no solution with non trivial topological charge has been explicitly found for this model. The existence of soliton solutions has been corroborated by numerical calculations using variational methods to find configurations which minimize the energy. [6]

We now discuss the integrability properties of the Skyrme-Faddeev model. Notice, that $\text{Im}\,(L_\mu \partial^\mu u^*) = 0$. Therefore, if one makes the correspondence

$$\mathcal{K}_\mu \rightarrow L_\mu \,, \tag{28}$$

one notices that (12) is satisfied, and the Skyrme-Faddeev model admits a zero curvature representation with the potentials being given by (7) and (8).

The conserved currents are given by (15), and they correspond to the three Noether currents associated to the invariance of the model under the $O(3)$ symmetry. The Skyrme-Faddeev model, however, does not admit an infinite number of conserved currents, obtainable through the procedures described in section (2), because the condition (16) is not satisfied, since

$$L_\mu \partial^\mu u = m^2 \left(\partial u\right)^2 . \tag{29}$$

However, the submodel obtained by imposing the constraint

$$\left(\partial u\right)^2 = 0 \,, \tag{30}$$

does possess an infinite number of conserved currents given by (17) (with the replacement (28)). In such case the equations of motion (25) become:

$$\partial^\mu \left(f\left(u\right) \partial_\mu u\right) = 0 \,, \tag{31}$$

where

$$f\left(u\right) \equiv m^2 - \eta_0 \frac{4}{e^2} \frac{\partial_\mu u \partial^\mu u^*}{\left(1+ \mid u \mid^2\right)^2} . \tag{32}$$

Since such submodel possesses an infinite number of local conserved currents, we believe that it should be easier to solve than the unconstrained Skyrme-Faddeev model.

3.2 A Solvable Model Possessing Toroidal Solitons

Using stereographic projection (1) the tensor (24) can be rewritten as

$$H_{\mu\nu} = \frac{-2i}{(1+ \mid u \mid^2)^2} \left(\partial_\mu u \partial_\nu u^* - \partial_\nu u \partial_\mu u^*\right) . \tag{33}$$

The model we consider here is defined by Lagrangian:[7]

$$\mathcal{L} \equiv -\eta_0 \left(H_{\mu\nu}^2\right)^{\frac{3}{4}} = -\eta_0 \, 8^{\frac{3}{4}} \frac{\left(K_\mu \partial^\mu u^*\right)^{\frac{3}{4}}}{(1+ \mid u \mid^2)^3} , \tag{34}$$

where K_μ is the same as in (19), and where $\eta_0 = \pm 1$, determines the choice of the signature of the Minkowski metric (see (23)).

The corresponding equations of motion are given by:

$$\partial^\mu \left(\frac{\left(K \partial u^*\right)^{-\frac{1}{4}} K_\mu}{1+ \mid u \mid^2} \right) = 0 , \tag{35}$$

and its complex conjugate.

This model possesses a representation in terms of the zero curvature (9) under substitution

$$\mathcal{K}_\mu = \frac{\left(K \partial u^*\right)^{-\frac{1}{4}} K_\mu}{1+ \mid u \mid^2} . \tag{36}$$

Indeed, such $\mathcal{K}_\mu$ satisfies (12) and (16) as a consequence of (20).

Consequently, the model (34) is integrable (or solvable) in the sense that it possesses an infinite number of conserved currents given by (17).

We are interested in constructing exact static finite energy solutions with non vanishing topological charges. The finite energy requirement imposes that the field $\vec{n}$ should be constant at spatial infinity. Therefore, for such purpose one can consider the three dimensional space as an S^3 where the spatial infinity is identified with the north pole. The relevant topological invariant is given by the Hopf map $S^3 \rightarrow S^2$, and the topological charge is

$$Q_h \equiv \frac{1}{4\pi^2} \int \epsilon_{ijk} H_{ij} A_k d^3x \quad ; \quad H_{ij} = \partial_i A_j - \partial_j A_i = \epsilon_{abc} n^a \partial_i n^b \partial_j n^c . \tag{37}$$

One of the main difficulties of constructing such type of solutions comes from scaling instabilities in the energy.[8–10] The choice of the Lagrangian density (34) allows to circumvent such problems. Indeed, the energy of static configurations is given by

$$E \equiv \int d^3x \, \Theta_{00} = 8^{\frac{3}{4}} \int d^3x \frac{\left(K_i \partial^i u^*\right)^{\frac{3}{4}}}{(1+ \mid u \mid^2)^3} , \tag{38}$$

with $i = 1, 2, 3$, and $\Theta_{\mu\nu}$ being the canonical energy-momentum tensor. Under a rescaling $x^i \to \lambda x^i$, one has $K_i \to \lambda^{-3} K_i$, and so energy is scale invariant.

We shall look for time-independent solutions to equations of motion (35) of the type[d]:

$$u(\eta, \xi, \varphi) \equiv f(\eta) e^{i(m\xi + n\varphi)}, \tag{39}$$

where we used toroidal coordinates on $\mathbb{R}^3$:

$$x = aq^{-1} \sinh\eta \cos\varphi\,,\ y = aq^{-1} \sinh\eta \sin\varphi\,,\ z = aq^{-1} \sin\xi\,,\ a > 0 \cos\xi\,, \tag{40}$$

where $q = \cosh\eta - \cos\xi$. The angles ξ, φ both vary from 0 to 2π and η varies from 0 to ∞. The surfaces of constant η are toroids that circle the z-axis, ξ=constant are spheres and φ=constant are half-planes. The corresponding gradient of u becomes in toroidal coordinates:

$$\vec{\nabla} u = (q/a) e^{i(m\xi + n\varphi)} \left(f'(\eta)\hat{e}_\eta + imf(\eta)\hat{e}_\xi + inf(\eta)\frac{\hat{e}_\varphi}{\sinh\eta} \right), \tag{41}$$

where we introduced the unit vectors spanning the orthogonal toroidal coordinate system having properties $\hat{e}_i \cdot \hat{e}_j = \delta_{ij}$, $i, j \equiv \eta, \xi, \varphi$.

In terms of the vector-field $\vec{\mathcal{K}}$ from (36) the equations of motion (35) become, in the static case, $\vec{\nabla} \cdot \vec{\mathcal{K}} = 0$, i.e. $\vec{\mathcal{K}}$ is solenoidal. Using (39) and (41) we obtain for the components of $\vec{\mathcal{K}}$:

$$\mathcal{K}_\eta = \mathcal{K}_0 \left(m^2 + \frac{n^2}{(\sinh\eta)^2} \right) f\,;\ \ \mathcal{K}_\xi = i\, m\mathcal{K}_0\, f'\,;\ \ \mathcal{K}_\varphi = i\, n\mathcal{K}_0\, \frac{f'}{\sinh\eta}\,,$$

with

$$\mathcal{K}_0 \equiv \sqrt{2}(q/a)^2 e^{i(m\xi + n\varphi)} \left(m^2 + \frac{n^2}{(\sinh\eta)^2} \right)^{-1/4} \frac{f^{1/2} f'^{1/2}}{(f^2 + 1)}\,, \tag{42}$$

where the components of the vector field are defined according to $\vec{V} = \hat{e}_\eta V_\eta + \hat{e}_\xi V_\xi + \hat{e}_\varphi V_\varphi$.

Plugging the components of $\vec{\mathcal{K}}$ into the expression for divergence of the vector field in the toroidal coordinates

$$\vec{\nabla} \cdot \vec{V} = \frac{q}{a} \left(\frac{\partial V_\eta}{\partial \eta} + \frac{\partial V_\xi}{\partial \xi} + \frac{1}{\sinh\eta}\frac{\partial V_\varphi}{\partial \varphi} - 2V_\xi \frac{\sin\xi}{q} + V_\eta \left(\frac{\cosh\eta}{\sinh\eta} - \frac{2\sinh\eta}{q} \right) \right),$$

and using

$$\frac{\partial q^2}{\partial \eta} = q^2 \frac{2\sinh\eta}{q}\,; \qquad \frac{\partial q^2}{\partial \xi} = q^2 \frac{2\sin\xi}{q}\,, \tag{43}$$

[d]We shall take m and n integers in order for u to be single valued.

we arrive at:

$$\frac{\partial}{\partial\eta}\ln\frac{ff'}{(f^2+1)^2} = -\frac{2m^2(\sinh\eta)^2 - n^2\cosh\eta}{m^2(\sinh\eta)^2 + n^2}\frac{\cosh\eta}{\sinh\eta}. \tag{44}$$

We take $m^2 > n^2$ in Eq.(44). The integration yields:

$$\frac{1}{f^2+1} = \frac{2k_1}{|m|(m^2-n^2)}\frac{\cosh\eta}{(\frac{n^2-m^2}{m^2}+\cosh^2\eta)^{1/2}} + k_2, \tag{45}$$

with k_2 being the integration constant of the last integration. Imposing boundary conditions:

$$\begin{array}{llllllll} \vec{n} & \to & (0,0,1) & \text{or} & |u| \to \infty & \text{or} & f \to \infty & \text{as} \quad \eta \to 0, \\ \vec{n} & \to & (0,0,-1) & \text{or} & u \to 0 & \text{or} & f \to 0 & \text{as} \quad \eta \to \infty, \end{array}$$

one gets

$$f^2 = \frac{\cosh\eta - \sqrt{(n^2/m^2) + \sinh^2\eta}}{\sqrt{1 + (m^2/n^2)\sinh^2\eta} - \cosh\eta}. \tag{46}$$

One observes that f depends only on the ratio m^2/n^2, and that $f^2 \geq 0$ for any value of m^2/n^2. Note, that taking the limit $m \to n$ in (46) and using L'Hospital rule yields:

$$\lim_{m\to n} f^2 = \frac{1}{\sinh^2\eta}, \tag{47}$$

which gives for $m = n$ a solution:

$$u = \frac{e^{im(\xi+\varphi)}}{\sinh\eta}. \tag{48}$$

In the special case of $m = n = 1$ Eq.(48) reproduces the standard soliton solution.[9,11] Expression (48) can be written as a composite of the Hopf map together with the stereographic map : $\mathbb{R}^3 \to S^3$ of degree 1.

Inserting our solution into (38) we obtain an expression for the energy $E = E_{m,n}$ of the soliton configuration

$$E_{m,n} = (2\pi)^2 8\cdot 2^{3/4}\int_0^\infty \frac{d\eta\sinh\eta}{(1+f^2)^3}\left(m^2 + \frac{n^2}{(\sinh\eta)^2}\right)^{3/4} f^{3/2}f'^{\,3/2}, \tag{49}$$

which after the η-integration yields

$$E_{m,n} = (2\pi)^2 4\cdot 2^{1/4}\sqrt{|n||m|(|n|+|m|)}. \tag{50}$$

We now turn to calculation of the Hopf numbers. Define functions $\Phi_i, i = 1, \ldots, 4$ as follows:

$$\Phi_{\left\{\begin{array}{c}1\\2\end{array}\right\}} = \left(\frac{\cosh\eta - \left(n^2/m^2 + \sinh^2\eta\right)^{1/2}}{\left(|m/n| - 1\right)\left(n^2/m^2 + \sinh^2\eta\right)^{1/2}}\right)^{1/2} \left\{\begin{array}{c}\cos m\xi\\ \sin m\xi\end{array}\right\}, \quad (51)$$

$$\Phi_{\left\{\begin{array}{c}3\\4\end{array}\right\}} = \left(\frac{\left(1 + (m^2/n^2)\sinh^2\eta\right)^{1/2} - \cosh\eta}{\left(|m/n| - 1\right)\left(n^2/m^2 + \sinh^2\eta\right)^{1/2}}\right)^{1/2} \left\{\begin{array}{c}\cos n\varphi\\ -\sin n\varphi\end{array}\right\}. \quad (52)$$

They provide parameterization of S^3 and satisfy:

$$u = \frac{Z_1}{Z_2} = \frac{\Phi_1 + i\Phi_2}{\Phi_3 + i\Phi_4}, \qquad\qquad |\, Z_1 \,|^2 + |\, Z_2 \,|^2 = 1, \quad (53)$$

or equivalently, for $\vec{n}$ defined from u via relation (1):

$$n_i = Z^\dagger \sigma_i Z\,; \quad Z^\dagger = (Z_1^*, Z_2^*)\,; \quad Z = \begin{pmatrix} Z_1 \\ Z_2 \end{pmatrix}. \quad (54)$$

Let, furthermore

$$A_i = \frac{i}{2}\left(Z^\dagger \partial_i Z - \partial_i Z^\dagger\, Z\right), \quad (55)$$

be a vector potential for the two-form $H_{ij} = \partial_i A_j - \partial_j A_i$. In terms of the toroidal coordinates the components of $\vec{A}$ are

$$A_\eta = 0\,; \quad A_\xi = -mq\left(\Phi_1^2 + \Phi_2^2\right); \quad A_\varphi = -\frac{nq}{\sinh\eta}\left(\Phi_3^2 + \Phi_4^2\right). \quad (56)$$

Next, we calculate the vector function $\vec{B} = \vec{\nabla} \times \vec{A}$ (such that $B_i = \epsilon_{ijk} H_{jk}/2$), using expression for the curl operator in toroidal coordinates and plugging it into the Hopf index defined as:

$$Q_H = \frac{1}{4\pi^2}\int d^3x\, \vec{A}\cdot\vec{B}, \quad (57)$$

for which we now find

$$Q_H = \frac{nm}{2}\left(\left(\Phi_1^2 + \Phi_2^2\right)^2\Big|_0^\infty - \left(\Phi_3^2 + \Phi_4^2\right)^2\Big|_0^\infty\right) = -nm\,. \quad (58)$$

It is always possible to choose the Hopf index such that it is positive $Q_H \geq 0$. This amounts to the right choice of orientation, which determines the sign of Q_H.

Note, that the following inequality:

$$|m| + |n| \geq 2\sqrt{|n||m|} \tag{59}$$

always holds. Hence

$$\sqrt{|n||m|(|n|+|m|)} \geq \sqrt{2}(|n||m|)^{3/4} \;\rightarrow\; E_{m,n} \geq (2\pi)^2 4 \cdot 2^{3/4}\, |Q_H|^{3/4}\,, \tag{60}$$

which agrees with the lower bound result of Vakulenko and Kapitansky [12] for the Faddeev-Skyrme model.

Notice that the solution obtained above implies that the vector field $\vec{n}$ (see (1)) is given by

$$n_1 = \frac{2f}{f^2+1}\cos\left(m\xi + n\varphi\right)\;;\quad n_2 = \frac{2f}{f^2+1}\sin\left(m\xi + n\varphi\right)\;;\quad n_3 = \frac{f^2-1}{f^2+1}\,.$$

Since f is a function of η only, one observes that the surfaces of constant n_3 are tori. In addition, on those surfaces, the lines of constant n_1 and n_2 wind around the tori with frequencies in the ξ and φ-directions given by m and n, respectively. One can check that n_3 falls monotonically from $n_3 = 1$ at $\eta = 0$ to $n_3 = -1$ at $\eta = \infty$. In addition, the bigger the ratio m/n is, the faster if performs that flip. Therefore the size of our soliton decreases with the increase of m/n.

The Hopf index of our solution can alternatively be calculated in the following way. The solution provides a mapping of the spatial $\mathbb{R}^3$ into S^3 defined by $\sum_{i=1}^4 \Phi_i^2 = 1$. Then, (1) and (53) provide the Hopf map $S^3 \rightarrow S^2$. The Hopf index is given by the linking number of the pre-images, under the Hopf map, of any two points of S^2. Consider the point $\vec{n} = (0,0,-1)$ which corresponds in $\mathbb{R}^3$ to $\eta \rightarrow \infty$. For this value of η, Φ_1, Φ_2 go to zero and $\Phi_3 = \cos n\varphi$ and $\Phi_4 = -\sin n\varphi$. On S^3 we find the circle $\Phi_3^2 + \Phi_4^2 = \cos^2 n\varphi + \sin^2 n\varphi = 1$ of radius 1 and as φ varies between 0 and 2π the preimage wraps $\mid n \mid$ times around the φ direction. Similarly the preimage of $\vec{n} = (0,0,1)$ corresponds to $\eta = 0$. For this value of η, Φ_3, Φ_4 go to zero and $\Phi_1 = \cos m\xi$ and $\Phi_2 = \sin m\xi$. On S^3 we find the circle $\Phi_1^2 + \Phi_2^2 = \cos^2 m\xi + \sin^2 m\xi = 1$ of radius 1 and as ξ varies between 0 and 2π the preimage wraps $\mid m \mid$ times around the ξ direction. Since these two circles intersect the linking number is $\mid nm \mid$. That is indeed the Hopf number calculated above.

Finally, let us mention that other models circumventing Derrick's theorem have been proposed previously in the Deser et al. and Nicole. [8,9] A common feature is that, like the model considered in the present paper, the Lagragians are non-polynomial functions of the fields and their derivatives. The corresponding solutions have been constructed and in the special case of Nicole [9] a soliton with the unit Hopf charge was obtained. Also, in Kundu [13] the Lagrangian (34) appeared on basis of dimensional arguments and it was observed that it possesses the solution (48) with $m = n = 1$. To our knowledge, an infinite number of soliton solutions to the field theoretical equations of motion with the Hopf charges bigger than unity were not obtained previously in the literature in an exact form.

References

1. L. A. Ferreira and E. E. Leite, *Nucl. Phys.* **B547**, 471 (1999), hep-th/9810067.
2. O. Alvarez, L. A. Ferreira and J. Sánchez Guillén, *Nucl. Phys.* **B529**, 689 (1998), hep-th/9710147.
3. D. Gianzo, J.O. Madsen, and J. Sánchez Guillén, *Nucl. Phys.* **B537**, 586 (1999), hep-th/9805094.
4. K. Fujii and T. Suzuki, *Lett. Math. Phys.* **46**, 49 (1998), hep-th/9802105; K. Fujii, Y. Homma, and T. Suzuki, *Phys. Lett.* **438B**, 290 (1998), hep-th/9806084; hep-th/9809149.
5. L. D. Faddeev and A. Niemi, *Phys. Rev. Lett.* **82**, 1624 (1999), hep-th/9807069.
6. J. Gladikowski and M. Hellmund, *Phys. Rev.* **D 56**, 5194 (1997), hep-th/9609035; L. Faddeev and A. Niemi, *Nature* **387**, 58 (1997), hep-th/9610193; R.A. Battye, P.M. Sutcliffe, *Phys. Rev. Lett.* **81**, 4798 (1998), hep-th/9811077; J. Hietarinta and P. Salo, *Phys. Lett.* **451B**, 60 (1999), hep-th/9811053.
7. H. Aratyn, L. A. Ferreira and A.H. Zimerman, *Phys. Rev. Lett.* **83**, 1723 (1999), hep-th/9905079.
8. S. Deser, M.J. Duff and C.J. Isham, *Nucl. Phys.* **B114**, 29 (1976).
9. D.A. Nicole, *J. Phys.* **G4**, 1363 (1978).
10. L. Faddeev, in *40 Years in Mathematical Physics*, (World Scientific, 1995).
11. H. Aratyn, L. A. Ferreira and A.H. Zimerman, *Phys. Lett.* **456B**, 162 (1999), hep-th/9902142.
12. A. F. Vakulenko and L.V. Kapitansky, *Sov. Phys. Dokl.* **24**, 432 (1979).
13. A. Kundu, *Annals of Physics (USA)* **139**, 36 (1982); see also *Acta Phys. Austriaca* **54**, 7 (1982).

REAL FORMS OF THE N=4 SUPERSYMMETRIC TODA CHAIN HIERARCHY[a]

F. DELDUC
Laboratoire de Physique,
Groupe de Physique Théorique ENS Lyon
46, allée d'Italie, F - 69364 - Lyon CEDEX 07

A. SORIN
Bogoliubov Laboratory of Theoretical Physics, JINR,
141 980, Dubna, Moscow Region, Russian Federation

Three inequivalent real forms of the complex N=4 supersymmetric Toda chain hierarchy in the real $N = 2$ superspace with one even and two odd real coordinates are presented. It is demonstrated that the first of them possesses a global $N = 4$ supersymmetry, while the other two admit a twisted $N = 4$ supersymmetry. A new superfield basis in which supersymmetry transformations are local is discussed and a manifest $N = 4$ supersymmetric representation of the $N = 4$ Toda chain in terms of a chiral and an anti-chiral $N = 4$ superfield is constructed. Its relation to the complex $N = 4$ supersymmetric KdV hierarchy is established. Darboux-Backlund symmetries and a new real form of this last hierarchy possessing a twisted $N = 4$ supersymmetry are derived.

1 Introduction

Recently the Lax pair representation of the even and odd flows of the complex $N = 4$ supersymmetric Toda chain hierarchy in $N = 2$ superspace were constructed in Delduc et al.[1] The corresponding local and nonlocal Hamiltonians, the finite and infinite discrete symmetries, the first two Hamiltonian structures and the recursion operator connecting all evolution equations and the Hamiltonian structures were also studied. The goal of the present work is first to analyse the possible real forms of the $N = 4$ Toda chain hierarchy in $N = 2$ superspace, second to derive a manifest $N = 4$ supersymmetric representation of its first nontrivial even flows in the real $N = 4$ superspace, and third to clarify its relation (if any) with the $N = 4$ supersymmetric KdV hierarchy.

Let us start with a short summary of the results that we shall need concerning the complex $N = 4$ supersymmetric Toda chain hierarchy (see Delduc et al., Leznov et al. and Lechtenfeld et al.[1–4] for more details).

The complex $N = 4$ supersymmetric Toda chain hierarchy in the complex $N = 2$ superspace comprises an infinite set of even and odd flows for two complex even $N = 2$ superfields $u(z, \theta^+, \theta^-)$ and $v(z, \theta^+, \theta^-)$, where z and $\theta^\pm$

---
[a]Talk given by A. Sorin

are complex even and odd coordinates, respectively. The flows are generated by complex even and odd evolution derivatives $\{\frac{\partial}{\partial t_k}, U_k^{\pm}, U_k, \overline{U}_k\}$ and $\{D_k^{\pm}, Q_k^{\pm}\}$ $(k \in \mathbb{N})$, respectively, with the following length dimensions:

$$[\tfrac{\partial}{\partial t_k}] = [U_k^{\pm}] = [U_k] = [\overline{U}_k] = -k, \quad [D_k^{\pm}] = [Q_k^{\pm}] = -k + \frac{1}{2}\,. \tag{1}$$

The first few of these flows are :

$$\frac{\partial}{\partial t_1}\begin{pmatrix} v \\ u \end{pmatrix} = \partial \begin{pmatrix} v \\ u \end{pmatrix}, \tag{2}$$

$$\begin{aligned}
\frac{\partial}{\partial t_2} v &= +v\,'' + 2uv(D_+D_-v) - (D_+D_-v^2u) \\
&\quad -v^2(D_+D_-u) + 2v(uv)^2\,, \\
\frac{\partial}{\partial t_2} u &= -u\,'' + 2uv(D_+D_-u) - (D_+D_-u^2v) \\
&\quad -u^2(D_+D_-v) - 2u(uv)^2\,,
\end{aligned} \tag{3}$$

$$\begin{aligned}
\frac{\partial}{\partial t_3} v &= v\,''' - 3(D_+v)\,'(D_-uv) + 3(D_-v)\,'(D_+uv) \\
&\quad -3v\,'(D_+u)(D_-v) + 3v\,'(D_-u)(D_+v) \\
&\quad -6vv\,'(D_+D_-u) + 6(uv)^2v\,'\,, \\
\frac{\partial}{\partial t_3} u &= u\,''' - 3(D_-u)\,'(D_+uv) + 3(D_+u)\,'(D_-uv) \\
&\quad -3u\,'(D_-v)(D_+u) + 3u\,'(D_+v)(D_-u) \\
&\quad -6uu\,'(D_-D_+v) + 6(uv)^2u\,'\,,
\end{aligned} \tag{4}$$

$$D_1^{\pm} v = -D_{\pm} v \pm 2vD_{\mp}^{-1}(uv)\,, \quad D_1^{\pm} u = -D_{\pm} u \mp 2uD_{\mp}^{-1}(uv)\,, \tag{5}$$

$$Q_1^{\pm}\begin{pmatrix} v \\ u \end{pmatrix} = Q_{\pm}\begin{pmatrix} v \\ u \end{pmatrix}, \tag{6}$$

$$\begin{aligned}
U_0^{\pm} v &= \frac{1}{2}v - \theta^{\pm}(D_{\pm}v \mp 2vD_{\mp}^{-1}(uv))\,, \\
U_0^{\pm} u &= \frac{1}{2}u - \theta^{\pm}(D_{\pm}u \pm 2uD_{\mp}^{-1}(uv))\,,
\end{aligned} \tag{7}$$

$$\begin{aligned}
U_0 v &= -\theta^{+}(D_-v + 2vD_{+}^{-1}(uv)) - \theta^{-}(D_+v - 2vD_{-}^{-1}(uv))\,, \\
U_0 u &= -\theta^{+}(D_-u - 2uD_{+}^{-1}(uv)) - \theta^{-}(D_+u + 2uD_{-}^{-1}(uv))\,,
\end{aligned} \tag{8}$$

$$\overline{U}_0 \begin{pmatrix} v \\ u \end{pmatrix} = \frac{1}{2} \left(\theta^- (D_+ + Q_+) - \theta^+ (D_- + Q_-) \right) \begin{pmatrix} v \\ u \end{pmatrix} . \tag{9}$$

Throughout this letter, we shall use the notations $u' = \partial u = \frac{\partial}{\partial z} u$. $D_\pm$ and $Q_\pm$ are odd covariant derivatives and supersymmetry generators,

$$D_\pm \equiv \frac{\partial}{\partial \theta^\pm} + \theta^\pm \partial \, , \quad Q_\pm \equiv \frac{\partial}{\partial \theta^\pm} - \theta^\pm \partial \, . \tag{10}$$

They form the algebra[b]

$$\{D_\pm, D_\pm\} = +2\partial \, , \qquad \{Q_\pm, Q_\pm\} = -2\partial \, . \tag{11}$$

Using the explicit expressions of the flows (2–9), one can calculate their algebra which has the following nonzero brackets:

$$\left\{ D_1^\pm \, , \, D_1^\mp \right\} = -2 \, \frac{\partial}{\partial t_1} \, , \quad \left\{ Q_1^\pm \, , \, Q_1^\mp \right\} = +2 \, \frac{\partial}{\partial t_1} \, , \tag{12}$$

$$\begin{gathered} \left[U_0 \, , \, U_0^\pm \right] = \pm \overline{U}_0 \, , \quad \left[\overline{U}_0 \, , \, U_0^\pm \right] = \pm U_0 \, , \\ \left[U_0 \, , \, \overline{U}_0 \right] = 2(U_0^+ - U_0^-) \, , \end{gathered} \tag{13}$$

$$\begin{gathered} \left[U_0^\pm \, , \, D_1^\pm \right] = -Q_1^\pm \, , \quad \left[U_0^\pm \, , \, Q_1^\pm \right] = -D_1^\pm \, , \\ \left[U_0 \, , \, D_1^\pm \right] = +Q_1^\mp \, , \quad \left[U_0 \, , \, Q_1^\pm \right] = +D_1^\mp \, , \\ \left[\overline{U}_0 \, , \, D_1^\pm \right] = \pm D_1^\mp \, , \quad \left[\overline{U}_0 \, , \, Q_1^\pm \right] = \pm Q_1^\mp \, . \end{gathered} \tag{14}$$

This algebra reproduces the algebra of the global complex $N = 4$ supersymmetry, together with its $gl(2, \mathbb{C})$ automorphisms. It is the algebra of symmetries of the nonlinear even flows (3–4). It may be realized in the superspace $\{t_k, \theta_k^\pm, \rho_k^\pm, h_k^\pm, h_k, \overline{h}_k\}$, where $t_k, h_k^\pm, h_k, \overline{h}_k$ $(\theta_k^\pm, \rho_k^\pm)$ are complex even (odd) abelian evolution times with the length dimensions

$$[t_k] = [h_k^\pm] = [h_k] = [\overline{h}_k] = k \, , \qquad [\theta_k^\pm] = [\rho_k^\pm] = k - \frac{1}{2} \, , \tag{15}$$

which are in one-to-one correspondence with the length dimensions (1) of the corresponding evolution derivatives.

[b]Hereafter, we explicitly present only non-zero brackets.

2 Real Forms of the N=4 Toda Chain Hierarchy

It is well known that different real forms derived from the same complex integrable hierarchy are inequivalent in general. Keeping this in mind it seems important to find as many different real forms of the $N = 4$ Toda chain hierarchy as possible.

With this aim let us discuss various inequivalent complex conjugations of the superfields $u(z,\theta^+,\theta^-)$ and $v(z,\theta^+,\theta^-)$, of the superspace coordinates $\{z,\ \theta^{\pm}\}$, and of the evolution derivatives $\{\frac{\partial}{\partial t_k},\ U_k^{\pm},\ U_k,\ \overline{U}_k,\ D_k^{\pm},\ Q_k^{\pm}\}$ which should be consistent with the flows (2–9). We restrict our considerations to the case when iz and $\theta^{\pm}$ are coordinates of the real $N = 2$ superspace which satisfy the following standard complex conjugation properties:

$$(iz,\theta^{\pm})^* = (iz,\theta^{\pm}), \tag{16}$$

where i is the imaginary unity. We will also use the standard convention regarding complex conjugation of products involving odd operators and functions (see, e.g., the books Gates et al.[5]). In particular, if $\mathcal{O}$ is some even differential operator acting on a superfield F, we define the complex conjugate of $\mathcal{O}$ by $(\mathcal{O}F)^* = \mathcal{O}^*F^*$. Then, in the case under consideration one can derive, for example, the following relations

$$\begin{aligned} &\partial^* = -\partial\,, \qquad (\epsilon^{\pm})^* = \epsilon^{\pm}\,, \qquad (\epsilon^+\epsilon^-)^* = -\epsilon^+\epsilon^-\,, \\ &(\epsilon^{\pm}D_{\pm})^* = \epsilon^{\pm}D_{\pm}\,, \qquad (D_+D_-)^* = -D_+D_- \end{aligned} \tag{17}$$

which we use in what follows. Here, $\epsilon^{\pm}$ are constant odd real parameters.

Direct verification shows that the flows (2–9) admit the three inequivalent complex conjugations:

$$\begin{aligned} &(v,u)^* = (-v,u)\,, \qquad (z,\theta^{\pm})^* = (-z,\theta^{\pm})\,, \\ &(t_p,U_p^{\pm},U_p,\overline{U}_p,\epsilon_p^{\pm}D_p^{\pm},\varepsilon_p^{\pm}Q_p^{\pm})^* \\ &= (-1)^p(t_p,U_p^{\pm},U_p,\overline{U}_p,-\epsilon_p^{\pm}D_p^{\pm},-\varepsilon_p^{\pm}Q_p^{\pm})\,, \end{aligned} \tag{18}$$

$$\begin{aligned} &(v,u)^{\bullet} = (u,v)\,, \qquad (z,\theta^{\pm})^{\bullet} = (-z,\theta^{\pm}), \\ &(t_p,U_p^{\pm},U_p,\overline{U}_p,\epsilon_p^{\pm}D_p^{\pm},\varepsilon_p^{\pm}Q_p^{\pm})^{\bullet} \\ &= (-t_p,U_p^{\pm},U_p,\overline{U}_p,\epsilon_p^{\pm}D_p^{\pm},\varepsilon_p^{\pm}Q_p^{\pm})\,, \end{aligned} \tag{19}$$

$$\begin{aligned} &(v,\ u)^{\star} = (\,-u(D_-D_+\ln u + uv),\frac{1}{u}\,)\,, \qquad (z,\theta^{\pm})^{\star} = (-z,\theta^{\pm})\,, \\ &(t_p,U_p^{\pm},U_p,\overline{U}_p,\epsilon_p^{\pm}D_p^{\pm},\varepsilon_p^{\pm}Q_p^{\pm})^{\star} \\ &= (-t_p,-U_p^{\pm},-U_p,\overline{U}_p,-\epsilon_p^{\pm}D_p^{\pm},\varepsilon_p^{\pm}Q_p^{\pm})\,, \end{aligned} \tag{20}$$

where $\epsilon_p^{\pm}$ and $\varepsilon_p^{\pm}$ are constant odd real parameters. We would like to underline that the complex conjugations of the evolution derivatives (the second lines of Eqs. (18–20)) are defined and fixed completely by the explicit expressions (2–9) for the flows. These complex conjugations extract different real forms of the algebra (12–14). The real forms of the algebra (12–14) with the involutions (18–19) correspond to a twisted real $N = 4$ supersymmetry, while the real form corresponding to the involution (20) reproduces the algebra of the real $N = 4$ supersymmetry. This last fact becomes more obvious if one uses the $N = 2$ basis of the algebra with the generators

$$\begin{aligned}
&\Sigma_1 \equiv U_0, \quad \Sigma_2 \equiv -i\overline{U}_0, \quad \Sigma_3 \equiv U_0^- - U_0^+, \quad \Sigma \equiv U_0^- + U_0^+, \\
&\mathcal{D}_1 \equiv Q_1^+ + D_1^+, \quad \mathcal{D}_2 \equiv Q_1^- + D_1^-, \\
&\overline{\mathcal{D}}^1 \equiv Q_1^+ - D_1^+, \quad \overline{\mathcal{D}}^2 \equiv Q_1^- - D_1^- .
\end{aligned} \tag{21}$$

Then, the nonzero algebra brackets (12–14) and the complex conjugation rule (20) are the standard ones for the real $N = 4$ supersymmetry algebra together with its $u(2)$ automorphisms

$$\begin{aligned}
&\left\{\mathcal{D}_\alpha , \overline{\mathcal{D}}^\beta\right\} = 4\delta_\alpha{}^\beta \frac{\partial}{\partial t_1}, && \left[\Sigma_a , \Sigma_b\right] = 2i\epsilon_{abc}\Sigma_c , \\
&\left[\Sigma_a , \mathcal{D}_\alpha\right] = (\sigma_a)_\alpha{}^\beta \mathcal{D}_\beta, && \left[\Sigma_a , \overline{\mathcal{D}}^\alpha\right] = -\overline{\mathcal{D}}^\beta (\sigma_a)_\beta{}^\alpha , \\
&\left[\Sigma , \mathcal{D}_\alpha\right] = -\mathcal{D}_\alpha, && \left[\Sigma , \overline{\mathcal{D}}_\alpha\right] = \overline{\mathcal{D}}_\alpha ,
\end{aligned} \tag{22}$$

$$\begin{aligned}
&(\tfrac{\partial}{\partial t_1}, \Sigma_a, \Sigma, \epsilon^\alpha \mathcal{D}_\alpha, \overline{\epsilon}_\alpha \overline{\mathcal{D}}^\alpha)^\star = -(\tfrac{\partial}{\partial t_1}, \Sigma_a, \Sigma, -\overline{\epsilon}_\alpha \overline{\mathcal{D}}^\alpha, -\epsilon^\alpha \mathcal{D}_\alpha) , \\
&(\epsilon^\alpha, \overline{\epsilon}_\alpha)^\star = (\overline{\epsilon}_\alpha, \epsilon^\alpha) .
\end{aligned} \tag{23}$$

Here, $\epsilon^\alpha, \overline{\epsilon}_\alpha$ are constant odd parameters, a summation over repeated indices $\alpha, \beta = 1, 2$ and $a, b, c = 1, 2, 3$ is understood in Eqs. (22), and σ_a are the Pauli matrices

$$\begin{aligned}
&\sigma_1 \equiv \begin{pmatrix} 0 & 1 \\ 1 & 0 \end{pmatrix}, \quad \sigma_2 \equiv \begin{pmatrix} 0 & -i \\ i & 0 \end{pmatrix}, \quad \sigma_3 \equiv \begin{pmatrix} 1 & 0 \\ 0 & -1 \end{pmatrix}, \\
&\sigma_a \sigma_b = \delta_{ab} + i\epsilon_{abc}\sigma_c ,
\end{aligned} \tag{24}$$

and ϵ_{abc} is a totally antisymmetric tensor ($\epsilon_{123} = 1$). Therefore, we conclude that the complex $N = 4$ supersymmetric Toda chain hierarchy with the complex conjugation (20) possesses a real $N = 4$ supersymmetry, and due to this remarkable fact it can be called the $N = 4$ supersymmetric Toda chain hierarchy (for some examples of $N = 4$ supersymmetric integrable systems, see

Delduc et al., Ivanov et al., Bonora et al., and Popowicz [6–12] and references therein).

Let us remark that a combination of the two involutions (20) and (19) generates the infinite-dimensional group of discrete Darboux transformations [2,3,1]

$$(v,\ u)^{\star\bullet} = (\ v(D_-D_+\ln v - uv),\ \frac{1}{v}\ ),\quad (z,\theta^{\pm})^{\star\bullet} = (z,\theta^{\pm}),$$
$$(t_p, U_p^{\pm}, U_p, \overline{U}_p, D_p^{\pm}, Q_p^{\pm})^{\star\bullet} = (t_p, -U_p^{\pm}, -U_p, \overline{U}_p, -D_p^{\pm}, Q_p^{\pm}). \quad (25)$$

This way of deriving discrete symmetries was proposed by Sorin [13] and applied to the construction of discrete symmetry transformations of the $N = 2$ supersymmetric GNLS hierarchies.

3 A KdV–Like Basis with Locally Realized Supersymmetries

The third complex conjugation (20) looks rather complicated when compared to the first two ones (18–19). However, in another superfield basis defined as

$$J \equiv uv + D_-D_+\ln u\,, \qquad \overline{J} \equiv -uv\,, \quad (26)$$

where J and $\overline{J}$ ($[J] = [\overline{J}] = -1$) are unconstrained even $N = 2$ superfields, it drastically simplifies. In this basis the complex conjugations (18–20) are given by

$$(J,\ \overline{J})^* = -(J,\ \overline{J})\,, \quad (27)$$

$$(J,\ \overline{J})^\bullet = (\ J - D_-D_+\ln\overline{J},\ \overline{J}\ )\,, \quad (28)$$

$$(J,\ \overline{J})^\star = (\overline{J},\ J)\,, \quad (29)$$

and the equations (3–9) become simpler as well,

$$\begin{aligned}
\frac{\partial}{\partial t_2}J &= -J\,'' - 2(JD_+^{-1}D_-\overline{J})\,' + D_+D_-(D_+^{-1}D_-J)^2\,,\\
\frac{\partial}{\partial t_2}\overline{J} &= +\overline{J}\,'' - 2(\overline{J}D_+^{-1}D_-J)\,' + D_+D_-(D_+^{-1}D_-\overline{J})^2\,,
\end{aligned} \quad (30)$$

$$\begin{aligned}
\frac{\partial}{\partial t_3}J &= J\,''' + 3\Big[J\,'D_+^{-1}D_-(J+\overline{J}) + J(D_+^{-1}D_-(J+\overline{J}))^2\\
&\quad + (D_-J)D_+J - \overline{J}J^2 - \frac{1}{3}J^3\Big]\,' - 3D_+D_-\Big[J^2D_+^{-1}D_-(J+\overline{J})\Big]\,,\\
\frac{\partial}{\partial t_3}\overline{J} &= \overline{J}\,''' - 3\Big[\overline{J}\,'D_+^{-1}D_-(J+\overline{J}) - \overline{J}(D_+^{-1}D_-(J+\overline{J}))^2\\
&\quad + (D_-\overline{J})D_+\overline{J} + J\overline{J}^2 + \frac{1}{3}\overline{J}^3\Big]\,' - 3D_+D_-\Big[\overline{J}^2D_+^{-1}D_-(J+\overline{J})\Big]\,,
\end{aligned} \quad (31)$$

$$D_1^{\pm}\begin{pmatrix} J \\ \bar{J}\end{pmatrix} = D_{\pm}\begin{pmatrix} +J \\ -\bar{J}\end{pmatrix}, \quad Q_1^{\pm}\begin{pmatrix} J \\ \bar{J}\end{pmatrix} = Q_{\pm}\begin{pmatrix} J \\ \bar{J}\end{pmatrix}, \tag{32}$$

$$\begin{aligned} U_0^{\pm}\begin{pmatrix} J \\ \bar{J}\end{pmatrix} &= -D_{\pm}\theta^{\pm}\begin{pmatrix} +J \\ -\bar{J}\end{pmatrix}, \\ U_0\begin{pmatrix} J \\ \bar{J}\end{pmatrix} &= (\theta^+ D_- + \theta^- D_+)\begin{pmatrix} +J \\ -\bar{J}\end{pmatrix}, \end{aligned} \tag{33}$$

$$\overline{U}_0\begin{pmatrix} J \\ \bar{J}\end{pmatrix} = \frac{1}{2}\left(\,\theta^-(D_+ + Q_+) - \theta^+(D_- + Q_-)\,\right)\begin{pmatrix} J \\ \bar{J}\end{pmatrix}. \tag{34}$$

Notice that the supersymmetry and $u(2)$ transformations (32–34) of the superfields J, $\bar{J}$ are local functions of the superfields, while the evolution equations (30–31) become nonlocal.

4 A Manifest $N = 4$ Supersymmetric Representation

The equations (30–31) admit a manifestly $N = 4$ supersymmetric representation

$$\begin{aligned} \tfrac{\partial}{\partial t_2}\mathcal{J} &= -\mathcal{J}\,'' - \mathcal{D}_+\mathcal{D}_-\left[2(\mathcal{J}\partial^{-1}\overline{\mathcal{J}})\,' - (\overline{\mathcal{D}}^+\overline{\mathcal{D}}^-\,\partial^{-1}\mathcal{J})^2\right], \\ \tfrac{\partial}{\partial t_2}\overline{\mathcal{J}} &= +\overline{\mathcal{J}}\,'' - \overline{\mathcal{D}}^+\overline{\mathcal{D}}^-\left[2(\overline{\mathcal{J}}\partial^{-1}\mathcal{J})\,' - (\mathcal{D}_+\mathcal{D}_-\partial^{-1}\overline{\mathcal{J}})^2\right], \end{aligned} \tag{35}$$

$$\begin{aligned} \tfrac{\partial}{\partial t_3}\mathcal{J} \;=\;& \mathcal{J}\,''' + \mathcal{D}_+\mathcal{D}_-\Big\{3\Big[\mathcal{J}\,'\partial^{-1}\overline{\mathcal{J}} + (\mathcal{J}\partial^{-1}\overline{\mathcal{J}})\mathcal{D}_+\mathcal{D}_-\partial^{-1}\overline{\mathcal{J}} \\ & - \frac{1}{2}(\overline{\mathcal{D}}^+\overline{\mathcal{D}}^-\,\partial^{-1}\mathcal{J})^2\Big]\,' - (\overline{\mathcal{D}}^+\overline{\mathcal{D}}^-\,\partial^{-1}\mathcal{J})^3 \\ & - 3(\overline{\mathcal{D}}^+\overline{\mathcal{D}}^-\,\partial^{-1}\mathcal{J})^2\mathcal{D}_+\mathcal{D}_-\partial^{-1}\overline{\mathcal{J}} + 6\mathcal{J}\overline{\mathcal{J}}\,\overline{\mathcal{D}}^+\overline{\mathcal{D}}^-\,\partial^{-1}\mathcal{J}\Big\}, \\ \tfrac{\partial}{\partial t_3}\overline{\mathcal{J}} \;=\;& \overline{\mathcal{J}}\,''' + \overline{\mathcal{D}}^+\overline{\mathcal{D}}^-\Big\{3\Big[-\overline{\mathcal{J}}\,'\partial^{-1}\mathcal{J} + (\overline{\mathcal{J}}\partial^{-1}\mathcal{J})\overline{\mathcal{D}}^+\overline{\mathcal{D}}^-\,\partial^{-1}\mathcal{J} \\ & + \frac{1}{2}(\mathcal{D}_+\mathcal{D}_-\partial^{-1}\overline{\mathcal{J}})^2\Big]\,' - (\mathcal{D}_+\mathcal{D}_-\partial^{-1}\overline{\mathcal{J}})^3 \\ & - 3(\mathcal{D}_+\mathcal{D}_-\partial^{-1}\overline{\mathcal{J}})^2\overline{\mathcal{D}}^+\overline{\mathcal{D}}^-\,\partial^{-1}\mathcal{J} + 6\mathcal{J}\overline{\mathcal{J}}\,\mathcal{D}_+\mathcal{D}_-\partial^{-1}\overline{\mathcal{J}}\Big\}, \end{aligned} \tag{36}$$

in terms of one chiral $\mathcal{J}(z,\theta^+,\theta^-,\eta^+,\eta^-)$ and one antichiral $\overline{\mathcal{J}}(z,\theta^+,\theta^-,\eta^+,\eta^-)$ even $N = 4$ superfield,

$$\mathcal{D}_{\pm}\mathcal{J} = 0\,, \qquad \overline{\mathcal{D}}^{\pm}\,\overline{\mathcal{J}} = 0\,. \tag{37}$$

Here $\mathcal{D}_\pm, \overline{\mathcal{D}}_\pm$ are $N=4$ odd covariant derivatives,

$$\mathcal{D}_\pm \equiv \frac{1}{2}\left(\frac{\partial}{\partial\theta^\pm} + i\frac{\partial}{\partial\eta^\pm} + (\theta^\pm + i\eta^\pm)\partial\right),$$
$$\overline{\mathcal{D}}^\pm \equiv \frac{1}{2}\left(\frac{\partial}{\partial\theta^\pm} - i\frac{\partial}{\partial\eta^\pm} + (\theta^\pm - i\eta^\pm)\partial\right), \qquad k,m=\pm,$$
$$\left\{\mathcal{D}_k\,,\overline{\mathcal{D}}^m\right\} = \delta_k{}^m\partial, \quad \left\{\mathcal{D}_k\,,\mathcal{D}_m\right\} = \left\{\overline{\mathcal{D}}^k\,,\overline{\mathcal{D}}^m\right\} = 0\,, \tag{38}$$

and $\eta^\pm$ are two additional real odd coordinates. The relations between the independent components of the $N=2$ superfields $\{J(z,\theta^+,\theta^-),\ \overline{J}(z,\theta^+,\theta^-)\}$ and those of the $N=4$ superfields $\{\mathcal{J}(z,\theta^+,\theta^-,\eta^+,\eta^-),\ \overline{\mathcal{J}}(z,\theta^+,\theta^-,\eta^+,\eta^-)\}$ are the following:

$$\mathcal{J}|_{\eta\pm=0} = J, \quad \overline{\mathcal{D}}^\pm\mathcal{J}|_{\eta\pm=0} = D_\pm J, \quad \overline{\mathcal{D}}^+\,\overline{\mathcal{D}}^-\mathcal{J}|_{\eta\pm=0} = D_+D_-J\,,$$
$$\overline{\mathcal{J}}|_{\eta\pm=0} = \overline{J}, \quad \mathcal{D}_\pm\overline{\mathcal{J}}|_{\eta\pm=0} = D_\pm\overline{J}, \quad \mathcal{D}_+\,\mathcal{D}_-\overline{\mathcal{J}}|_{\eta\pm=0} = D_+D_-\overline{J}\,. \tag{39}$$

Let us also present a manifestly $N=4$ supersymmetric form of the discrete Darboux transformations

$$\mathcal{J}^{\star\bullet\star\bullet} = \mathcal{J} - \mathcal{D}_-\mathcal{D}_+\ln\overline{\mathcal{J}}, \quad \overline{\mathcal{J}}^{\star\bullet\star\bullet} = \overline{\mathcal{J}} - \overline{\mathcal{D}}^-\overline{\mathcal{D}}^+\ln\mathcal{J}^{\star\bullet\star\bullet}\,, \tag{40}$$

which can easily be derived using Eqs. (28–29) and (39). They are discrete symmetries of the even and odd flows of the $N=4$ supersymmetric Toda chain hierarchy. In other words, if the set $\{\mathcal{J},\overline{\mathcal{J}}\}$ is a solution of the $N=4$ Toda chain hierarchy, then the set $\{\mathcal{J}^{\star\bullet\star\bullet},\overline{\mathcal{J}}^{\star\bullet\star\bullet}\}$, related to the former by Eqs. (40), is a solution of the hierarchy as well. The equations (40) reproduce the one-dimensional reduction of the two-dimensional $N=(2|2)$ superconformal Toda lattice.[14,15]

Finally, we would like to remark that the equations (35–36) can be rewritten in a local form, if one introduces a new superfield basis defined by the following invertible transformations:

$$\mathcal{J} \equiv \mathcal{D}_+\overline{\Psi}, \qquad \overline{\mathcal{J}} \equiv \overline{\mathcal{D}}^-\Psi\,,$$
$$\overline{\Psi} \equiv \overline{\mathcal{D}}^+\partial^{-1}\mathcal{J}, \quad \Psi \equiv \mathcal{D}_-\partial^{-1}\overline{\mathcal{J}}\,, \tag{41}$$

where $\Psi,\overline{\Psi}$ ($[\Psi]=[\overline{\Psi}]=-1/2$) are new constrained odd $N=2$ superfields

$$\mathcal{D}_-\Psi = \overline{\mathcal{D}}^+\Psi = 0\,, \qquad \mathcal{D}_-\overline{\Psi} = \overline{\mathcal{D}}^+\overline{\Psi} = 0\,, \tag{42}$$

with the reality conditions which can be derived from Eqs. (27–29) and (41). Then, these equations become

$$\frac{\partial}{\partial t_2}\Psi = +\Psi\,'' + 2\mathcal{D}_-\overline{\mathcal{D}}^-(\overline{\Psi}\,\overline{\mathcal{D}}^-\Psi) - \overline{\mathcal{D}}^+(\mathcal{D}_+\Psi)^2\,,$$
$$\frac{\partial}{\partial t_2}\overline{\Psi} = -\overline{\Psi}\,'' - 2\overline{\mathcal{D}}^+\mathcal{D}_+(\Psi\,\mathcal{D}_+\overline{\Psi}) + \mathcal{D}_-(\overline{\mathcal{D}}^-\overline{\Psi})^2\,, \qquad (43)$$

$$\begin{aligned}
\frac{\partial}{\partial t_3}\Psi &= \Psi\,''' + 3\mathcal{D}_-\Big[(\overline{\mathcal{D}}^-\Psi)\,'\,\overline{\mathcal{D}}^-\overline{\Psi} + (\overline{\mathcal{D}}^-\Psi)(\overline{\mathcal{D}}^-\overline{\Psi})^2 \\
&+ \frac{1}{2}\overline{\mathcal{D}}^+\overline{\mathcal{D}}^-(\mathcal{D}_+\Psi)^2\Big] + \overline{\mathcal{D}}^+\Big[(\mathcal{D}_+\Psi)^3 - 3(\mathcal{D}_+\Psi)^2\overline{\mathcal{D}}^-\overline{\Psi} \\
&- 6(\mathcal{D}_+\overline{\Psi})(\overline{\mathcal{D}}^-\Psi)\mathcal{D}_+\Psi\Big]\,, \\
\frac{\partial}{\partial t_3}\overline{\Psi} &= \overline{\Psi}\,''' + 3\overline{\mathcal{D}}^+\Big[(\mathcal{D}_+\overline{\Psi})\,'\,\mathcal{D}_+\Psi + (\mathcal{D}_+\overline{\Psi})(\mathcal{D}_+\Psi)^2 \\
&- \frac{1}{2}\mathcal{D}_+\mathcal{D}_-(\overline{\mathcal{D}}^-\overline{\Psi})^2\Big] + \mathcal{D}_-\Big[(\overline{\mathcal{D}}^-\overline{\Psi})^3 - 3(\overline{\mathcal{D}}^-\overline{\Psi})^2\mathcal{D}_+\Psi \\
&- 6(\mathcal{D}_+\overline{\Psi})(\overline{\mathcal{D}}^-\Psi)\overline{\mathcal{D}}^-\overline{\Psi}\Big]\,.
\end{aligned} \qquad (44)$$

5 Relation with the $N = 4$ Supersymmetric KdV Hierarchy

It is well known that there are often hidden relationships between a priori unrelated hierarchies. Some examples are the $N = 2$ NLS and $N = 2$ $\alpha = 4$ KdV, [16] the "quasi" $N = 4$ KdV and $N = 2$ $\alpha = -2$ Boussinesq, [9] the $N = 2$ (1,1)-GNLS and $N = 4$ KdV. [13,11] These relationships may lead to a deeper understanding of the hierarchies. They may help to obtain a more complete description and to derive solutions.

The existence of a manifestly $N = 4$ supersymmetric, local form (43–44) of equations belonging to the $N = 4$ supersymmetric Toda chain hierarchy, gives an additional evidence in favour of the existence of a hidden relationship with the $N = 4$ supersymmetric KdV hierarchy. [6,7]

It turns out that such a relationship indeed exists. Let us present it at the level of the second flow equations (30) which in a new superfield basis $\{\widetilde{J}, \Phi, \overline{\Phi}\}$ take the following local form:

$$-i\frac{\partial}{\partial t_2}\widetilde{J} = -\frac{1}{2}(\Phi + \overline{\Phi})\,'' - 2(\widetilde{J}(\Phi - \overline{\Phi}))\,' + [D, \overline{D}](\widetilde{J}(\Phi + \overline{\Phi}))\,,$$
$$-i\frac{\partial}{\partial t_2}\Phi = 2D\overline{D}(\widetilde{J}\,' - \widetilde{J}^2 - \frac{3}{4}\Phi^2 + \frac{1}{2}\Phi\overline{\Phi})\,,$$
$$-i\frac{\partial}{\partial t_2}\overline{\Phi} = 2\overline{D}D(\widetilde{J}\,' + \widetilde{J}^2 + \frac{3}{4}\overline{\Phi}^2 - \frac{1}{2}\Phi\overline{\Phi})\,, \qquad (45)$$

where $\widetilde{J}, \Phi, \overline{\Phi}$ ($[\widetilde{J}] = [\Phi] = [\overline{\Phi}] = -1$) are new unconstrained, chiral ($D\,\Phi = 0$) and antichiral ($\overline{D}\,\overline{\Phi} = 0$) even $N = 2$ superfields, respectively, related to the superfields $J, \overline{J}$ (26) by the following invertible transformations:

$$J \equiv \frac{1}{2}(\Phi + \overline{\Phi}) - i\widetilde{J}\,, \qquad \overline{J} \equiv \frac{1}{2}(\Phi + \overline{\Phi}) + i\widetilde{J}\,,$$
$$\widetilde{J} \equiv \frac{i}{2}(J - \overline{J})\,, \quad \Phi \equiv D\overline{D}\partial^{-1}(J + \overline{J}), \quad \overline{\Phi} \equiv \overline{D}D\partial^{-1}(J + \overline{J})\,, \tag{46}$$

and $D, \overline{D}$ are $N = 2$ odd covariant derivatives,

$$D \equiv \frac{1}{2}(D_{+} + iD_{-})\,, \qquad \overline{D} \equiv \frac{1}{2}(D_{+} - iD_{-})\,,$$
$$\left\{D\,, \overline{D}\right\} = \partial\,, \qquad \left\{D\,, D\right\} = \left\{\overline{D}\,, \overline{D}\right\} = 0\,. \tag{47}$$

Now, one can easily recognize that Eq. (45) is the second flow of the $N = 4$ KdV hierarchy in a particular "$SU(2)$ frame" (compare Eqs. (45) with Eqs. (4.5) and (4.3c) from Delduc et al. [7]).

Moreover, in this basis the second Hamiltonian structure of the $N = 4$ Toda chain hierarchy [1] reproduces the $N = 4$ $SU(2)$ superconformal algebra and the flow (45) is generated by the Hamiltonian H^{t}_{2} [1]

$$\begin{aligned} H^{t}_{2} &= \int dz d\theta^{+} d\theta^{-} uv\,' \equiv i \int dz d\theta^{+} d\theta^{-} \overline{J}[D, \overline{D}]\partial^{-1}(J + \overline{J}) \\ &\equiv \int dz d\theta^{+} d\theta^{-} \widetilde{J}(\overline{\Phi} - \Phi)\,. \end{aligned} \tag{48}$$

The same relationship is certainly valid for any other flow of the $N = 4$ Toda and $N = 4$ KdV hierarchies both in the $N = 2$ and $N = 4$ superspaces.

The relationship just established allows to apply the formalism of Delduc et al.,[1] developed for the case of the $N = 4$ Toda chain hierarchy, for a more complete description of the $N = 4$ KdV hierarchy. It can be used to construct new bosonic and fermionic flows and Hamiltonians, new finite and infinite discrete symmetries, the tau function, etc.. Let us present as an example the three involutions

$$(\,\Phi,\ \overline{\Phi},\ \widetilde{J}\,)^{*} = (\,-\overline{\Phi},\ -\Phi,\ \widetilde{J}\,)\,, \tag{49}$$

$$\begin{aligned} \Phi^{\bullet} &= \overline{\Phi} - i\overline{D}D \ln(\Phi + \overline{\Phi} + 2i\widetilde{J})\,, \\ \overline{\Phi}^{\bullet} &= \Phi + iD\overline{D} \ln(\Phi + \overline{\Phi} + 2i\widetilde{J})\,, \\ \widetilde{J}^{\bullet} &= -\widetilde{J} + \frac{1}{2}[D, \overline{D}] \ln(\Phi + \overline{\Phi} + 2i\widetilde{J})\,, \end{aligned} \tag{50}$$

$$(\Phi, \overline{\Phi}, \widetilde{J})^{\star} = (\overline{\Phi}, \Phi, \widetilde{J}), \tag{51}$$

and the Darboux-Backlund symmetries

$$\begin{aligned} \Phi^{\star\bullet} &= \Phi + iD\overline{D}\ln(\Phi + \overline{\Phi} + 2i\widetilde{J}), \\ \overline{\Phi}^{\star\bullet} &= \overline{\Phi} - i\overline{D}D\ln(\Phi + \overline{\Phi} + 2i\widetilde{J}), \\ \widetilde{J}^{\star\bullet} &= -\widetilde{J} + \frac{1}{2}[D, \overline{D}]\ln(\Phi + \overline{\Phi} + 2i\widetilde{J}), \end{aligned} \tag{52}$$

derived using the relationship (46) and Eqs. (27–29). The involution (50) looks rather complicated, however let us remember that in the original superfield basis $\{u, v\}$ it has a very simple form (19). The involutions (49) and (51) were discussed in Ivanov and Krivonos, [8] to our knowledge but the involution (50) and the corresponding real form of the $N = 4$ KdV hierarchy as well as its Darboux-Backlund symmetries (52) are presented here for the first time.

6 Conclusion

In this letter, we have described three distinct real forms of the $N = 4$ Toda chain hierarchy introduced in Delduc et al. [1] Then it has been shown that the symmetry algebra of one of these real forms contains the usual (untwisted) real $N = 4$ supersymmetry algebra. A set of $N = 2$ superfields with simple conjugation properties in the untwisted case have been introduced. It has then been shown how to extend these superfields to superfields in $N = 4$ superspace, and write all flows and conjugation rules directly in $N = 4$ superspace. Finally, a change of basis in $N = 4$ superspace has allowed us to eliminate all nonlocalities in the flows. As a byproduct, a relationship between the complex $N = 4$ Toda chain and $N = 4$ KdV hierarchies has been established, which allows to derive Darboux-Backlund symmetries and a new real form of the last hierarchy possessing a twisted $N = 4$ supersymmetry.

It is obvious that there remains a lot of work to do in order to improve our understanding of the hierarchy in $N = 4$ superspace. A first step in this direction would be to derive a Lax formulation of the flows in terms of $N = 4$ operators.

Acknowledgments

A.S. would like to thank L. Freidel for useful discussions and to Organizers of XIV-th Max Born Symposium for the invitation to present this work as well as for kind hospitality. This work was partially supported by the PICS Project No. 593, RFBR-CNRS Grant No. 98-02-22034, RFBR Grant No. 99-02-18417, Nato Grant No. PST.CLG 974874 and INTAS Grant INTAS-96-0538.

References

1. F. Delduc, L. Gallot and A. Sorin, *Nucl. Phys.* **B558**, 545 (1999), solv-int/9907004.
2. A.N. Leznov and A.S. Sorin, *Phys. Lett.* **B389**, 494 (1996), hep-th/9608166; *Nucl. Phys.* (Proc. Suppl.) **B56**, 258 (1997).
3. O. Lechtenfeld and A. Sorin, *Nucl. Phys.* **B557**, 535 (1999), solv-int/9810009.
4. O. Lechtenfeld and A. Sorin, *Supersymmetric KP hierarchy in $N = 1$ superspace and its $N = 2$ reductions*, ITP-UH-14/99, JINR E2-99-216, solv-int/9907021, *Nucl. Phys.* **B**, to appear.
5. S.J. Gates, Jr., M.T. Grisaru, M. Roček and W. Siegel, *Superspace or one thousand and one lessons in supersymmetry*, the Benjamin/Cummings Publishing Company, Inc, 1983, pgs. 58-59;
   P. West, *Introduction to supersymmetry and supergravity*, extended second edition, World Scientific, 1990, pgs. 393-394.
6. F. Delduc and E. Ivanov, *Phys. Lett.* **B309**, 312 (1993), hep-th/9301024.
7. F. Delduc, E. Ivanov and S. Krivonos, *J. Math. Phys.* **37**, 1356 (1996); Erratum-ibid. **38**, 1224 (1997), hep-th/9510033.
8. E. Ivanov and S. Krivonos, *Phys. Lett.* **A231**, 75 (1997), hep-th/9609191.
9. F. Delduc, L. Gallot and E. Ivanov, *Phys. Lett.* **B396**, 122 (1997), hep-th/9611033.
10. E. Ivanov, S. Krivonos and F. Toppan, *Phys. Lett.* **B405**, 85 (1997), hep-th/9703224.
11. L. Bonora and A. Sorin, *Phys. Lett.* **B407**, 131 (1997), hep-th/9704130.
12. Z. Popowicz, *Phys. Lett.* **B459**, 150 (1999), hep-th/9903198.
13. A. Sorin, *Phys. Lett.* **B395**, 218 (1997), hep-th/9611148; *Phys. Atom. Nucl.* **61**, 1768 (1998), solv-int/9701020.
14. J. Evans, T. Hollowood, *Nucl. Phys.* **B352**, 723 (1991).
15. V.B. Derjagin, A.N. Leznov and A. Sorin, *Nucl. Phys.* **B527**, 643 (1998), solv-int/9803010.
16. S. Krivonos and A. Sorin, *Phys. Lett.* **B357**, 94 (1995), hep-th/9504084; S. Krivonos, A. Sorin and F. Toppan, *Phys. Lett.* **A206**, 146 (1995), hep-th/9504138.

$SU(N)$ SKYRMIONS AND TWO DIMENSIONAL CP^N RATIONAL MAPS[a]

T. IOANNIDOU[b]
Institute of Mathematics, University of Kent at Canterbury, Canterbury CT2 7NF, UK

B. PIETTE[c] AND W.J. ZAKRZEWSKI[d]
Department of Mathematical Sciences, University of Durham, Durham DH1 3LE, UK

We show that classical solutions of the $SU(N)$ Skyrme models can be approximated using an ansatz involving two dimensional CP^N rational maps. We study various properties of these maps and show that, in general, their energies are only marginally higher than the energies of the corresponding $SU(2)$ embeddings. We also show that a generalisation of the rational maps ansatz leads to exact solutions of the $SU(N)$ model.

1 Introduction

The Skyrme model was first introduced by Skyrme[1] in an attempt to describes nuclei using topological ideas. Since then, QCD and the standard model have become well established as the theory of elementary particles and as such should describe nuclear phenomena. However, these theories of nuclei are hard to solve at low energies; thus it is difficult to calculate energies of bound states, *i.e.* nuclei. Witten, however, showed[2] that QCD, in the limit of a large number of colours, gives the Skyrme model as an effective low energy theory of baryons. To describe real (physical) baryons, one should take the classical solutions of the Skyrme model and quantise some of their degrees of freedom (a problem not fully solved yet).

The Skyrme model is described by the Lagrangian

$$\mathcal{L} = \frac{F^2}{16}\,\mathrm{tr}\left(\partial_\mu U\,\partial^\mu U^{-1}\right) + \frac{1}{32a^2}\,\mathrm{tr}\left[\partial_\mu U\,U^{-1}, \partial_\nu U\,U^{-1}\right]^2 + \frac{F^2}{16}\,M_\pi^2\,\mathrm{tr}\left(U^{-1} + U - 2\,I\right), \tag{1}$$

where $U(\vec{x},t)$ is an $SU(N)$ valued scalar field, $F \approx 189$ MeV is the pion decay constant and a is a dimensional constant. The second term, called the Skyrme

[a]talk given by W.J. Zakrzewski
[b]e-mail address: T.Ioannidou@ukc.ac.uk
[c]e-mail address: B.M.A.G.Piette@durham.ac.uk
[d]e-mail address:W.J.Zakrzewski@durham.ac.uk

term, was introduced by Skyrme to have finite size static solutions. This term also introduces small interaction forces between baryons. The last term makes the pion field $U(\vec{x},t)$ massive and one usually considers the cases $M_\pi = 0$ or $M_\pi = 138 MeV$.

To have finite energy configurations, one must require that the field $U(\vec{x},t)$ goes to a constant matrix, say I, at spatial infinity: $U \to I$ as $|\vec{x}| \to \infty$. This effectively compactifies the three dimensional Euclidean space into S^3 and hence implies that the field configurations of the Skyrme model can be considered as mappings from S^3 into $SU(N)$. As the third homotopy class of $SU(N)$ is Z every field configuration is characterised by an integer:

$$B = \frac{1}{24\pi^2} \int_{R^3} \varepsilon_{ijk} \, \mathrm{tr}\, (\partial_i U\, U^{-1} \partial_j U\, U^{-1} \partial_k U\, U^{-1})\, d^3\vec{x}\,. \tag{2}$$

It was Skyrme's main idea to interpret this quantity as the baryon number. [2,3] The lowest energy state in the $B = 1$ sector is then naturally identified with the (classical) nucleon and static solutions with larger values for B then describe nuclei.

In this paper, we concentrate our attention on studying static properties of the model and we consider $U(\vec{x})$ fields, which are stationary points of the energy functional.

To study classical solutions of the model, it is convenient to perform a change of variables to express the Lagrangian in dimensionless units. To do this we rescale the space-time coordinates [4,5] $\vec{x} \to 2\vec{x}/aF$ and, defining $m_\pi = M_\pi 2/aF$, we write the static energy corresponding to the Lagrangian (1) as

$$\begin{aligned} E \;=\; & \frac{F}{4a} \int_{R^3} \left\{ -\frac{1}{2}\, \mathrm{tr}\, (\partial_i U\, U^{-1})^2 - \frac{1}{16}\, \mathrm{tr}\, [\partial_i U\, U^{-1}, \partial_j U\, U^{-1}]^2 \right. \\ & \left. - \frac{m_\pi^2}{2} \mathrm{tr}(U^{-1} + U - 2\, I) \right\} d^3\vec{x}\,. \end{aligned} \tag{3}$$

From now on, we take $F/4a = 1/12\pi^2$ so that the energy is expressed in the same units as the baryon number. In this case, the static fields U obey the equation

$$\partial_i \left(\partial_i U\, U^{-1} - \frac{1}{4}\, [\partial_j U\, U^{-1}, [\partial_j U\, U^{-1}, \partial_i U\, U^{-1}]] \right) - \frac{m_\pi^2}{4}\, (U - U^{-1}) = 0\,. \tag{4}$$

The simplest nontrivial solution of (4) is the so-called hedgehog ansatz for the configuration corresponding to a single Skyrmion ($B = 1$). In polar coordinates, the angular dependence of the ansatz is explicit while the radial dependence reduces the problem to a second order ordinary differential equation for a profile function.

Finding solutions with higher baryon numbers is much more difficult and, so far, most of the work has been done numerically. For example, Battye and Sutcliffe [5] have computed numerically solutions up to $B = 9$ and have shown that these solutions are nearly radially symmetric *i.e.* that the energy density is concentrated on the edges and the vertices of various polyhedra. This observation led to the rational map ansatz [6] which provides a good approximation to the $SU(2)$ solutions. The energy of such configurations is only slightly higher than that of the solutions, and, when taken as initial conditions for a (numerical) time evolution of the Skyrme model, the rational maps configurations relax to the corresponding static solutions without changing much.

Most of the work performed so far has involved $SU(2)$ skyrmions and very little has been done for the $SU(N)$ model when $N > 2$. An interesting question then arises as to whether there are any finite energy solutions of the $SU(N)$ ($N > 2$) model which are not embeddings of the $SU(2)$ model and, if they exist, whether they have lower energies than their $SU(2)$ counterparts.

The first example of such a *non-embedding* configuration for a higher group was the $SO(3)$ soliton, which corresponds to a bound system of two skyrmions, and which was found using the chiral field ansatz by Balachandran et al. [7] Another configuration, with a large $SU(3)$ strangeness content, was found by Kopeliovich et al. [8] However, all other known skyrmion configurations seem to have been the embeddings of the solutions of the $SU(2)$ model.

In the following sections, we will show how the idea of Houghton et al. can be generalised and how this leads to configurations with a larger group of symmetry. We will also show that the rational map ansatz leads, in some cases, to exact solutions of the model. We describe here our work reported previously in Ioannidou et al. [9,10]

2 CP^N Rational Map Ansatz

To write down the rational map ansatz for the $SU(N)$ model we use the polar coordinates (r, θ, ϕ) in $\mathbf{R}^3$, and express the $SU(N)$ field as

$$\begin{aligned} U(r,\theta,\phi) &= e^{2ig(r)(P-I/N)} \\ &= e^{-2ig(r)/N}\left(I + (e^{2ig} - 1)P\right), \end{aligned} \tag{5}$$

where P is a $N \times N$ hermitian projector which depends only on the angular variables (θ, ϕ) and $g(r)$ is a radial profile function. The overall phase $e^{-2ig(r)/N}$ is required to ensure that the field U belongs to $SU(N)$. We will only consider the case where P is a rank one projector, so the ansatz implies that the matrix P is a mapping from S^2 into $CP^{(N-1)}$. It is then convenient to map the sphere onto the complex plane via a stereographic projection and use the complex coordinate ξ and its conjugate instead of θ and ϕ. More explicitly, the relation between the two is given by

$$\xi = \tan\frac{\theta}{2}\, e^{i\phi}\,. \tag{6}$$

We can now write P as

$$P(V) = \frac{V \otimes V^\dagger}{|V|^2}\,, \tag{7}$$

where V is a N component complex vector dependent on ξ and $\bar{\xi}$.

For (5) to be well-defined at the origin, we have to impose that the profile function is a multiple of π at the origin and at infinity. We can always assume that $g(\infty) = 0$, while at the origin we will impose the condition that $g(0) = k\pi$, where k is an integer which most of the time will be 1.

If we insert the ansatz (5) into the expression for the baryon number (2) we get

$$\begin{aligned} B &= \frac{i}{\pi^2}\int d\xi\, d\bar{\xi}\, \mathrm{tr}\left(P\left[\partial_\xi P,\, \partial_{\bar{\xi}} P\right]\right) \int_0^\infty dr\, \sin^2 g\, g_r \\ &= \frac{i}{2\pi^2}\int d\xi\, d\bar{\xi}\, \mathrm{tr}\left(P\left[\partial_{\bar{\xi}} P,\, \partial_\xi P\right]\right)\left(g(0) - g(\infty)\right), \end{aligned} \tag{8}$$

which is the topological charge of the two-dimensional $CP^{(N-1)}$ sigma model multiplied by the value of g at the origin. We thus see that to have a topologically non-trivial configuration, the field g must be non-zero at the origin and the projector P must be a topologically non-trivial $CP^{(N-1)}$ field.

If we plug the ansatz (5) into the expression for the energy density (3) we get

$$E = \frac{1}{3\pi}\int dr\left(A_N\, g_r^2\, r^2 + 2\mathcal{N}\, \sin^2 g\, (1 + g_r^2) + \mathcal{I}\,\frac{\sin^4 g}{r^2} + 2m_\pi^2(1 - \cos g)r^2\right), \tag{9}$$

where

$$A_N = \frac{2}{N}(N-1)\,, \tag{10}$$

$$\mathcal{N} = \frac{i}{2\pi}\int d\xi\, d\bar{\xi}\, \mathrm{tr}\left(|\partial_\xi P|^2\right), \tag{11}$$

$$\mathcal{I} = \frac{i}{4\pi}\int d\xi\, d\bar{\xi}\, (1 + |\xi|^2)^2\, \mathrm{tr}\left(\left[\partial_\xi P,\, \partial_{\bar{\xi}} P\right]^2\right). \tag{12}$$

An attractive feature of the ansatz (5) is that it leads to a simple expression for the energy density which can be successively minimized with respect to the parameters of the projector P and then with respect to the shape of the profile function $g(r)$. This is then expected to give good approximations to multi-skyrmion field configurations. As the integrals $\mathcal{N}$ and $\mathcal{I}$ in (9) are independent of r, we can minimise (9) by first minimising $\mathcal{N}$ and $\mathcal{I}$ as functions of P and then with respect to the profile function g.

However, since $\mathcal{N}$ is the expression for the energy of the two dimensional Euclidean CP^2 sigma model, all classical solutions contain the so-called self-dual solutions, instantons or holomorphic maps from S^2 into $CP^{(N-1)}$, first given in D'Adda et al.,[11] which are given by the projector P of the form (7) with $V = f(\xi)$. In this case, the energy $\mathcal{N}$ is given by the degree of f, *ie* the degree of the highest order polynomial in ξ among the components of f after all their common factors have been cancelled. Moreover, if we impose the boundary condition $g(0) = \pi$, the baryon number will be given by the degree of the rational map.

The procedure to compute the rational map ansatz can thus be summarised as follows: First we must choose which baryon number B the configuration must have. This fixes the value of $\mathcal{N} = B$ but also implies that the entries of $V = f(\xi)$ are polynomials of degree B (or lower but at least one entry must be of degree B). We then determine the coefficients of the polynomials by minimising $\mathcal{I}$. Finally, considering $\mathcal{I}$, $\mathcal{N}$ as constants, we can compute g by solving the Euler-Lagrange equations for (9) and compute the energy of the resultant configurations.

3 Computing the Rational Map

To compute the rational map ansatz, we must minimise $\mathcal{I}$ (10) over all the polynomial functions $f(z)$ of a chosen degree B. To do this it is helpful to get rid of all gauge degrees of freedom and reduce the number of parameters to a minimum.

First of all it is obvious that the relation between the projector P and the vector $V = f$ (7) implies that f is defined up to an overall coefficient which can be any function of ξ. Moreover, under the gauge transformation $U \to AUA^\dagger$, f becomes Af. We can thus "mix" the components of f using any $SU(N)$ constant matrix. We can use this invariance to make the first component of f a polynomial of degree B, the second one a polynomial of degree $B-1$ and in general make the nth component a polynomial of degree $B-n+1$. In other words, we can write the most general expression for f as

$$f = \begin{pmatrix} \xi^B + a^{(1)}_{B-1}\,\xi^{B-1} + \ldots + a^{(1)}_1\,\xi \\ a^{(2)}_{B-1}\,\xi^{B-1} + a^{(2)}_{B-2}\,\xi^{B-2} + \ldots + a^{(2)}_1\,\xi + a^{(2)}_0 \\ a^{(3)}_{B-2}\,\xi^{B-2} + a^{(3)}_{B-3}\,\xi^{B-3} + \ldots + a^{(3)}_1\,\xi + a^{(3)}_0 \\ \vdots \\ a^{(N)}_{B-N+1}\,\xi^{B-N+1} + a^{(N)}_{B-N}\,\xi^{B-N} + \ldots + a^{(N)}_1\,\xi + a^{(N)}_0 \end{pmatrix}, \qquad (13)$$

where we have used the overall coefficient of f to set the coefficient of ξ^B to one. We have also used the translational invariance of the model to set $a^{(1)}_1$ to 0. Moreover, all the coefficients are complex except the highest coefficient of each polynomial $(a^{(r)}_{B-r+1})$, which can be taken to be real; this comes from the fact the energy (9) is invariant under the transformation $f_r \to e^{i\alpha_r} f_r$ where α_r is any constant.

One consequence of this is that any rational map ansatz constructed out of a holomorphic vector f of degree $n < N$ can be reduced to an $SU(n+1)$ ansatz embedding. This means that the $B = 1$ ansatz, which is nothing but the hedgehog solution of Skyrme, is the same for all $SU(N)$ models. For $B = 2$, we will have only two types of ansatz, one for $SU(2)$ and one for $SU(3)$. For $B = 3$, we will have ansätze for $SU(2)$, $SU(3)$ and $SU(4)$.

To minimise $\mathcal{I}$ we can use two different methods. The method used by Houghton et al.[6] consists in looking for rational maps which are invariant under some subgroup of $SU(2)$. The sphere is indeed invariant under the $SO(3) \equiv SU(2)$ group, so for any rotation $s = \begin{pmatrix} a & b \\ -\bar{b} & \bar{a} \end{pmatrix}$ such that $|a|^2 + |b|^2 = 1$, we have $s(\xi) = (a\xi + b)/(-\bar{\xi} + \bar{a})$. A rational map ansatz is then invariant under the subgroup G of the $SU(2)$ group if for any $g \in G$ we have

$$P(f(g(z))) = P(A_g(f(\xi)))\,, \qquad (14)$$

where A_g belongs to $SU(N)$. The set of A_g will then form an embedding of G in $SU(N)$. Solving (14) for a rational map of a given degree reduces the number of parameters and makes the minimisation of $\mathcal{I}$ somewhat easier, but this assumes that the minimal configuration has a certain symmetry (this happens to be true for $B < 10$).

Another way to determine the coefficients of the rational map f (13) is to insert the general expression (13) for f in the expression for $\mathcal{I}$ (10) and minimise it numerically. This is somewhat more straightforward and does not make any assumption, but on the other hand, it is a rather tedious operation when B is large.

In what follows, we will need to use some properties of the analytic projectors P, *ie* projectors of the form

$$P_0 = P(f) = \frac{f\,f^\dagger}{|f|^2}\,, \tag{15}$$

where f is a holomorphic vector (*ie* whose entries are functions only of ξ). First of all, it is easy to show that P_0 satisfies the equation

$$P_0\,\partial_\xi P_0 = 0\,, \qquad \partial_\xi P_0\,P_0 = \partial_\xi P_0\,, \tag{16}$$

which is nothing but the self-dual equations of the two dimensional $CP^{(N-1)}$ sigma models.[12]

Following Din and Zakrzewski,[13] we define the operator P_+ by its action on any vector $v \in C^N$ as

$$P_+ v = \partial_\xi v - \frac{v\,(v^\dagger\,\partial_\xi v)}{|v|^2}\,, \tag{17}$$

and then define further vectors $P_+^k v$ by induction: $P_+^k v = P_+(P_+^{k-1} v)$.

If f is holomorphic, then the following relations follow directly from the definition of the P_+ operator or are very easy to prove[12]

$$(P_+^k f)^\dagger\,P_+^l f = 0\,, \qquad k \neq l\,, \tag{18}$$

$$\partial_{\bar\xi}\,(P_+^k f) = -P_+^{k-1} f\,\frac{|P_+^k f|^2}{|P_+^{k-1} f|^2}\,, \quad \partial_\xi\left(\frac{P_+^{k-1} f}{|P_+^{k-1} f|^2}\right) = \frac{P_+^k f}{|P_+^{k-1} f|^2}\,. \tag{19}$$

4 $SU(N)$ Radially Symmetric Configurations

The largest symmetry we can impose on our configuration is the radial symmetry. This can be done by using the following ansatz:

$$N = 2\,, \qquad f = (1, \xi)^t\,, \tag{20}$$

$$N = 3\,, \qquad f = (1, \sqrt{2}\xi, \xi^2)^t\,, \tag{21}$$

$$N = 4\,, \qquad f = (1, \sqrt{3}\xi, \sqrt{3}\xi^2, \xi^3)^t\,, \tag{22}$$

$$N = n\,, \qquad f = (f_0, f_1, \ldots, f_{n-1})^t : \quad f_i = \xi^i \sqrt{C_{i+1}^{n-1}}\,, \tag{23}$$

where C_{i+1}^{n-1} denotes the binomial coefficients. It is easy to show that for these very special choices of f, $|f|^2 = (1+|\xi|^2)^{n-1}$ and

$$\frac{|P_+ f|^2}{|f|^2} \equiv \frac{n}{(1+|\xi|^2)^2} \,. \tag{24}$$

Using the condition (24) the integrals involving P in the energy (9) can be evaluated analytically,

$$\mathcal{N} = n\,, \qquad \mathcal{I} = (N-1)^2 = n^2\,,. \tag{25}$$

Using the analyticity of the projector P_0, it is straightforward to verify that the baryon number of this field is $B = n$, *ie* the degree of f.

When $m_\pi \neq 0$ the energy (9) for the ansatz (5) becomes

$$E = \frac{1}{3\pi}\int dr \Bigg\{ A_N\, g_r^2\, r^2 + 2\mathcal{N}\sin^2 g\,(1+g_r^2) + \mathcal{I}\,\frac{\sin^4 g}{r^2}$$
$$+ m_\pi^2\, r^2 \left[(N-1)\left(1-\cos(\frac{2g}{N})\right) + 1 - \cos\left(\frac{(N-1)\,2g}{N}\right)\right]\Bigg\}. \tag{26}$$

Minimising (26) given (25) leads to the following equation for the profile function

$$g_{rr}\left(A_N + 2n\,\frac{\sin^2 g}{r^2}\right) + 2\,A_N\,\frac{g_r}{r} + \frac{\sin 2g}{r^2}\left(n\,(g_r^2-1) - n^2\,\frac{\sin^2 g}{r^2}\right) \tag{27}$$
$$m_\pi^2\left(\frac{N-1}{N}\right)\left[\sin\left(\frac{2g}{N}\right) + \sin\left(\frac{(N-1)2g}{N}\right)\right] = 0\,,$$

where A_N is given by (10).

We have looked at field configurations corresponding to $B = 2-4$ for the $SU(2)$ embeddings and for the $SU(N)$ spherical symmetric fields (21)-(23) where $N = B+1$ (*ie* $SU(3)$ for $B = 2$) and studied the dependence of their energies on m_π. Solving (28) to determine g and then calculating the energy of the configuration we find that, in all cases at low values of the mass, the embeddings have lower energies while as the mass increases the energies increase. However, as the embedding energies increase faster for all low B there is a value of m_π above which the embedding energy is higher. Unfortunately, this value of m_π is quite large and it increases with the increase of B.

These results are summarised in Table 1, which gives values of the energy for different values of the mass. Note that the energy per skyrmion of the harmonic ansatz configuration is always lower than the energy of a single skyrmion.

m_π	SU(2) $E_{B=1}$	SU(2) $E_{B=2}$	SU(3) $E_{B=2}$	SU(2) $E_{B=3}$	SU(4) $E_{B=3}$	SU(2) $E_{B=3}$	SU(5) $E_{B=4}$
0	1.232	2.416	2.444	3.553	3.644	4.546	4.838
0.2	1.247	2.444	2.472	3.594	3.683	4.597	4.886
1	1.416	2.795	2.808	4.125	4.172	5.270	5.520
2.23	1.693	3.381	3.370	5.021	5.006	6.419	6.615
7	2.510	5.101	5.030	7.634	7.478	9.776	9.880
30	4.783	9.836	9.633	14.793	14.339	18.971	18.948

Table 1: Mass dependence of the energy for the radially symmetric configurations in the $SU(2) - SU(5)$ models.

5 Low Energy $SU(N)$ Configurations

Various rational maps have been computed for the $SU(2)$ [6] and $SU(N)$ Skyrme models. [9] In Table 2 we summarise the properties of a few rational maps for the lowest baryon numbers when $m_\pi = 0$.

In Table 3 we list the shapes and symmetries of these configurations. The shape and symmetries of the $SU(2)$ rational maps are the same as the symmetries and shape of the exact $SU(2)$ solutions.

	B	Rat. Map En./Sk	$\mathcal{I}$	True En./Sk	f^t
$SU(2)$	1	1.232	1	1.232	$(1, \xi)$
$SU(2)$	2	1.208	5.81	1.171	$(1, \xi^2)$
$SU(3)$	2	1.222	4	1.171	$(1, \sqrt{2}\xi, \xi^2)$
$SU(2)$	3	1.184	13.58	1.143	$(i\sqrt{3}\xi^2 - 1, \xi(\xi^2 - i\sqrt{3}))$
$SU(3)$	3	1.215	10.654	1.143	$(1/\sqrt{(2)}, 1.576\xi, \xi^3)$
$SU(4)$	3	1.215	9	1.143	$(1, \sqrt{3}\xi, \sqrt{3}\xi^2, \xi^3)$
$SU(2)$	4	1.137	20.65	1.116	$(\xi^4 + 2i\sqrt{3}\xi^2 + 1, \xi^4 - 2i\sqrt{3}\xi^2 + 1)$
$SU(3)$	4	1.184	18.045	1.116	$(1, 2.7191\xi^2, \xi^4)$
$SU(5)$	4	1.2095	16	1.116	$(1, \sqrt{4}\xi, \sqrt{6}\xi^2, \sqrt{4}\xi^3, \xi^4)$

Table 2: Properties of various $SU(N)$ rational maps configurations when $m_\pi = 0$

When comparing the rational maps of Table 2 we see that the symmetries of these maps for a given baryon number increase with N. This higher symmetry has the effect of decreasing $\mathcal{I}$ but this effect is more than compensated by the increase of $A_N = \frac{2}{N}(N-1)$ in (10) and, as a result, the more symmetrical configurations have a higher energy than the $SU(2)$ ones. Nevertheless, we can conclude that the $SU(N)$ model has solutions with the symmetries listed in Table 3 and that these solutions have an energy most probably higher than the corresponding $SU(2)$ embedding but lower that the energy of the corresponding ansatz. Moreover, each of these solutions has an energy per skyrmion lower than the energy of a single Skyrmion.

	B	Symmetry	Shape
$SU(2)$	1	Radial	Sphere
$SU(2)$	2	Torus	Axial
$SU(3)$	2	Radial	Sphere
$SU(2)$	3	Tetrahedral	Tetrahedron
$SU(3)$	3	Axial	Torus + sphere on top of it
$SU(4)$	3	Radial	Sphere
$SU(2)$	4	Cubic	Cube
$SU(3)$	4	Axial	2 tori on top of each other
$SU(5)$	4	Radial	Sphere

Table 3: Properties of various $SU(N)$ rational maps configurations when $m_\pi = 0$

6 Radially Symmetric Solution of the $SU(N)$ Model

To generalise the hedgehog ansatz and construct radially symmetric solutions of the $SU(N)$ model we have to generalise the rational map ansatz (5) and use the set of mutually orthogonal projectors constructed from the family of $P_+^k f$ vectors as follows:

$$P_0 = P(f), \qquad P_1 = P(P_+ f), \qquad \ldots, \qquad P_i = P(P_+^i f). \qquad (28)$$

The multi-projector ansatz is given by

$$\begin{aligned} U=&\exp\{ig_0(P_0 - \frac{I}{N}) + ig_1(P_1 - \frac{I}{N}) - \ldots + ig_{N-2}(P_{N-2} - \frac{I}{N})\} \qquad (29)\\ =&e^{-ig_0/N}(I + A_0 P_0)e^{-ig_1/N}(I + A_1 P_1)\ldots e^{-ig_{N-2}/N}(I + A_{N-1}P_{N-2}), \end{aligned}$$

where $g_k = g_k(r)$, for $k = 0, \ldots, N-2$, are the profile functions and $A_k = e^{ig_k} - 1$.

Taking for f the harmonic function given in (23), one can show that

$$|P_+^k f|^2 = \alpha(1 + |\xi|^2)^{N-2k-1}, \qquad (30)$$

where α depends on N and k. From this it is possible to prove that (30) is a solution of the $SU(N)$ model if, defining $F_k = g_k - g_{k+1}$, $(k = 0, ..N-2)$ and

$F_{N-2} = g_{N-2}$, the following equations are satisfied

$$\begin{aligned}
&-\frac{2(l+1)}{N}\sum_{i=0}^{N-2}(i+1)\ddot{F}_i + 2\sum_{k=0}^{l}\sum_{i=k}^{N-2}\ddot{F}_i + \frac{1}{r^2}\ddot{F}_l(l+1)(N-l-1)(1-\cos F_l)\\
&+\frac{1}{2r^2}\dot{F}_l^2(l+1)(N-l-1)\sin F_l\\
&+\frac{2}{r}\left(-\frac{2(l+1)}{N}\sum_{i=0}^{N-2}(i+1)\dot{F}_i + 2\sum_{k=0}^{l}\left(\sum_{i=k}^{N-2}\dot{F}_i\right)\right)\\
&-\frac{2}{r^2}(l+1)(N-l-1)\sin F_l\\
&-\frac{1}{r^4}(l+1)^2(N-l-1)^2(1-\cos F_l)\sin F_l\\
&+\frac{1}{2r^4}(l+1)(N-l-1)\sin F_l\big[l(N-l)(1-\cos F_{l-1})\\
&+(l+2)(N-l-2)(1-\cos F_{l+1})\big] = 0\,.
\end{aligned} \tag{31}$$

The topological charge of these solutions is then given by

$$B = \frac{1}{2\pi}\sum_{i=0}^{N-2}(i+1)(N-i-1)\,(F_i - \sin F_i)_{r=0}^{r=\infty}\,, \tag{32}$$

and as $g_i(\infty) = 0$ (required for the finiteness of the energy) the only contributions to the topological charge come from $F_i(0)$. By choosing various boundary condition for g_i at the origin, one can construct various families of solutions, most of which are topologically non-trivial, but with a relatively large energy (cfr. Ioannidou et al.[10] for more details).

7 Conclusions

In this paper we have summarised our previous work[9,10] which has shown that one can use the rational map of the two-dimensional CP^N model to construct some low energy configurations of the $SU(N)$ Skyrme model. This construction leads to some low energy configurations which are not $SU(2)$ embeddings into an $SU(N)$, but they have larger energies than the known embedded $SU(2)$ solutions.

We have also shown that the multi projector rational map ansatz leads to some radially symmetric solutions of the $SU(N)$ model.

Acknowledgments

We would like to thank C. J. Houghton, N. S. Manton and P.M. Sutcliffe for their interest and correspondence about the symmetries of our field configurations. WJZ thanks Profs. Lukierski and Popowicz for their invitation to present this work at the XIV Max Born Symposium and their hospitality in Karpacz.

References

1. T. H. R. Skyrme, *Nucl. Phys.* **31**, 556 (1962).
2. E. Witten, *Nucl. Phys.* B **223**, 422 (1983).
3. T. H. R. Skyrme, *Proc. R. Soc.* A **260**, 127 (1961).
4. N. S. Manton, *Phys. Lett.* B **192**, 177 (1987).
5. R. A. Battye and P. M Sutcliffe, *Phys. Rev. Lett.* **79**, 363 (1997).
6. C. J. Houghton, N. S. Manton and P. M. Sutcliffe, *Nucl. Phys.* B **510**, 507 (1998).
7. A. P. Balachandran, A. Barducci, F. Lizzi, V. G. J. Rodgers and A. Stern, *Phys. Rev. Lett.* **52**, 887 (1984).
8. V.B. Kopeliovich, B.E. Schwesinger and B.E. Stern, *JETP Lett.* **62**, 185-90 (1995).
9. T. Ioannidou, B. Piette and W. J. Zakrzewski *Low Energy States in the SU(N) Skyrme Model*: Proceedings of ISPM-98, Eds. M. Eliashvili et al (1998), 91-123
10. T. Ioannidou, B. Piette and W.J. Zakrzewski, *Jour. Math. Phys.* **40**, 6223 (1999).
11. A. D'Adda, P. Di Vecchia and M. Luscher, *Nucl. Phys.* B **146**, 63 (1998).
12. W. J. Zakrzewski, Low dimensional sigma models (IOP, 1989).
13. A. Din and W. J. Zakrzewski, *Nucl. Phys.* **B 174**, 397 (1980). (1980).

DIVERSE PBGS PATTERNS AND SUPERBRANES

E. IVANOV

Bogoliubov Laboratory of Theoretical Physics, JINR,
141 980, Dubna, Moscow Region, Russian Federation

This is a brief account of the approach to superbranes based upon the concept of Partial Breaking of Global Supersymmetry (PBGS).

1 Introduction

The view of superbranes as theories of partial spontaneous breaking of global SUSY (PBGS) has received a considerable attention (see, e.g., References[1–12]). The salient feature of this approach is that it deals with the Goldstone superfields living on the worldvolume superspace of unbroken SUSY and accommodating the superbrane physical degrees of freedom.

It is well known that the superbranes in the Green-Schwarz (GS) formulation break half of the target SUSY (see, e.g., Duff[13]). Choosing the static gauge with respect to the worldvolume diffeomorphisms and killing half of the target θ-coordinates by κ-symmetry, one ends up with the physical multiplet which comprises transverse target bosonic coordinates and the rest of θ-coordinates (in the case of D-branes and their generalizations, the physical multiplets can include additional fields). The remaining θ's are shifted under half of the target SUSY, which suggests to interpret them as the relevant Goldstone fermions. Thus one half of the original SUSY, with respect to which the physical degrees of freedom form a supermultiplet, is unbroken, while the other is spontaneously broken, with the physical fermions as the Goldstone ones.

One can reverse the argument: adopt PBGS as the guiding principle and deduce superbranes just from it. In doing so, one can take advantage of the nonlinear realizations method [14–16] which provides the universal framework for treating spontaneously broken symmetries. In this method, the Goldstone (super)fields are identified with the parameters of the coset space of the full (super)symmetry group over its unbroken symmetry subgroup. The invariant actions of the Goldstone superfields constructed as integrals over the worldvolume superspace are reduced to the corresponding GS-type actions after passing to components and eliminating the auxiliary fields.

The nonlinear realizations method was worked out for constructing nonlinear sigma models of internal symmetries. It is less known that the bosonic p-branes in the physical (or "static") gauge can also be treated in the nonlinear realizations language. [17] To describe in this way some p-brane moving in

D-dimensional Minkowski space, one should consider a nonlinear realization of the relevant Poincaré group $\mathcal{P}_D = ISO(1, D-1)$, such that the residual unbroken (vacuum stability) symmetry group is the product of the Poincaré group of $p+1$-dimensional brane worldvolume and the rotation group of the transverse brane coordinates. The relevant coset manifold is $\mathcal{P}_D/SO(1,p)\otimes SO(D-p-1)$. One splits the full D space translation generator P_M, $M = 0, 1, \ldots D-1$, as

$$P_M \quad \Rightarrow \quad (P_m, \, P_\mu) \, , \quad m = 0, 1, \ldots p; \;\; \mu = p+1, \ldots D \, , \tag{1}$$

and associates with P_m, P_μ the worldvolume coordinate x^m and the Goldstone field $X^\mu(x)$ as the coset parameters (X^μ becomes the transverse p-brane coordinate). Also, one introduces the Goldstone fields $\Lambda^\mu_m(x)$ parametrizing the spontaneously broken part of the Lorentz group $SO(1, D-1)$ (with the generators L^m_μ):

$$P_m \Rightarrow x^m \, , \quad P_\mu \Rightarrow X^\mu(x) \, , \quad L^m_\mu \Rightarrow \Lambda^\mu_m(x) \, . \tag{2}$$

The application of the Cartan forms techniques augmented with some extra covariant constraints (the inverse Higgs [18]) gives rise to the following minimal invariant action

$$S_{br} \sim \int d^{p+1}x \left(\sqrt{(-1)^p \, g} - 1 \right) , \; g = \det \left(\eta_{mn} - \partial_m X^\mu \partial_n X^\mu \right) . \tag{3}$$

It is the static gauge form of the p-brane Nambu-Goto (NG) action.

As an illustration, let us consider the simplest example of the massive particle (0-brane) in $D = 2$. The $D = 2$ Poincaré group $\mathcal{P}_{(2)}$ involves two translation generators P_0, P_1 and the $SO(1,1)$ Lorentz generator L, with the only non-vanishing commutators

$$[\, L, P_0 \,] = iP_1 \, , \quad [\, L, P_1 \,] = iP_0 \, . \tag{4}$$

In accord with the said above, we should construct a nonlinear realization of $\mathcal{P}_{(2)}$, with the one-dimensional "Poincaré group" generated by the generator P_0 as the stability subgroup. Thus we are led to place all generators into the coset (in this particular case it coincides with the full group):

$$G = \mathrm{e}^{itP_0} \, \mathrm{e}^{iX(t)P_1} \, \mathrm{e}^{i\Lambda(t)L} \, . \tag{5}$$

Here the wordline evolution parameter t (time) is the coset coordinate associated with P_0, and the Goldstone fields are associated with the rest of generators. The group $\mathcal{P}_{(2)}$ acts as left shifts of G. The Cartan forms

$$G^{-1} \, dG = i \, \omega_t \, P_0 + i \, \omega_1 \, P_1 + i\omega_L \, L \, , \tag{6}$$

$$\omega_t = \sqrt{1+\Sigma^2}\, dt + \Sigma\, dX \ , \quad \omega_1 = \sqrt{1+\Sigma^2}\, dX + \Sigma\, dt \ ,$$
$$\omega_L = \frac{1}{\sqrt{1+\Sigma^2}}\, d\Sigma \ , \qquad \Sigma \equiv \mathrm{sh}\, \Lambda \ , \tag{7}$$

by construction are invariant under this left group action. Next, we observe that the Lorentz Goldstone field $\Sigma(t)$ can be traded for $\dot{X}(t)$ by the inverse Higgs [18] constraint

$$\omega_1 = 0 \quad \Rightarrow \quad \Sigma = -\frac{\dot{X}}{\sqrt{1-\dot{X}^2}} \ . \tag{8}$$

This constraint is covariant since ω_1 is the group invariant (in the generic case, the coset Cartan forms undergo homogeneous rotation in their stability subgroup indices). Thus the obtained expression for Σ possesses correct transformation properties. Substituting it into the remaining Cartan forms we find

$$\omega_t = \sqrt{1-\dot{X}^2}\, dt \ , \quad \omega_L = \sqrt{1-\dot{X}^2}\, \frac{d}{dt}\left(\frac{\dot{X}}{\sqrt{1-\dot{X}^2}}\right) dt \ . \tag{9}$$

The simplest invariant action, the covariant length

$$S = \int \omega_t = \int dt \sqrt{1-\dot{X}^2} \ , \tag{10}$$

is recognized, up to a renormalization factor of the dimension of mass, as the action of $D=2$ massive particle in the static gauge $X^0(t) = t$. The equation of motion for $X(t)$ can also be given the manifestly covariant form

$$\omega_L = 0 \ . \tag{11}$$

Actually, we could start from the action (10) with the original expression (7) for ω_t and reproduce (8) as the algebraic equation of motion for $\Sigma(t)$. It is important to realize that the correct form of the action is recovered just because we have included the Lorentz Goldstone field into the coset.

In the generic case of p-brane in D-dimensional Minkowski space the construction is analogous. One equates to zero the Cartan forms ω^μ associated with the transverse translation generators, which is once again the manifestly covariant constraint, and in this way eliminates the Lorentz Goldstone fields $\Lambda^{\mu m}(x)$:

$$\Lambda^\mu_m \sim \partial_m X^\mu \ . \tag{12}$$

Substituting these expressions into the Cartan forms ω^m which are covariant differentials of x^m, one finds that the invariant volume of x-space, i.e. the integral of the external product of these 1-forms, is just the p-brane static gauge NG action (3).

2 Superbranes from PBGS

The PBGS approach is the generalization of the above nonlinear realization view of branes to the superbranes. Though originally [1] the phenomenon of partial breaking of global SUSY ($N = 2$ down to $N = 1$ in $D = 4$) was studied without any reference to superbranes, the subsequent study [2–12] revealed the profound relationship between both concepts. It was shown in Ref. [2] that $N = 1, D = 4$ superstring in the static gauge can be understood as the theory of partial breaking of $N = 1\,, D = 4$ SUSY to its $N = (2,0)\,, d = 2$ subgroup. The very existence of self-consistent GS type actions for superbranes, with the appropriate fermionic κ-symmetry, was inferred in the pioneering paper [3] from the study of the partial breaking $N = (1,0)\,, D = 6 \;\Rightarrow\; N = 1\,, d = 4$. The same PBGS pattern was treated in Ref. [5] from the $D = 4$ perspective as the partial breaking of $N = 2$ SUSY with two central charges down to $N = 1$ one, with the systematic use of the nonlinear realizations and inverse Higgs effect techniques. It was shown that the self-consistent theory can be constructed in terms of the (covariantly) chiral and antichiral bosonic $N = 1$ superfields which are the Goldstone ones corresponding to the central charges (that is, to the translation operators in fifth and sixth directions from the $D = 6$ viewpoint). The fermionic Goldstone superfields associated with the spontaneously broken supertranslation generators are covariantly expressed in terms of these basic Goldstone superfields by the covariant constraints of the type (8). The relevant Goldstone superfields action, a nonlinear generalization of the standard free action of $N = 1$ chiral superfields, was constructed in Refs. [7,8] In components, after eliminating the auxiliary fields, it yields just the static gauge form of the GS action for the $N = (1,0)\,, D = 6$ 3-brane in a flat background. An interesting new phenomenon was discovered by Bagger and Galperin. [6] It turned out that the same SUSY admits several different PBGS options depending on into which multiplet of unbroken SUSY one embeds the Goldstone fermion. In the $N = 2 \;\Rightarrow\; N = 1\,, D = 4$ case, instead of placing this field into the chiral $N = 1$ multiplet, one can place it into the abelian vector $N = 1$ multiplet as a "photino", by imposing the appropriate covariant constraints on the Goldstone fermionic superfield which generalize the Bianchi identities for the flat $N = 1$ Maxwell superfield strength. The relevant invariant action, on the one hand, is $N = 2$ extension of the Born-Infeld (BI) action with the hidden, nonlinearly realized half of supersymmetry, and, on the other, is (in components) a gauge-fixed form of the GS action of the "space-time filling" D3-brane. One more option is to embed the Goldstone fermion into the $N = 1$ tensor (or linear) multiplet. [7,8,9] The emerging brane is the super "L3-brane" in $D = 5$ (in terminology of Howe et al., [19]) it is related to the $N = (1,0)\,, D = 6$

3-brane via the familiar duality between $N = 1$ tensor and chiral superfields.

The study of the PBGS patterns corresponding to partial breaking of SUSY with 16 supercharges ($N = 1, D = 10$; $N = 2, D = 6$; $N = 4, D = 4$; ...) was initiated in our works.[10,11] Let us dwell on this and related subjects in some detail.

3 Hypermultiplet as a Goldstone Superfield

To describe the 1/2 breaking of Poincaré SUSY with 16 supercharges, it is natural to start from the maximally symmetric situation where the broken and unbroken SUSY "live" as simple ones. This amounts to considering the PBGS option $N = 1$, $D = 10 \Rightarrow N = (1,0)$, $d = 6$ (we could equally choose $N = (0,1)$). From the $d = 6$ perspective, $N = 1$, $D = 10$ is a central extension of $N = (1,1)$:

$$N = 1,\ D = 10 \qquad \propto \quad \left\{Q^i_\alpha, P_{\alpha\beta}, S^{\beta a}, Z^{ia}\right\} \ , \tag{13}$$

where $\alpha, \beta = 1, ..., 4$; $i = 1, 2$; $a = 1, 2$ are, respectively, the $Spin(1,5)$ indices and the $SU(2)$ doublet indices. The basic anticommutation relations read

$$\left\{Q^i_\alpha, Q^j_\beta\right\} = \epsilon^{ij} P_{\alpha\beta} \ , \ \left\{Q^i_\alpha, S^{a\beta}\right\} = \delta^\beta_\alpha Z^{ia} \ , \ \left\{S^{a\alpha}, S^{b\beta}\right\} = \epsilon^{ab} P^{\alpha\beta} \ . \tag{14}$$

One also must add generators of the $D = 10$ Lorentz group $SO(1,9) \propto \left\{M_{\alpha\beta\,\gamma\delta},\ T^{ij},\ T^{ab},\ K^{\alpha\beta}_{ia}\right\}$, with M and T generating the $d = 6$ Lorentz group $SO(1,5)$ and R-symmetry $SU(2) \times SU(2)$.

We wish $N = (1,0), d = 6$ SUSY $\propto \left\{Q^i_\alpha, P_{\alpha\beta}, T^{ij}, T^{ab}, M_{\alpha\beta\,\gamma\delta}\right\}$ to be unbroken. Like in the bosonic case of Sect.1, we are led to place the generators $Q^i_\alpha, P_{\alpha\beta}, S^{\alpha a}, Z^{ia}, K^{ia}_{\alpha\beta}$ into the coset and to treat the coset parameters associated with two first generators as the coordinates of $N = (1,0)$, $d = 6$ superspace, while the remaining ones as the Goldstone superfields:

$$\begin{aligned} &Q^i_\alpha \Rightarrow \theta^\alpha_i \ , \quad P_{\alpha\beta} \Rightarrow x^{\alpha\beta} \ . \\ &S^{\alpha a} \Rightarrow \Psi_{\alpha a}(x,\theta) \ , \ Z^{ia} \Rightarrow q_{ia}(x,\theta) \ , \ K^{ia}_{\alpha\beta} \Rightarrow \Lambda^{\alpha\beta}_{ia}(x,\theta) \ . \end{aligned} \tag{15}$$

The coset element G in the exponential parametrization is constructed like in eq. (5). The Cartan forms are defined by

$$G^{-1}dG = \Omega_Q + \Omega_P + \Omega_Z + \Omega_S + \Omega_K + \dots \ , \tag{16}$$

where the subscripts denote the relevant generators. The forms $\Omega_{Q,P}$ are the covariant differentials of the superspace coordinates, $\Omega_{Z,S,K}$ are those of Goldstone superfields, they are homogeneously transformed under the supergroup left shifts. We shall be interested in the linearized structure of $\Omega_Z = \Omega_Z^{ia}\, Z_{ia}$:[10,11]

$$\Omega_Z^{ia} = dq^{ia} + 2\Lambda^{ia}_{\alpha\beta}\, dx^{\alpha\beta} + \psi^a_\alpha d\theta^{i\alpha} + \ldots . \tag{17}$$

Comparing it with the form ω_1, eq. (7) of the toy example of Sect. 1, we observe that both forms have a similar structure. Hence, the superfields Λ, Ψ can be expressed through the basic Goldstone superfield q^{ia} from the inverse Higgs constraint analogous to (8)

$$\begin{aligned} \Omega_Z = 0 \Rightarrow \Lambda^{ia}_{\alpha\beta} &= \nabla_{\alpha\beta} q^{ia} = \partial_{\alpha\beta} q^{ia} + \ldots , \\ \Psi^a_\alpha &= \frac{1}{2}\nabla^k_\alpha q^a_k = \frac{1}{2} D^k_\alpha q^a_k + \ldots . \end{aligned} \tag{18}$$

Here $D^i_\alpha = \partial / \partial\theta^\alpha_i - \frac{1}{2}\theta^{i\beta}\partial_{\beta\alpha}$ and dots stand for nonlinear terms. Thus, q^{ia} is the essential Goldstone superfield, analogue of $X(t)$ in the $D = 2$ example. Its first bosonic component $q^{ia}(x)$ parametrizes the transverse directions in the $D = 10$ Minkowski space, while the physical fermionic component is the Goldstino related to the spontaneously broken supertranslations (with the generator $S^{a\alpha}$). This field content is that of the scalar $N = 1\, , D = 10$ 5-brane.

The constraint (18) differs in a few important aspects from (8) and the similar inverse Higgs constraint considered in Ref. [5] It not only eliminates the redundant Goldstone superfields, but also implies the differential constraint for $q^{ia}(x, \theta)$

$$\nabla^{(k}_\alpha q^{i)a} = D^{(k}_\alpha q^{i)a} + \ldots = 0 . \tag{19}$$

It is just the nonlinear, "brane" generalization of the standard hypermultiplet constraint. [20] The latter reduces the field content of q^{ia} to the set of (8+8) components and puts them on shell:

$$q^{ia}(x, \theta) \;\Rightarrow\; \phi^{ia}(x) + \theta^{\alpha i}\psi^a_\alpha(x) + x\text{-derivatives} , \tag{20}$$

$$\Box\phi^{ia}(x) = 0\, , \quad \partial^{\alpha\beta}\psi^a_\beta = 0 . \tag{21}$$

Thus eq. (19) describes the on-shell dynamics of $N = 1, D = 10$ 5-brane as the natural generalization of the free hypermultiplet dynamics. Since the off-shell superfield action for the latter can be constructed only in harmonic superspace,[21] it is reasonable to expect that there exists a brane generalization of the hypermultiplet harmonic superspace action. It would be $N = (1, 0), d = 6$ (or $N = 2, d = 4$ after dimensional reduction) analogue of the $N = 1$ Goldstone superfield action of Refs. [7,11] No systematic recipes are known so far for constructing PBGS actions. At present we are aware only of the appropriate fourth-order correction to the free hypermultiplet action. Even for the dimensionally-reduced cases, including the simplest case of $N = 2\,, D = 5$ superparticle, the full action is rather difficult to find. For the superparticle case it is known up to the sixth order in fields: [11]

$$S^q_{br} = \int d\zeta^{(-4)} q^+_a D^{++} q^{+a} + \alpha \int dZ (A^2 + 2\alpha\, AB^{++}B^{--} + \ldots) \qquad (22)$$

where

$$A = q^+_a D^{--} q^{+a} \ , \quad B^{++} = q^+_a \partial_t q^{+a} \ ,$$

$q^{+a}, D^{\pm\pm}$ are the $d = 1$ reduction of the analytic hypermultiplet superfield and harmonic derivatives, [21] B^{--} is defined by

$$D^{++}B^{--} - D^{--}B^{++} = 0$$

and α is a coupling constant. It would be tempting to find out the geometric principle allowing to restore the whole action. Let us now apply to a simpler PBGS case where the complete off-shell action can be constructed. [12]

4 N=1, D=4 Supermembrane

This case corresponds to the partial breaking of $N = 1, D = 4$ SUSY to $N = 1, d = 3$ one. The $N = 1, D = 4$ superalgebra in the $d = 3$ notation reads

$$\{Q_a, Q_b\} = P_{ab}, \{Q_a, S_b\} = \epsilon_{ab} Z, \{S_a, S_b\} = P_{ab},\ a, b = 1, 2\ . \qquad (23)$$

The $d = 3$ momentum operator P_{ab} together with the central charge Z form the $D = 4$ translation operator. The vacuum stability subalgebra consists of the $N = 1, d = 3$ superalgebra (generators Q_a and P_{ab}) and that of the $d = 3$ Lorentz group $SO(1,2)$. The second SUSY, the central charge Z and the generators K_{ab} of the coset $SO(1,3)/SO(1,2)$ define spontaneously broken symmetries. The appropriate coset parameters are introduced as

$$Q_a \Rightarrow \theta^a, P_{ab} \Rightarrow x^{ab}, S_a \Rightarrow \psi^a(x,\theta), Z \Rightarrow q(x,\theta), K_{ab} \Rightarrow \lambda^{ab}(x,\theta). \qquad (24)$$

After construction of the relevant Cartan forms and covariant elimination of the Goldstone superfields $\psi^a(x,\theta), \lambda^{ab}(x,\theta)$ by the appropriate inverse Higgs constraints, one is left with $q(x,\theta)$ as the only unremovable Goldstone superfield. Its physical fields (one boson and two fermions) parametrize one transverse bosonic and two fermionic directions in $N = 1, D = 4$ superspace, the auxiliary bosonic field also admits a nice geometric interpretation of the Goldstone field for the spontaneously broken $U(1)$ automorphism group of $N=1, D=4$ superalgebra ("γ^5 invariance"). The physical field content of q coincides with that of $N=1, D=4$ supermembrane.

The inverse Higgs conditions in this case do not imply the equation of motion for q. However, the dynamical equation for q turns out to admit, like in the $D=2$ particle example, a covariant representation as the vanishing of the covariant $d\theta$ projection of the coset Cartan form ω_S^a associated with the spontaneously broken SUSY generator S_a.[12] We have no place to present details. Let us explain how to construct the off-shell action for this case.

Like in other PBGS cases, the nonlinear realizations approach on its own provides no clear recipe how to construct such an action. This becomes possible using the trick similar to the one exploited in.[6] Namely, let us start from a *linear* realization of $N=1, D=4$ SUSY in terms of $N=1, d=3$ superfields $\Phi, \xi_a \equiv D_a\rho$ with the following transformation rules under the second SUSY:

$$\delta\rho = \theta^a\eta_a - 2D^a\Phi\eta_a \;,\; \delta\Phi = \frac{1}{2}\eta^a D_a\rho \;, \tag{25}$$

where η_a is the transformation parameter and D_a is the flat $N=1, d=3$ spinor covariant derivative, $\{D_a, D_b\} = \partial_{ab}$. It is easy to check that the closure of these transformations and those of manifest $N=1, d=3$ SUSY is just the superalgebra (23), with Z realized as a pure shift of ρ. The transformation of $\xi_a = D_a\rho$ starts with η_a, suggesting the interpretation of this superfield as the linear realization Goldstone fermion. After some work one finds that Φ can be covariantly expressed in terms of ξ_a as follows

$$\Phi = \frac{1}{2}\frac{\xi^2}{1+\sqrt{1+D^2\xi^2}} \,.$$

Recalling the transformation law of Φ, one finds that the integral

$$S = \int d^3x d^2\theta \Phi \equiv \frac{1}{2}\int d^3x d^2\theta \frac{\xi^2}{1+\sqrt{1+D^2\xi^2}}, \quad \xi^a = D^a\rho \;, \tag{26}$$

is invariant under the hidden SUSY transformations as well as $D=4$ Poincaré translations and so it can be identified with the sought $d=3$ worldvolume superspace action of $N=1, D=4$ supermembrane. Indeed, it is easy to find that the bosonic core of this action is just the static gauge membrane NG action:

$$S = \int d^3x \left(1 - \sqrt{1 - \frac{1}{2}\partial q \cdot \partial q}\right) \,. \tag{27}$$

One can find the equivalence field redfinition relating ξ_a to the nonlinear realization Goldstone fermion ψ_a and ρ to q. Also, the equations of motion following from the action (26) can be shown to be equivalent to those conjectured at the level of the Cartan forms. This implies the presence of hidden $SO(1,3)$ Lorentz symmetry in the action. It still remains to prove the precise equivalence of this action to the GS one (e.g., along the lines of Achucarro et al.[4]) and the action proposed in the superembedding approach.[19] Note an interesting peculiarity: one can add to the lagrangian in (26) the "cosmological" term $\sim \rho$. This term is invariant under the second SUSY (up to surface terms) and the shifts $\rho \rightarrow \rho + const$. Adding it changes the equation of motion for the auxiliary field and so can influence the structure of the component action (without this term, the auxiliary field is vanishing on shell). The pure physical boson part of the action is always given by (27).

Besides a scalar multiplet ρ (or q), we can choose a vector $N=1, d=3$ multiplet as the Goldstone one. It is represented by $N=1$ spinor superfield strength μ_a obeying the constraint:

$$D^a \mu_a = 0 \,. \tag{28}$$

It leaves in μ_a the first fermionic component together with the divergenceless vector $F_{ab} \equiv D_a \mu_b|_{\theta=0}$ (just the gauge field strength). Due to the vector-scalar $d=3$ duality, the superfield μ_a is expected to describe a D2-brane which is dual to the supermembrane.

The relevant action can be found using the previous trick. One can extend μ^a to the $N=1, D=4$ multiplet (μ^a, ϕ) with the following transformation rules under the second SUSY

$$\delta\mu_a = \eta_a - D^2\phi\eta_a + \partial_{ab}\phi\eta^b \,, \qquad \delta\phi = \frac{1}{2}\eta^a\mu_a \,. \tag{29}$$

Thus μ^a can also be interpreted as the linear realization Goldstone fermionic superfield. The following expression for ϕ

$$\phi = \frac{1}{2}\,\frac{\mu^2}{1+\sqrt{1-D^2\mu^2}} \tag{30}$$

can be checked to be consistent with (29). Then the action

$$S = -\int d^3x d^2\theta\, \phi = -\frac{1}{2}\int d^3x d^2\theta\, \frac{\mu^2}{1+\sqrt{1-D^2\mu^2}} \tag{31}$$

is invariant under the second SUSY in virtue of the transformation rule of ϕ (29) and the constraint (28). This nonlinear generalization of the standard

$N=1, d=3$ abelian vector multiplet action $\sim \mu^2$ involves as its bosonic core the $d=3$ BI action

$$S = \int d^3x \left(\sqrt{1+2F^2} - 1 \right) , \tag{32}$$

$$\partial^{ab} F_{ab} = 0 \quad \rightarrow \quad F_{ab} = \partial_{ac} G^c_b + \partial_{bc} G^c_a \,. \tag{33}$$

So it is $N=2$ extension of the $d=3$ BI action with nonlinearly realized second SUSY. It can be regarded as the worldvolume superfield action of the space-time filling D2-superbrane in a flat background. It is easy to prove its dual equivalence to the action (26) by inserting the constraint (28) in it with the Lagrange multiplier ρ and integrating out μ^a.[12]

Finally, let us show how the duality between the bosonic NG and BI actions (27), (32) can be recovered in the nonlinear realizations approach of Sect.1. It is easy to find that in the case of nonlinear realization of $\mathcal{P}_{(4)}$ in the coset $\mathcal{P}_{(4)}/SO(1,2)$, which corresponds just to membrane in $D=4$, the covariant differentials of x^m read

$$\omega^m = dx^m + 2\frac{\lambda^m}{1-2\lambda^2}(2\lambda_n + \partial_n q)\, dx^n \equiv \omega^m_n\, dx^n \,, \tag{34}$$

where $\lambda^m(x)$ is the Lorentz $SO(1,3)/SO(1,2)$ Goldstone field in the appropriate parametrization and $q(x)$ is the transverse coordinate of membrane. We could eliminate λ^m by the inverse Higgs constraint but in the present case it is instructive to reproduce it as the equation of motion for λ^m. The minimal invariant action constructed as the covariant worldvolume

$$S_{mem} = \int d^3x \det \omega^m_n \,, \tag{35}$$

up to a normalization factor and constant shift, is

$$S_{mem} = \int d^3x \frac{1}{1-2\lambda^2}\left(2\lambda^2 + \lambda\, \partial q\right) . \tag{36}$$

Varying λ^m yields the inverse Higgs expression for it

$$\lambda_m = -\frac{1}{2}\frac{\partial_m q}{1+\sqrt{1-\frac{1}{2}(\partial q)^2}} \,. \tag{37}$$

After this (36) takes the standard NG form (27). On the other hand, one can treat q in (36) as the Lagrange multiplier for the differential constraint on λ^m:

$$\partial_m F^m = 0 \,, \quad F^m \equiv 2\frac{\lambda^m}{1-2\lambda^2} \,. \tag{38}$$

Expressing λ^2 through F^m

$$\lambda^2 = \frac{1}{4F^2}\left(1 - \sqrt{1+2F^2}\right)^2 ,$$

one reduces (36) just to the BI form (32)

$$S_{mem} \sim \int d^3x \left(\sqrt{1+2F^2} - 1\right) . \tag{39}$$

This simple consideration shows that within the nonlinear realization approach the $d = 3$ Maxwell field strength entering the $d = 3$ BI action acquires the nice geometric interpretation as the Goldstone field representing the Lorentz coset $SO(1,3)/SO(1,2)$, while the BI action itself can be algorithmically derived as the action dual to the static gauge membrane NG action.

5 Concluding Remarks

The geometric PBGS approach can be thought of as a useful and viable alternative to the standard GS description of superbranes. Its main merit is that it gives *manifestly* worldsurface supersymmetric off-shell superfield actions. Leaving aside such important conceptual questions as whether it can be helpful for quantization, etc, we list here a few more modest problems solving which could extend its range of applicability.

First of all, it is desirable to develop convenient general recipes of constructing superfield PBGS actions similar to those provided by the nonlinear realizations for the internal symmetries sigma models. At present, constructing such actions is an art to some extent.

It is important to learn how to construct self-consistent PBGS actions on non-trivial bacgrounds involving the worldvolume and target supergravity and super Yang-Mills fields.

At last, it seems interesting to set up PBGS actions for the systems with 1/4 and other exotic partial breaking options. Some progress in this direction is reported in the contribution by Delduc, Krivonos and myself in this Volume. [22]

After this contribution has been sent to the Editors, I became aware of preprint [23] where, in particular, the same interpretation of bosonic branes as in Sect. 1 is discussed.

Acknowledgments

I thank Jerzy Lukierski for inviting me to give this contribution. This work was supported in part by grants RFBR-CNRS 98-02-22034, RFBR 99-02-18417, INTAS-96-0538, INTAS-96-0308 and NATO Grant No.PST.CLG 974874.

References

1. J. Bagger, J. Wess, *Phys. Lett.* **B 138**, 105 (1984).
2. J. Hughes, J. Polchinski, *Nucl. Phys.* **B 278**, 147 (1986).
3. J. Hughes, J. Liu, J. Polchinski, *Phys. Lett.* **B 180**, 370 (1986).
4. A. Achucarro, J. Gauntlett, K. Itoh, P.K. Townsend, *Nucl. Phys.* **B 314**, 129 (1989).
5. J. Bagger, A. Galperin, *Phys. Lett.* **B 336**, 25 (1994).
6. J. Bagger, A. Galperin, *Phys. Rev.* **D 55**, 1091 (1997).
7. J. Bagger, A. Galperin, *Phys. Lett.* **B 412**, 296 (1997).
8. M. Roček, A. Tseytlin, *Phys. Rev.* **D 59**, 106001 (1999).
9. F. Gonzalez-Rey, I.Y. Park, M. Roček, *Nucl. Phys.* **B 544**, 243 (1999).
10. S. Bellucci, E. Ivanov, S. Krivonos, *Fortschr. f. Phys.* **48**, 19 (2000).
11. S. Bellucci, E. Ivanov, S. Krivonos, *Phys. Lett.* **B 460**, 348 (1999).
12. E. Ivanov, S. Krivonos, *Phys. Lett.* **B 453**, 237 (1999).
13. M.J. Duff, "Supermembranes", hep-th/9611203.
14. S. Coleman, J. Wess, B. Zumino, *Phys. Rev.* **177**, 2239 (1969); C. Callan, S. Coleman, J. Wess, B. Zumino, *Phys. Rev.* **177**, 2247 (1969).
15. D.V. Volkov, *Sov. J. Part. Nucl.* **4**, 3 (1973).
16. V.I. Ogievetsky, Proceedings of X-th Winter School of Theoretical Physics in Karpacz, Vol.1. p. 227 (Wroclaw, 1974).
17. E. Ivanov, unpublished.
18. E.A. Ivanov, V.I. Ogievetsky, *Teor. Mat. Fiz.* **25**, 164 (1975).
19. P.S. Howe, O. Raetzel, E. Sezgin, JHEP **9808** (1998) 011; P.S. Howe, O. Raetzel, I. Rudychev, E. Sezgin, *Class. Quant. Grav.* **16**, 705 (1999).
20. M.F. Sohnius, *Nucl. Phys.* **B 138**, 109 (1978).
21. A. Galperin, E. Ivanov, S. Kalitzin, V. Ogievetsky, E. Sokatchev, *Class. Quant. Grav.* **1**, 469 (1984).
22. F. Delduc, E. Ivanov and S. Krovonos, *1/4 PBGS and Superparticle Actions*; hep-th/9912292.
23. P. West, *Automorphisms, Non-linear Realizations and Branes*, hep-th/0001216.

SUPEREXTENSION $n = (2,2)$ OF THE COMPLEX LIOUVILLE EQUATION AND ITS SOLUTION

A. A. KAPUSTNIKOV

Department of Physics, Dnepropetrovsk University,
49050, Dnepropetrovsk, Ukraine
E-mail: kpstnkv@creator.dp.ua

It is shown that the method of the nonlinear realization of local supersymmetry previously developed in the framework of supergravity, being applied to the $n = (2,2)$ superconformal symmetry, allows one to get the new form of the exactly solvable $n = (2,2)$ super-Liouville equation. The general advantage of this version as compared to the conventional one is that its bosonic part includes the complex Liouville equation. We obtain the suitable supercovariant constraints imposed on the corresponding superfields which provide the set of the resulting system of component equations be the same as that in model of $N = 2$, $D = 4$ Green-Schwarz superstring. The general solution of this system is derived from the corresponding solution of the bosonic string equation.

Introduction

It is well-known that a doubly supersymmetric generalization of the geometrical approach to superstring [1,2] in the case of $N = 2$, $D = 3$ Green-Schwarz superstring leads to the new version of the Liouville equation known in literature as $n = (1,1)$.[3] The latter as one can expect include the real Liouville equation in its bosonic part. The problem, however, arise when we try to extend this result to the case of $N = 2$, $D = 4$ superstring. It turns out that in this case the well-known form of the suitable super-Liouville equation [4] is not appropriate in virtue of absence of the complex Liouville equation in the corresponding bosonic part. Thus, the equation proposed in Ref. [4] can not be applied to the description of the $N = 2$, $D = 4$ Green-Schwarz superstring, which to the ordinary complex Liouville equation when all reduces fermionic component fields are neglected.

In this paper we would like to propose a new version of the $n = (2,2)$ super-Liouville equation which appears to be in agreement with the equations of motion of the $N = 2$, $D = 4$ Green-Schwarz superstring. Our approach is based on the method of the nonlinear realization of local supersymmetries developed by Ivanov and Kapustnikov in frame of supergravity.[5] It will be shown that when applied to the $n = (2,2)$ superconformal symmetry this method gives a possibility to impose the supercovariant constraints on the superfields in such a way that all unphysical degrees of freedom which occur in the original equation will get removed from the residual set of the equation of motion. The latter amounts to the complex Liouville equation for the bosonic worldsheet variable $\tilde{u}(\tilde{\xi}^{++},\ \tilde{\xi}^{--})$ supplemented with two first order free equations $\tilde{\partial}_{--}\lambda^{+}(\tilde{\xi}^{++},\ \tilde{\xi}^{--}) = \tilde{\partial}_{++}\lambda^{-}(\tilde{\xi}^{++},\ \tilde{\xi}^{--}) = 0$ for the fermions of the opposite chirality.

In Section 3 we present the general solution of this equation in terms of the restricted Lorentz harmonic variables,[6] which in a proper way extends the corresponding bosonic string solution obtained in Ref.[7]

1 New Version of the $n = (2,2)$ Super-Liouville Equation

1.1 Linear Realization

We begin with the linear realization of two copies of the one dimensional superconformal group acting separately on the light-cone complex coordinates of $N = 2$, $D = 4$ superstring $\mathbf{C}^{(2|2)} = (\xi_L^{++} = \xi^{++} + i\eta^+\overline{\eta}^+, \eta^+; \xi_L^{--} = \xi^{--} + i\eta^-\overline{\eta}^-, \eta^-)$:

$$\begin{aligned}
\xi_L^{\pm\pm\prime} &= \Lambda^{\pm\pm} - \overline{\eta}^{\pm}\overline{D}_{\pm}\Lambda^{\pm\pm} \qquad (1)\\
&= a_L^{\pm\pm}(\xi_L^{\pm\pm}) + 2i\eta^{\pm}\overline{\epsilon}^{\pm}(\xi_L^{\pm\pm})g^{(\pm\pm)}(\xi_L^{\pm\pm})e^{i\rho^{(\pm\pm)}(\xi_L^{\pm\pm})},\\
\eta^{\pm\prime} &= -\frac{i}{2}\overline{D}_{\perp}\Lambda^{\pm\pm} = \epsilon^{\pm}(\xi_L^{\pm\pm}) + \eta^{\pm}g^{(\pm\pm)}(\xi_L^{\pm\pm})e^{i\rho^{(\pm\pm)}(\xi_L^{\pm\pm})},\\
a_L^{\pm\pm}(\xi_L^{\pm\pm}) &= \xi_L^{\pm\pm} + a^{\pm\pm}(\xi_L^{\pm\pm}) + i\epsilon^{\pm}(\xi_L^{\pm\pm})\overline{\epsilon}^{\pm}(\xi_L^{\pm\pm}),\\
g^{(\pm\pm)} &= \sqrt{1 + \partial_{\pm\pm}a^{\pm\pm} + i(\epsilon^{\pm}\partial_{\pm\pm}\overline{\epsilon}^{\pm} + \overline{\epsilon}^{\pm}\partial_{\pm\pm}\epsilon^{\pm})}\,.
\end{aligned}$$

In Eq. (1) the general superfield (SF)

$$\begin{aligned}
\Lambda^{\pm\pm}(\xi_L^{\pm\pm}, \eta^{\pm}, \overline{\eta}^{\pm}) = {}& a_L^{\pm\pm}(\xi_L^{\pm\pm}) + 2i\eta^{\pm}\overline{\epsilon}^{\pm}(\xi_L^{++})g^{(\pm\pm)}(\xi_L^{\pm\pm})e^{i\rho^{(\pm\pm)}(\xi_L^{++})} \qquad (2)\\
& + 2i\overline{\eta}^{\pm}\epsilon^{\pm}(\xi_L^{\pm\pm}) - 2i\eta^{\pm}\overline{\eta}^{\pm}g^{(\pm\pm)}(\xi_L^{\pm\pm})e^{i\rho^{(\pm\pm)}(\xi_L^{\pm\pm})}\,,
\end{aligned}$$

is composed out from parameters $\epsilon^{+}(\xi_L^{++})$, $\epsilon^{-}(\xi_L^{--})$ of local supertranslations; two real parameters $a^{++}(\xi_L^{++})$, $a^{--}(\xi_L^{--})$ of $D1$-reparametrizations and two real parameters $\rho^{++}(\xi_L^{++})$, $\rho^{--}(\xi_L^{--})$ describing local $U(1) \times U(1)$-rotations. The spinor covariant derivatives are defined as

$$\begin{aligned}
D_{\pm} &= \partial_{\pm} + 2i\overline{\eta}^{\pm}\partial_{\pm\pm}\,, \qquad (3)\\
\overline{D}_{\pm} &= \overline{\partial}_{\pm}\,.
\end{aligned}$$

It is worth to mention that since the parameters $\xi_L^{\pm\pm\prime}$ and $\eta^{\pm\prime}$ in Eqs. (1) are subjects to the constraints

$$D_{\pm}\xi_L^{\pm\pm\prime} - 2i\overline{\eta}^{\pm\prime}D_{\pm}\eta^{\pm\prime} = 0 \qquad (4)$$

the *flat* spinor covariant derivatives (3) are transformed homogeneously with respect to (1)

$$D_{\pm} = (D_{\pm}\eta^{\pm\prime})D'_{\pm}\,. \qquad (5)$$

Therefore, the following superconformal-covariant equation can be proposed as a natural candidate for $n = (2,2)$ superextension of the corresponding $n = (1,1)$ super-Liouville equation [3]

$$D_{-}D_{+}W = e^{2W}\,\Psi_{+}^{--}\Psi_{-}^{++}\,. \qquad (6)$$

In Eq. (6) one double-analytical SF

$$W(\xi_L^{\pm\pm},\eta^{\pm}) = u(\xi_L^{\pm\pm}) + \eta^{+}\psi^{-}(\xi_L^{\pm\pm}) + \eta^{-}\psi^{+}(\xi_L^{\pm\pm}) + \eta^{-}\eta^{+}F(\xi_L^{\pm\pm})\,, \qquad (7)$$

and two general SFs $\Psi_{+}(\xi_L^{++},\eta^{+},\overline{\eta}^{+})$, $\Psi_{-}(\xi_L^{--},\eta^{-},\overline{\eta}^{-})$, depending separately on the $(2,0)$ and $(0,2)$ light-cone variables, are introduced. [a] Eq. (6) is invariant under the following gauge transformations

$$\begin{aligned} W'(\xi_L^{\pm\pm\prime},\eta^{\pm\prime}) &= W(\xi_L^{\pm\pm},\eta^{\pm}) - \frac{1}{2}ln(\overline{D}_{+}\overline{\eta}^{+\prime}) - \frac{1}{2}ln(\overline{D}_{-}\overline{\eta}^{-\prime})\,, \qquad (8)\\ \Psi'_{+}(\xi_L^{++\prime},\eta^{+\prime},\overline{\eta}^{+\prime}) &= (D_{+}\eta^{+\prime})^{-1}(\overline{D}_{+}\overline{\eta}^{+\prime})\Psi_{+}(\xi_L^{++},\eta^{+},\overline{\eta}^{+})\,,\\ \Psi'_{-}(\xi_L^{--\prime},\eta^{-\prime},\overline{\eta}^{-\prime}) &= (D_{-}\eta^{-\prime})^{-1}(\overline{D}_{-}\overline{\eta}^{-\prime})\Psi_{-}(\xi_L^{--},\eta^{-},\overline{\eta}^{-})\,. \end{aligned}$$

Note that due to the nilpotence of the covariant derivatives ($D_{\pm}^2 = 0$) the SFs W and $\Psi_{\pm}$ included in the Eq. (6) appear restricted

$$D_{\pm}\Psi_{\pm} + 2(D_{\pm}W)\Psi_{\pm} = 0\,. \qquad (9)$$

A particular property we shall encounter with here is, however, that the constraints (9) can be solved explicitly in terms of the *unrestricted* Fs

$$\Psi_{+} = D_{+}M + 2(D_{+}W)M, \qquad \Psi_{-} = D_{-}N + 2(D_{-}W)N\,. \qquad (10)$$

In Eq. (10) the general SFs [b]

$$\begin{aligned} M(\xi_L^{++},\eta^{+},\overline{\eta}^{+}) &= f(\xi_L^{++}) + \eta^{+}\omega^{-}(\xi_L^{++}) \qquad (11)\\ &\quad + \overline{\eta}^{+}\chi^{-}(\xi_L^{++}) + \eta^{+}\overline{\eta}^{+}m^{--}(\xi_L^{++})\,,\\ N(\xi_L^{--},\eta^{-},\overline{\eta}^{-}) &= g(\xi_L^{--}) + \eta^{-}\omega^{+}(\xi_L^{--})\\ &\quad + \overline{\eta}^{-}\chi^{+}(\xi_L^{--}) + \eta^{-}\overline{\eta}^{-}n^{++}(\xi_L^{--})\,, \end{aligned}$$

[a] We omit temporarily the upper indices of SFs Ψ and M for the enlightening of formulas but we shall come back to them in Section 3.

[b] It can be shown that in the case of chiral SFs $M(\xi_L^{++},\eta^{+}), N(\xi_L^{--},\eta^{-})$ Eq. (6) is reduced to free one $D_{+}D_{-}\tilde{W} = 0$ for the SF $\tilde{W} = W + \frac{1}{2}ln(MN)$.

are supposed transform as a superconformal densities

$$\begin{aligned} M'(\xi_L^{++\prime}, \eta^{+\prime}, \overline{\eta}^{+\prime}) &= (\overline{D}_+ \overline{\eta}^{+\prime}) M(\xi_L^{++}, \eta^+, \overline{\eta}^+), \\ N'(\xi_L^{--\prime}, \eta^{-\prime}, \overline{\eta}^{-\prime}) &= (\overline{D}_- \overline{\eta}^{-\prime}) N(\xi_L^{--}, \eta^-, \overline{\eta}^-). \end{aligned} \tag{12}$$

Although the component content of SFs W, M, N even upon the gauge fixing is still too large to be related with the $N = 2$, $D = 4$ superstring there is very important feature of Eq. (6). It contains *complex* Liouville equation in its bosonic part

$$\partial_{++}\partial_{--}u(\xi_L^{\pm\pm}) = \frac{1}{4}e^{2u(\xi_L^{\pm\pm})}m^{--}(\xi_L^{++})n^{++}(\xi_L^{--}) + \dots, \tag{13}$$

where all the unessential terms in the r.h.s. are omitted. It is clear, however, that to be connected with the superstring theory the SFs we have considered here must be covariantly constrained. In the next Section we are going to show that the desirable constraints could be imposed in frame of the nonlinear realization of $n = (2, 2)$ superconformal symmetry in which the original SFs becomes reducible.

1.2 Nonlinear Realization

To see this let us suppose that the v.e.v. of the component fields $m^{--}(\xi_L^{++})$ and $n^{++}(\xi_L^{--})$ in (11) are not equal to zero and as consequence of this the local supersymmetry (1) is actually spontaneously broken. In this case the fermionic components $\chi^{\pm}$ acquire the sense of the corresponding Goldstone fermions and one can exploit them for the singling out of the complex Liouville equation from the system (6) in a manifestly covariant manner. Indeed, it is well-known that in the models with spontaneously broken supersymmetry all the SFs becomes reducible. [10,8] Their irreducible parts are transformed, however, universally with respect to the action of the original supergroups, as the linear representations of the underlying unbroken subgroups but with the parameters depending nonlinearly on the Goldstone fermions. It makes the possibility to impose generally on the SFs in question some absolutely covariant restrictions providing to remove out from the model under consideration undesirable degrees of freedom. Here we can to avail oneself of the opportunity to restrict the SFs enter the Eq. (6) with the help of this approach.

For the beginning let us derive the nonlinear realization of the superconformal symmetry in superspace. Following closely to the general method developed in Ref. [8,8] we need firstly splits the general finite element of the group (1)

$$G(\zeta_L) \equiv \zeta_L', \tag{14}$$

where $\zeta_L = \{\xi_L^{\pm\pm}, \eta^\pm\}$, onto the product of two successive transformations

$$G(\zeta_L) = K(G_0(\zeta_L))\,. \tag{15}$$

In Eq. (15) the following standard notations are used. As before the $G_0(\zeta_L)$ refer to the "primes" coordinates ζ'_L but index *zero* means that they referring now only to the stability subgroup

$$\begin{aligned} \xi_L^{\pm\pm\prime} &= \xi_L^{\pm\pm} + a^{\pm\pm}(\xi_L^{\pm\pm})\,, \\ \eta^{\pm\prime} &= \eta^\pm e^{i\rho^{(\pm\pm)}(\xi_L^{\pm\pm})}\sqrt{1+\partial_{\pm\pm}a^{\pm\pm}}\,. \end{aligned} \tag{16}$$

The latter include only the ordinary conformal transformations (parameters $a^{\pm\pm}(\xi_L^{\pm\pm})$) supplemented with the local $U(1)\times U(1)$-rotations (parameters $\rho^{(\pm\pm)}(\xi_L^{\pm\pm})$). Note, that the first multiplier in the decomposition (15) is easily recognized as the representatives of the left coset space G/G_0 [c]

$$\begin{aligned} K^{\pm\pm}(\zeta_L) &= \xi_L^{\pm\pm} + i\epsilon^\pm(\xi_L^{\pm\pm})\overline{\epsilon}^\pm(\xi_L^{\pm\pm}) \\ &\quad +2i\eta^\pm\overline{\epsilon}^\pm(\xi_L^{\pm\pm})\sqrt{1+i(\epsilon^\pm\partial_{\pm\pm}\overline{\epsilon}^\pm + \overline{\epsilon}^\pm\partial_{\pm\pm}\epsilon^\pm)}\,, \\ K^\pm(\zeta_L) &= \epsilon^\pm(\xi_L^{\pm\pm}) + \eta^\pm\sqrt{1+i(\epsilon^\pm\partial_{\pm\pm}\overline{\epsilon}^\pm + \overline{\epsilon}^\pm\partial_{\pm\pm}\epsilon^\pm)}\,. \end{aligned} \tag{17}$$

It deserves to mention that in the decomposition (15) the comultipliers K and G_0 are chosen in such a way that the irreduciblity constraint (3) is satisfied separately for both of them. The prescription for constructing the corresponding nonlinear realization is as follows. [5] Let us identify the local parameters $\epsilon^\pm(\xi_L^{\pm\pm})$, $\overline{\epsilon}^\pm(\xi_L^{\pm\pm})$ in (17) with the Goldstone fields $\lambda^\pm(\xi_L^{\pm\pm})$, $\overline{\lambda}^\pm(\xi_L^{\pm\pm})$

$$\begin{aligned} \tilde{K}^{\pm\pm}(\tilde{\zeta}_L) &= \tilde{\xi}_L^{\pm\pm} + i\lambda^\pm(\tilde{\xi}_L^{\pm\pm})\overline{\lambda}^\pm(\tilde{\xi}_L^{\pm\pm}) \\ &\quad +2i\tilde{\eta}^\pm\overline{\lambda}^\pm(\tilde{\xi}_L^{\pm\pm})\sqrt{1+i(\lambda^\pm\tilde{\partial}_{\pm\pm}\overline{\lambda}^\pm + \overline{\lambda}^\pm\tilde{\partial}_{\pm\pm}\lambda^\pm)}\,, \\ \tilde{K}^\pm(\tilde{\zeta}_L) &= \lambda^\pm(\tilde{\xi}_L^{\pm\pm}) + \tilde{\eta}^\pm\sqrt{1+i(\lambda^\pm\tilde{\partial}_{\pm\pm}\overline{\lambda}^\pm + \overline{\lambda}^\pm\tilde{\partial}_{\pm\pm}\lambda^\pm)} \end{aligned} \tag{18}$$

and take for $\tilde{K}(\tilde{\zeta}_L)$ the transformation law associated to (15)

$$G(\tilde{K}(\tilde{\zeta}_L)) = \tilde{K}'(\tilde{G}_0(\tilde{\zeta}_L))\,. \tag{19}$$

In Eq. (19) the newly introduced coordinates $\tilde{\zeta}_L = \{\tilde{\xi}_L^{\pm\pm}, \tilde{\eta}^\pm\}$ are transformed differently as compared with $\zeta_L = \{\xi_L^{\pm\pm}, \eta^\pm\}$ in (1). Indeed, in accordance

[c] In virtue of (15) all the parameters in (1) should be regarded as composite ones which are composed out from the parameters of transformations (16) and (17).

with (16) they change only under the vacuum stability subgroup

$$\begin{aligned} \tilde{\xi}_L^{--\prime} &= \tilde{\xi}_L^{\pm\pm} + \tilde{a}^{\pm\pm}(\tilde{\xi}_L^{\pm\pm}), \\ \tilde{\eta}^{\pm\prime} &= \tilde{\eta}^{\pm} e^{i\tilde{\rho}^{(\pm\pm)}(\tilde{\xi}_L^{\pm\pm})}\sqrt{1 + \tilde{\partial}_{\pm\pm}\tilde{a}^{\pm\pm}}\,, \end{aligned} \tag{20}$$

where the parameters $\tilde{a}^{\pm\pm}(\tilde{\xi}_L^{\pm\pm})$ and $\tilde{\rho}^{(\pm\pm)}(\tilde{\xi}_L^{\pm\pm})$ turns out to be dependent nonlinearly on the fields $\lambda^{\pm}(\xi_L^{\pm\pm})$, $\overline{\lambda}^{\pm}(\xi_L^{\pm\pm})$. Eqs. (19) and (20) determine the transformation properties of the Goldstone fermions $\lambda^{\pm}(\xi_L^{\pm\pm})$, $\overline{\lambda}^{\pm}(\xi_L^{\pm\pm})$ with respect to the nonlinear realization of the superconformal group G in coset space (18).

2 Splitting Superspace and Irreducible Form of SFs

Up to now we have dealt with only formal prescription of construction of the nonlinear realization of superconformal group G without any relation of this procedure to the original equation (6). Nevertheless, there is the simple possibility to gain a more deeper insight into the model we started with if we compare two Eqs. (14) and (19). We find that $\tilde{K}(\tilde{\zeta}_L)$ transform under G in precisely the same manner as the initial coordinates ζ_L of superspace $\mathbf{C}^{(2|2)}$. Thus we have the unique possibility to identify them

$$\zeta_L = \tilde{K}(\tilde{\zeta}_L)\,. \tag{21}$$

Eq. (21) establish the relationship between two forms of the realization of superconformal symmetries in superspace. One of the remarkable futures of the transformations (21) is that superspace of the nonlinear realization $\tilde{\mathbf{C}}^{(2|2)} = \tilde{\zeta}_L$ turns out to be completely "splitting" in virtue of the transformations (20) which are not mixed the bosonic and fermionic variables. Due to this very important fact the SFs of the nonlinear realization becomes actually reducible. Indeed, let us perform the change of variables (21) in the Eq. (6)

$$\tilde{D}_-\tilde{D}_+\tilde{W} = e^{2\tilde{W}}\,\tilde{\Psi}_+\tilde{\Psi}_-\,, \tag{22}$$

where the SFs and covariant derivatives of the nonlinear realization (19), (20) and (21) are introduced

$$W = \tilde{W} - \frac{1}{2}ln(\tilde{\overline{D}}_+\overline{\eta}^+) - \frac{1}{2}ln(\tilde{\overline{D}}_-\overline{\eta}^-), \qquad D_\pm = (\tilde{D}_\pm\eta^\pm)^{-1}\tilde{D}_\pm\,, \tag{23}$$

$$\begin{gathered} \tilde{\Psi}_+ = \tilde{D}_+\tilde{M} + 2(\tilde{D}_+\tilde{W})\tilde{M}, \qquad \tilde{\Psi}_- = \tilde{D}_-\tilde{N} + 2(\tilde{D}_-\tilde{W})\tilde{N}\,, \\ M(\xi_L^{++},\eta^+,\overline{\eta}^+) = (\tilde{\overline{D}}_+\overline{\eta}^+)\tilde{M}(\tilde{\xi}_L^{++},\tilde{\eta}^+,\tilde{\overline{\eta}}^+), \end{gathered} \tag{24}$$

$$N(\xi_L^{--},\eta^-,\overline{\eta}^-) = (\overline{\tilde{D}}_-\overline{\eta}^-)\tilde{N}(\tilde{\xi}_L^{--},\tilde{\eta}^-,\overline{\tilde{\eta}}^-)\,. \tag{25}$$

Note should be taken that the covariant derivatives $\tilde{D}_\pm$ in (22) have the same structure as those of linear realization (3). This follows from the structure of the coset space representatives (17) which are defined in such a way that the irreducibility conditions (4) are fulfilled for them automatically.

Although the form of the Eq. (22) is precisely the same as the original one (6) the SFs of the nonlinear realization appearing in (22) are distinguished drastically from the SFs of linear realization. As it follows from (20) and (5) the SFs $\tilde{W}$ and $\tilde{\Psi}$ are transformed under the action of G only with respect to their stability subgroup (20)

$$\begin{aligned}
\tilde{W}'(\tilde{\xi}_L^{--\prime},\tilde{\eta}^{\pm\prime}) &= \tilde{W}(\tilde{\xi}_L^{\pm\pm},\tilde{\eta}^{\pm}) - \frac{1}{2}ln(\overline{\tilde{D}}_+\overline{\tilde{\eta}}^{+\prime}) - \frac{1}{2}ln(\overline{\tilde{D}}_-\overline{\tilde{\eta}}^{-\prime})\,, \quad (26)\\
\tilde{M}'(\tilde{\xi}_L^{++\prime},\tilde{\eta}^{+\prime},\overline{\tilde{\eta}}^{+\prime}) &= (\overline{\tilde{D}}_+\overline{\tilde{\eta}}^{+\prime})\tilde{M}(\tilde{\xi}_L^{++},\tilde{\eta}^{+},\overline{\tilde{\eta}}^{+})\,,\\
\tilde{N}'(\tilde{\xi}_L^{--\prime},\tilde{\eta}^{-\prime},\overline{\tilde{\eta}}^{-\prime}) &= (\overline{\tilde{D}}_-\overline{\tilde{\eta}}^{-\prime})\tilde{N}(\tilde{\xi}_L^{--},\tilde{\eta}^{-},\overline{\tilde{\eta}}^{-})\,.
\end{aligned}$$

Substituting here the explicit form of gauge parameters deduced from the transformations (20)

$$\overline{\tilde{D}}_\pm\overline{\tilde{\eta}}^{\pm\prime} = e^{i\tilde{\rho}^{(\pm\pm)}(\tilde{\xi}_L^{\pm\pm})}\sqrt{1+\tilde{\partial}_{\pm\pm}\tilde{a}^{\pm\pm}(\tilde{\xi}_L^{\pm\pm})}\,, \tag{27}$$

one concludes that all the component fields of the SFs $\tilde{W}$ and $\tilde{M},\tilde{N}$ are transformed *independently* of each other. Thus we can put down the following manifestly covariant constraints

$$\tilde{W}(\tilde{\xi}_L^{\pm\pm},\tilde{\eta}^{\pm}) = \tilde{u}(\tilde{\xi}_L^{\pm\pm})\,, \tag{28}$$

$$\begin{aligned}
\tilde{M}(\tilde{\xi}_L^{++},\tilde{\eta}^{+},\overline{\tilde{\eta}}^{+}) &= \tilde{\eta}^{+}\overline{\tilde{\eta}}^{+}\,\tilde{m}^{--}(\tilde{\xi}_L^{++})\,,\\
\tilde{N}(\tilde{\xi}_L^{--},\tilde{\eta}^{-},\overline{\tilde{\eta}}^{-}) &= \tilde{\eta}^{-}\overline{\tilde{\eta}}^{-}\,\tilde{n}^{++}(\tilde{\xi}_L^{--})\,,
\end{aligned} \tag{29}$$

which leaves intact the G-invariance of theory. Returning these constraints back into the Eq. (22) we obtain the final component form of the Eq. (6)

$$\tilde{\partial}_{--}\tilde{\partial}_{++}\tilde{u} = \frac{1}{4}e^{2\tilde{u}}\tilde{m}^{--}(\tilde{\xi}_L^{++})\tilde{n}^{++}(\tilde{\xi}_L^{--})\,. \tag{30}$$

This Eq. together with chirality conditions of the Goldstone fermions $\lambda^\pm(\xi_L^{\pm\pm})$, $\overline{\lambda}^\pm(\xi_L^{\pm\pm})$ gives the whole system of Eqs. describing dynamics of the $N=2$, $D=4$ superstring in the component level. [9]

3 General Solution

Let us consider shortly the problem of construction of general solution of the Eq. (6). It is well-known that the Virasoro constraints simplifying significantly the string equations of motion can generally be solved in terms of two copies (left- and right-moving) of the Lorentz harmonic variables parameterizing the compact coset spaces isomorphic to the $(D-2)$-dimensional sphere [9]

$$S_{D-2} = \frac{SO(1,D-1)}{SO(1,1)\times SO(D-2)\times K_{D-2}}. \tag{31}$$

Moreover, it was shown in Ref. [7] that from these variables the particular Lorentz covariant combinations can be formed which resolve generally the corresponding nonlinear σ-model equations of motion inspired by the bosonic strings in the geometrical approach. [1] By the construction the number of two copies of chiral variables parameterizing the coset space (31) is apparently enough to recover the $2(D-2)$ physical degrees of freedom of D-dimensional bosonic strings. But in the case of superstrings these variables replaced by the world-sheet superfields must be properly restricted to provide the necessary balance between bosonic and fermionic degrees of freedom $(D-2)B \equiv (D-2)F$.

In this Section we shall show that the suitable constraints can be achieved within the method of the nonlinear realization of superconformal symmetry developed in Section 1.2.

Proceeding from [7] one can check that the general solution of the Liouville Eq. (13) can be written in form

$$\begin{aligned} e^{-2\tilde{u}(\tilde{\xi}_L^{\pm\pm})} &= \frac{1}{2}\tilde{r}_m^{++}(\tilde{\xi}_L^{--})\tilde{l}^{--m}(\tilde{\xi}_L^{++}), \\ \tilde{m}_{++}^{--}(\tilde{\xi}_L^{++}) &= \tilde{l}_m^{--}(\tilde{\xi}_L^{++})\tilde{\partial}_{++}\tilde{l}^m(\tilde{\xi}_L^{++}), \\ \tilde{n}_{--}^{++}(\tilde{\xi}_L^{--}) &= \tilde{r}_m^{++}(\tilde{\xi}_L^{--})\tilde{\partial}_{--}\tilde{r}^m(\tilde{\xi}_L^{--}), \end{aligned} \tag{32}$$

where the left(right)-moving Lorentz harmonics are normalized as follows

$$\tilde{l}_m^{++}\tilde{l}^{m++} = 0, \quad \tilde{l}_m^{--}\tilde{l}^{m--} = 0, \quad \tilde{l}_m\tilde{l}^{m\pm\pm} = 0, \tag{33}$$

$$\tilde{l}_m^{--}\tilde{l}^{m++} = 2, \quad \tilde{l}_m\tilde{l}^m = -1. \tag{34}$$

Substituting these solutions into the Eqs. (29) and taking account of the expressions (25) one finds

$$M \equiv M^{--}(\xi_L^{++},\eta^+,\overline{\eta}^+) = (\tilde{\overline{D}}_+\overline{\eta}^+)\tilde{\eta}^+\tilde{\overline{\eta}}^+\tilde{l}_m^{--}(\tilde{\xi}_L^{++})\tilde{\partial}_{++}\tilde{l}^m(\tilde{\xi}_L^{++}), \tag{35}$$

$$N \equiv M^{++}(\xi_L^{--}, \eta^-, \overline{\eta}^-) = (\tilde{\overline{D}}_- \overline{\eta}^-)\tilde{\eta}^- \tilde{\overline{\eta}}^- \tilde{r}_m^{++}(\tilde{\xi}_L^{--})\tilde{\partial}_{--}\tilde{r}^m(\tilde{\xi}_L^{--})\,. \quad (36)$$

Now comparing these SFs with explicit form of the general solution of the Eq. (6)

$$\begin{aligned} e^{-2W(\xi_L^{\pm\pm}, \eta^\pm)} &= \frac{1}{2} r_m^{++}(\xi_L^{--}, \eta^-) l^{--m}(\xi_L^{++}, \eta^+)\,, \quad (37) \\ \Psi_+^{--}(\xi_L^{++}, \eta^+, \overline{\eta}^+) &= \frac{1}{2i} l_m^{--}(\xi_L^{++}, \eta^+) D_+ l^m(\xi_L^{++}, \eta^+)\,, \\ \Psi_-^{++}(\xi_L^{--}, \eta^-, \overline{\eta}^-) &= \frac{1}{2i} r_m^{++}(\xi_L^{--}, \eta^-) D_- r^m(\xi_L^{--}, \eta^-)\,, \end{aligned}$$

one can derives the following expressions for the corresponding SFs of linear realization

$$M^{\pm\pm} = \Phi_\pm^\pm \Omega^\mp \Psi_\mp^{\pm\pm}\,, \qquad \overline{D}_\pm \Phi_\pm^\pm = 0\,, \qquad D_\pm \Omega^\pm = 1\,, \quad (38)$$

$$\Phi_\pm^\pm = \tilde{\overline{D}}_\pm \overline{\eta}^\pm, \qquad \Omega^\pm = (\tilde{D}_\pm \eta^\pm)\tilde{\eta}^\pm\,, \quad (39)$$

$$\begin{aligned} l_m^{\pm\pm,0}(\xi_L^{++}, \eta^+) &= \tilde{l}_m^{\pm\pm,0}(\tilde{\xi}_L^{++})\,, \quad (40) \\ r_m^{\pm\pm,0}(\xi_L^{--}, \eta^-) &= \tilde{l}_m^{\pm\pm,0}(\tilde{\xi}_L^{--})\,. \end{aligned}$$

4 Conclusion

Thus, we have established that the $n = (2,2)$ appriopriate generalization of complex Liouville equation related to the $N = 2, D = 4$ superstring is given by the Eq. (6) in which the auxiliary SFs Ψ are subjects to the constraints (10) and (38). Then the general solution of this equation can be given in terms of Lorentz harmonics (37) which in turn are also restricted by the conditions (40). Note, that in actually means that the Eq. (6) is exactly solvable as the corresponding bosonic string equation does,[7] but unlike to the bosonic case the corresponding harmonic SFs becomes *essentially restricted* by the constraints (40) which provide the supersymmetric balance between bosonic and fermionic degrees of freedom. [d] It is worth mentioning that the first constraint in Eqs. (38) implies that the SFs M are actually nilpotent $M^2 = 0$. From the theory of spontaneously broken supersymmetries we know that such a type of constraints leads directly to the nonlinear realizations of the underlaying symmetries,[10,11] in frame of which these constraints could be solved explicitly in terms of the corresponding Goldstone (super)fields. In the case under consideration we find

[d]It is obvious, that the same property of $n = (1,1)$ super-Liouville equation[3] can be derived from this one with the help of dimensional reduction from (37) and (40).

the suitable manifestly supercovariant solution (39) in terms of the Goldstone fermions of the nonlinear realization of $n = (2,2)$ superconformal symmetry $\lambda^{\pm}(\xi_L^{\pm\pm})$, $\overline{\lambda}^{\pm}(\xi_L^{\pm\pm})$.

We are convinced that this approach actually gives the universal way of deriving the equations of motion as well as their solutions for the superstrings in the cases of higher dimensions too, i.e. $D = 6, 10$. In particular, the $N = 2, D = 6$ superstring is expected should be described by the nonlinear realization of the $n = (4,4)$ supersymmetric WZNW σ-model in which W is replaced by the double-analytical SF $q^{(1,1)}$ representing twisted multiplet in the harmonic $(4,4)$ superspace. [4,12]

We hope to return to this question in a forthcoming publications.

Acknowledgments

It is a great pleasure for me to express thanks to I. Bandos, E. Ivanov, S. Krivonos, A. Pashnev and D. Sorokin for their interest to this work and valuable discussions.

References

1. F. Lund and T. Regge *Phys. Rev.* **D 14,** 1524 (1976);
   R. Omnes *Nucl. Phys.* **B 149,** 269 (1979);
   B.M. Barbashov and V.V. Nesterenko *Commun. Math. Phys.* **78,** 499 (1981).
   A. Zheltukhin *Sov. J. Nucl. Phys. (Yadern.Fiz.)* **33,** 1723 (1981); *Theor. Mat. Phys.* **52,** 73 (1982); *Phys. Lett.* **B 116,** 147 (1982); *Theor. Mat. Phys.* **56,** 230 (1983).
2. I. Bandos, P. Pasti, D. Sorokin, M. Tonin and D. Volkov, *Nucl. Phys.* **B 446,** 79-119 (1995)(hep-th/9501113).
3. I. Bandos, D. Sorokin and D. Volkov, *Phys. Lett.* **B 372,** 77-82 (1996).
4. E. Ivanov and S. Krivonos, *J. Physics* **A17,** L671 (1984).
5. E. Ivanov and A. Kapustnikov, *Nucl. Phys.* **B 333,** 439 (1990);
6. A.S. Galperin, P.S. Howe and K.S. Stelle, *Nucl. Phys.* **B 368,** 281 (1992).
7. I. Bandos, E. Ivanov, A. Kapustnikov and S. Ulanov, *J. Math. Phys.* **40,** 5203-5223 (1999)(hep-th/9810038).
8. S. Coleman, J. Wess and B. Zumino, *Phys. Rev.* **177,** 2239 (1969);
   C. Callan, S. Coleman, J. Wess and B. Zumino, *Phys. Rev.* **177,** 2247 (1969);
   D. Volkov, *J. Elem. Part. Atom. Nucl.* **4,** 3 (1973);

V. Ogievtsky, *in Proc. X Winter School of Theor. Physics* (Wroclaw, 1974), **vol. 1**, p. 117.

9. D.B. Fairlie and C.A. Manogue, *Phys. Rev.* **D 36,** 475-479 (1987); N. Berkovits, *Phys. Lett.* **B 241,** 497-502 (1990).
10. E. Ivanov, A.A. Kapustnikov, *J. Physics* **A11,** 2375 (1978); *J. Physics* **G8,** 167 (1982); *Int. J. Mod. Phys.* **A 7,** 2153 (1992).
11. M. Rocek, A.A. Tseytlin, ITP-SB-98-68, Imperial/TP/98-99/013, hep-th/9811232.
12. E. Ivanov, S. Krivonos and V. Leviant, *Nucl. Phys.* **B 304,** 601 (1988); O. Gorovoy and E. Ivanov, *Nucl. Phys.* **B 381,** 394 (1992);
    E. Ivanov and A. Sutulin, *Nucl. Phys.* **B 432,** 246-280 (1994); *Nucl. Phys.* **B 483,** 531 (1997)(E) ;
    E. Ivanov and A. Sutulin, *Class. Quantum Grav.* **14,** 843-863 (1997).

GAUGING OF 1d–SPACE TRANSLATIONS FOR NONRELATIVISTIC MATTER

P.C. STICHEL
An der Krebskuhle 21
D-33619 Bielefeld[a]

We develop in a systematic fashion the idea of gauging 1d-space translations with fixed Newtonian time for nonrelativistic matter (particles and fields). By starting with a nonrelativistic free theory we obtain its minimal gauge invariant extension by introducing two gauge fields with a Maxwellian self interaction. We fix the gauge so that the residual symmetry group is the Galilei group. The reduced N-particle Lagrangian describes geodesic motion in a $(N-1)$-dimensional (Pseudo-) Riemannian space. The singularity of the metric for negative gauge coupling leads in classical dynamics to the formation of geometric bags in the case of two or three particles. The ordering problem within the quantization scheme for N-particles is solved by canonical quantization of a pseudoclassical Schrödinger theory obtained by adding to the continuum generalization of the point-particle Lagrangian an appropriate quantum correction. We solve the two-particle bound state problem for both signs of the gauge coupling. At the end we speculate on the possible physical relevance of the new interaction induced by the gauge fields.

1 Introduction

Gauging a symmetry means to introduce space-time dependent functions instead of the constant parameters of a global symmetry group. The original action, invariant with respect to the global symmetry group, will in general not be gauge invariant. In order to cancel the terms which violate the invariance one has to introduce compensating fields as functions of space but also of time. These gauge fields have to obey suitable transformation laws with respect to the now local symmetry.

The most important interactions in current physics are gauge interactions. In elementary particle physics internal symmetry groups are gauged (Quantumelectrodynamics, Quantumchromodynamics, Standard Model). In general relativity, the space-time symmetry group of special relativity (Poincaré group) is gauged. Gauge interactions are also of importance in condensed matter physics (fractional quantum hall effect). Therefore the "gauge principle" may be called the leading principle for the construction of fundamental interactions in physics. [1]

Our subject will be 'gauging of the Galilei group' in nonrelativistic physics (cp. De Pietri et al. [2]). In contrast to De Pietri et al. [2] we start with the following leading principles:

- We ask for the smallest number of gauge fields doing the job with a minimal coupling Lagrangian,
- we don't start with the nonrelativistic limit of a relativistic theory.

In this talk we treat the simplest example: point particles and nonrelativistic fields moving on the line. The underlying global transformations are space translations whose gauging leads to general coordinate transformations at fixed Newtonian time. We will show that the minimal coupling for this example consists of only two gauge fields with a Maxwellian interaction term. We don't introduce an additional dilaton field as in (1+1)-gravity. Arbitrary gauge function will be chosen such, that the residual symmetry group of the action will be the Galilei group (cp. Stichel [3,4]).
We ask: What are the physical consequences of such a gauge interaction in classical resp. quantum physics?

2 Minimal-Coupling Lagrangian

A) Point Particles

We start with N nonrelativistic particles in free motion on a line ($\mathbf{R}^1$) described by the Lagrangian

$$L_0 = \frac{1}{2}\sum_{\alpha=1}^{N}(\dot{x}_\alpha(t))^2 \,. \tag{2.1}$$

The equations of motion (EOM) following from (2.1)

$$\ddot{x}_\alpha = 0 \tag{2.2}$$

are invariant with respect to global Galilei-transformations

$$(x,t) \rightarrow (x',t'), \tag{2.3}$$

with

$$x' = x + a + vt \,, \tag{2.4}$$

and

$$t' = t + b \,, \tag{2.5}$$

where the parameters a, v and b take values in $\mathbf{R}^1$. Now we generalize (2.4) to a local transformation, given in infinitesimal form by

$$\delta x = a(x,t) \,, \tag{2.6}$$

where $a(x,t)$ is an arbitrary, twice differentiable and bounded function of its arguments. Eq. (2.6) describes local space translations (including local boosts). We keep time translations with a constant b (cp. (2.5)).

Obviously the EOM (2.2) are not invariant with respect to the transformation (2.6). In order to repair that, we introduce two gauge fields $h(x,t)$ and $e(x,t)$ and replace $\dot{x}$ for each particle in (2.1) by the function ξ

$$\xi = h(x,t)\dot{x} + e(x,t)\,. \tag{2.7}$$

Invariance of ξ with respect to (2.6) requires the following transformation rules for the gauge fields

$$\delta h = -h\partial_x a\,, \qquad \delta e = -h\partial_t a\,, \tag{2.8}$$

where

$$\delta f(x,t) := f'(x+\delta x, t+\delta t)\,. \tag{2.9}$$

With (2.7) we supply the minimal gauge invariant extension for the particle velocities $\dot{x}_\alpha$.

We conclude from (2.7) and (2.8) that our gauge fields $e(x,t)$ and $h(x,t)$ are components of the Zweibein $\{E^b_\mu\}$ with components $\{E^0_\mu\}) = (1,0)$ and $\{E^1_\mu\} = (e,h)$. In our case the vector $\{E^0_\mu\}$ is trivial because time has been held fixed up to translations with a constant.

Now L_0 in (2.1) has to be replaced by

$$L_{\text{matter}} = \frac{1}{2}\sum_{\alpha=1}^{N}(\xi_\alpha(t))^2\,, \tag{2.10}$$

with

$$\xi_\alpha(t) := h_\alpha(t)\dot{x}_\alpha(t) + e_\alpha(t)\,, \tag{2.11}$$

where we defined

$$f_\alpha(t) := f(x_\alpha(t),t) \tag{2.12}$$

for an arbitrary field $f(x,t)$.

We must supplement (2.10) by an invariant (or quasi-invariant) Lagrangian L_{field} describing the self-interaction of the gauge fields. Let us define a field strength F

$$F := \frac{1}{h}(\partial_t h - \partial_x e)\,. \tag{2.13}$$

From (2.9) we obtain easily the commutator between δ and any partial differentiation $\partial \in (\partial_t, \partial_x)$

$$\delta\partial f = \partial\delta f - (\partial a)\partial_x f\,. \tag{2.14}$$

We infer from (2.8) and (2.14) that our field strength F is gauge invariant

$$\delta F = 0 \ . \tag{2.15}$$

Therefore, any integral of the form

$$\int_{\mathbf{R}^1} d\mu_t(x) K(F(x,t)) \,, \tag{2.16}$$

with the invariant measure

$$d\mu_t(x) := h(x,t)dx \tag{2.17}$$

is a candidate for $\mathrm{L}_{\text{field}}$. The simplest, nontrivial example for K is a quadratic

$$K(Z) = Z^2 \ . \tag{2.18}$$

With this Maxwellian choice for $\mathrm{L}_{\text{field}}$ our action takes the form

$$S = \int dt (\mathrm{L}_{\text{matter}} + \mathrm{L}_{\text{field}}) \,, \tag{2.19}$$

with

$$\mathrm{L}_{\text{field}} = \frac{1}{2\lambda} \int dx h(x,t) F^2(x,t) \,, \tag{2.20}$$

where λ is a coupling strength. With our Zweibein $\{E^b_\mu\}$ we identify hF as the only nonvanishing component of the torsion tensor and (2.20) as the related local quadratic Lagrangian.

By varying S with respect to x_α we get the particle-EOM

$$\dot{\xi}_\alpha + \xi_\alpha F_\alpha = 0 \ . \tag{2.21}$$

For the following it is advantageous to replace the Lagrangian (2.10) by its 1st-order form

$$\mathrm{L}_{\text{matter}} = \sum_\alpha \xi_\alpha (h_\alpha \dot{x}_\alpha + e_\alpha) - \frac{1}{2} \sum_\alpha \xi_\alpha^2 \ . \tag{2.22}$$

Variation of S with respect to ξ_α leads to (2.11) now as a constraint rather than a definition.

B) Fields

The continuum generalisation of the Lagrangian (2.22) will be given in terms of hydrodynamics variables $\rho(x,t)$ and $\theta(x,t)$ describing a density field and a potential field respectively. Therefore we look for a Lagrangian $\mathrm{L}_{\mathrm{matter}}$ such that for the particular choice

$$\rho(x,t) = \frac{1}{h(x,t)} \sum_\alpha \delta(x - x_\alpha(t)) \,, \tag{2.23}$$

and the identification

$$\xi(x,t) := \frac{1}{h(x,t)} \partial_x \theta(x,t) \,, \tag{2.24}$$

we obtain up to a total time-derivative the expression (2.22).

It can easily be shown that the desired expression for $\mathrm{L}_{\mathrm{matter}}$ is given by

$$\begin{aligned} \mathrm{L}_{\mathrm{matter}} &= -\int d\mu_t(x) \rho(x,t)(D_t\theta)(x,t) \\ &\quad -\frac{1}{2}\int d\mu_t(x)\rho(x,t)(D\theta)^2(x,t) \,, \end{aligned} \tag{2.25}$$

where the invariant derivatives D and D_t are defined by means of the inverse Zweibein $\{\tilde{E}^\mu_b\}$

$$D := \tilde{E}^1_1 \partial_x = \frac{1}{h(x,t)} \partial_x \,, \tag{2.26}$$

$$\text{and} \quad D_t := \tilde{E}^\mu_0 \partial_\mu = \partial_t - e(x,t) D \,. \tag{2.27}$$

Under the proviso that our hydrodynamic field e and θ are invariant with respect to local coordinate transformations (2.6)

$$\delta\rho = \delta\theta = 0 \,, \tag{2.28}$$

the invariance of $\mathrm{L}_{\mathrm{matter}}$

$$\delta \mathrm{L}_{\mathrm{matter}} = 0 \tag{2.29}$$

follows.

3 Gauge Fixing

In this section we will show, that the gauge field $h(x,t)$ is not a dynamical variable but a kinematical entity to be determined by fixing the class of physically admissible frames of reference.

Let us start with introducing a gauge function Λ by

$$h(x,t) = \partial_x \Lambda(x,t) . \tag{3.1}$$

With the local transformation properties (2.8) and (2.14) we infer from (3.1)

i) The most general form of $e(x,t)$ is given by

$$e(x,t) = \partial_t \Lambda(x,t) + \hat{e}(x,t) , \tag{3.2}$$

with $\delta \hat{e} = 0$ (but in general $\hat{e}$ depends on Λ!)

ii) $\delta\Lambda = 0$, or for finite transformations generated by $a(x,t)$

$$\Lambda'(x,t) = e^{-a(x,t)\partial_x} \Lambda(x,t) . \tag{3.3}$$

Physically admissible frames of reference in the nonrelativistic regime are inertial frames connected to each other by Galilei transformations. Therefore we have to fix a class of functions $\{\Lambda'(x,t)\}$ such that the remaining (residual) symmetry group is the Galilei group only. This means we must look for $\{\Lambda'\}$ invariant with respect to transformations generated by

$$\tilde{a}(x,t) = \alpha + \beta t \tag{3.4}$$

only. This is the case iff

$$\Lambda'(x,t) = x - (a + vt) \tag{3.5}$$

which is equivalent to fixing the gauge by

$$h(x,t) = 1 , \tag{3.6}$$

and demanding

$$\delta e = -\delta\beta . \tag{3.7}$$

It remains to show that for an arbitrary function $\Lambda(x,t)$ with $\partial_x \Lambda \neq 0$ it is always possible to find a transformation

$$x \to x'(x,t) , \tag{3.8}$$

such that $\Lambda'(x',t)$ is given by (3.5). Obviously this goal will be achieved with

$$x'(x,t) = \Lambda(x,t) + (a + vt) , \tag{3.9}$$

since (3.3) may be written as

$$\Lambda'(x',t) = \Lambda(x,t) . \tag{3.10}$$

4 Classical Dynamics

A) Reduced Matter Lagrangian

In order to express our Lagrangian L in terms of matter variables only, we will follow the symplectic Hamiltonian procedure of Faddeev and Jackiw. [5]

We enlarge our phase space by an additional field $\pi(x,t)$ and obtain for the canonical 1st order form of L

$$L = \int dx(-\rho\partial_t\theta - \mathcal{H}(x,t)), \tag{4.1}$$

with (after a partial integration)

$$\begin{aligned}\mathcal{H}(x,t): &= \frac{1}{2}\rho(\partial_x\theta)^2 + \frac{\lambda}{2}\pi^2 \\ &-e(\partial_x\pi + \rho\partial_x\theta)\,.\end{aligned} \tag{4.2}$$

In (4.2) $e(x,t)$ is a Lagrange-multiplier field. Variation of S with respect to e leads to the constraint

$$\partial_x\pi = -\rho\partial_x\theta\,, \tag{4.3}$$

which has the solution

$$\pi(x,t) = -\frac{1}{2}\int dy\epsilon(x-y)\rho(y,t)\partial_y\theta(y,t)\,, \tag{4.4}$$

where $\epsilon(x) := x/|x|$.

In order that the boundary term arising in the derivation of (4.1-2) vanishes and to obtain $|\int dx\pi^2| < \infty$ we have to require a boundary condition at spatial infinity:

$$\pi\,|_{x=\pm\infty} = 0\,. \tag{4.5}$$

From (4.4) we conclude that (4.5) implies the constraint

$$\int dx\rho(x,t)\partial_x\theta(x,t) = 0\,, \tag{4.6}$$

i.e. the total canonical momentum of matter has to vanish.

The constraint (4.6) Eq. can't be solved explicitly in terms of dynamical variables. Therefore we have to take into account (4.6) by means of a Lagrange multiplier v. With that and the insertion of (4.4) into (4.2) we obtain for the reduced matter Lagrangian L_{red}

$$\mathrm{L}_{red} = \int dx(-\rho\partial_t\theta - \mathrm{H}_{red}(x,t)\,, \tag{4.7}$$

with

$$\mathrm{H}_{red}(x,t) := \frac{1}{2}(\partial_x\theta)^2 + \frac{\lambda}{2}\pi^2 + v\rho\partial_x\theta\,, \tag{4.8}$$

where $\pi(x,t)$ is given by (4.4).

Comparison of the EOM's derived from L_{red} with the corresponding Poisson-bracket (PB) relations $\dot{A} = \{A, \mathrm{H}_{red}\}$ for any field A leads to the symplectic structure

$$\{\rho(x,t), \theta(y,t)\} = \delta(x-y)\,, \tag{4.9}$$

with all other PB's vanishing.

By means of (2.23-24) we may specialize (4.7) to the case of N point particles. We obtain after a partial integration and use of (4.6) up to a total time derivative

$$\mathrm{L}_{red} = \sum_\alpha \xi_\alpha(\dot{x}_\alpha - \frac{1}{2}\xi_\alpha - v) + \frac{\lambda}{4}\sum_{\alpha,\beta}|x_{\alpha\beta}|\xi_\alpha\xi_\beta\,, \tag{4.10}$$

where $x_{\alpha\beta} := x_\alpha - x_\beta$.

From (4.6) we read off the constraint

$$\sum_{\alpha=1}^{N}\xi_a = 0\,, \tag{4.11}$$

which may be solved for ξ_1 leading to (using the summation convention, indices running from 2 to N)

$$\mathrm{L}'_{red} = \xi_\alpha\dot{\eta}^\alpha - \mathrm{H}_{red}\,, \tag{4.12}$$

with

$$\mathrm{H}'_{red} = \frac{1}{2}g^{\alpha\beta}(\eta)\xi_\alpha\xi_\beta\,, \tag{4.13}$$

where

$$\eta^\alpha := x^\alpha - x^1\,,$$

and

$$g^{\alpha\beta}(\eta) := \delta^{\alpha\beta} + 1 + \frac{\lambda}{2}\left(|\eta_\alpha| + |\eta_\beta| - |\eta_\alpha - \eta_\beta|\right)\,. \tag{4.14}$$

L'_{red} is now independent of v. Galilei symmetry becomes a hidden symmetry.

We observe that H'_{red} looks like a free Hamiltonian describing the geodesic motion of one-'particle' in a $(N-1)$-dimensional (Pseudo-) Riemannian space M with metric $\{g_{\alpha\beta}\}$ (obtained as the inverse of (4.14)).

By varying the action $S' = \int dt \mathrm{L}'_{red}$ with respect to ξ_α we obtain the constraint

$$\dot{\eta}^\alpha = g^{\alpha\beta}(\eta)\xi_\beta \ . \tag{4.15}$$

If we express L'_{red} and H'_{red} by means of (4.18) in terms of $(\{\eta^\alpha\}, \{\dot{\eta}^\alpha\})$ we obtain

$$\mathrm{L}'_{red} = \mathrm{H}'_{red} = \frac{1}{2} g_{\alpha\beta}(\eta)\dot{\eta}^\alpha \dot{\eta}^\beta \ . \tag{4.16}$$

Therefore the Euler-Lagrange equations become the geodesic equations.

On the other hand the variations of S' with respect to η^α leads to the EOM

$$\dot{\xi}_\alpha = -\frac{1}{2}\partial_\alpha (g^{\beta\gamma})\xi_\beta \xi_\gamma \ . \tag{4.17}$$

Writing (4.15) and (4.17) in PB-form we infer from (4.13) the canonical symplectic structure

$$\{\eta^\alpha, \xi_\beta\} = \delta^\alpha_\beta \ , \qquad \{\eta^\alpha, \eta^\beta\} = \{\xi_\alpha, \xi_\beta\} = 0 \ . \tag{4.18}$$

B) String–Like Behaviour

In the point particle case the π-field (Eq. (4.4) takes the form

$$\pi(x,t) = -\frac{1}{2}\sum_{\alpha=1}^{N} \epsilon(x - x_\alpha(t))\xi_\alpha(t) \, , \tag{4.19}$$

with the constraint (4.11). Therefore $\pi(x,t)$ vanishes outside a closed x-intervall

$$\pi(x,t) = 0 \qquad \forall x \notin [\mu_1(t), \mu_2(t)] \ , \tag{4.20}$$

with $\mu_{\frac{1}{2}}(t) := \begin{smallmatrix}\min\\ \max\end{smallmatrix}\{x_\alpha\}$.

In particular for $N = 2$ $\pi(x,t)$ is string-like with point masses at its ends.

From (4.15,17) we obtain the following EOM for relative particle motion ($x := x_1 - x_2$)

$$\ddot{x} - \lambda E \epsilon(x) = 0 \, , \tag{4.21}$$

where the conserved energy E is given by

$$E := \frac{1}{4}\,\frac{\dot{x}^2}{1 + \frac{\lambda}{2}|x|} \ . \tag{4.22}$$

We conclude

i) For $E > 0$ but $\lambda < 0$ the 'string' is always of finite length ($|x| < 2/|\lambda|$). This is a geometric bag as a dynamical consequence of our gauge theory.

ii) The interparticle potential is proportional to their distance with strength proportional to the energy. This leads according to (4.21-22) to a constant positive (negative) acceleration for $x > 0$ ($x < 0$) for either sign of λ (for $\lambda < 0$ if $|x| > 2/|\lambda|$ only) resulting in arbitrary large velocities if time goes on. Therefore the model is unphysical for $\lambda > 0$ and for $\lambda < 0$ if $|x| > 2/|\lambda|$.

iii) For $N = 2$, M is a one-dimensional Riemannian space and therefore flat. This allows by means of the transformation

$$x \to y := \frac{4}{|\lambda|}(1 + \frac{\lambda}{2}|x|)^{1/2}, \tag{4.23}$$

the transition to a $1d$-Euclidean space E_1

$$E_1 = \begin{cases} [4/\lambda, \infty] & \text{for} \quad \lambda > 0\,, \\ [0, 4/|\lambda|] & \text{for} \quad \lambda < 0\,. \end{cases}$$

On E_1 we have the free EOM

$$\ddot{y} = 0\,, \tag{4.24}$$

where the finite end points of E_1 are points of reflection.

Remark: A careful analysis of the three-particle case shows for $\lambda < 0$ again the formation of a geometric bag as a consequence of a singularity of the metric!

5 Quantization

A) The Ordering Problem

In quantizing the N-particle system described by the Hamiltonian (4.13) and the symplectic structure (4.18) we have according to the canonical quantization recipe to substitute

$$\xi_\alpha \to p_\alpha (= \frac{\hbar}{i}\frac{\partial}{\partial \eta_\alpha} =: \frac{\hbar}{i}\partial_\alpha), \tag{5.1}$$

and

$$\{A, B\} \to \frac{[A, B]}{i\hbar}, \tag{5.2}$$

where [A,B] is the commutator of the operators A and B which denote any pair from the phase space set $\{\eta_2, \ldots, \eta_N, p_2, \ldots, p_N\}$. But (5.1) does not specify uniquely the quantum analogon of the Hamiltonian $g^{\alpha\beta}(\eta)\xi_\alpha\xi_\beta$, because the space variables $\{\eta^i\}$ and the momentum variables $\{p_i\}$ become non-commuting operators. This is the well known ordering problem. The only restriction we have is the requirement of hermiticity for the Hamilton operator. But this leaves a whole family of admissible Hamilton operators for a given classical Hamiltonian. Let us illustrate this for the simplest case, $N = 2$. From (4.13) we have

$$\mathrm{H_{class}} = (1 + \frac{\lambda}{2}|x|)\xi^2 . \tag{5.3}$$

From that we obtain two fundamental hermitian Hamilton operators (with respect to the usual inner product)

$$H_0 = p(1 + \frac{\lambda}{2}|x|)p, \tag{5.4}$$

$$H_1 = \frac{1}{2}(p^2(1 + \frac{\lambda}{2}|x|) + (1 + \frac{\lambda}{2}|x|)p^2) . \tag{5.5}$$

The most general hermitian Hamilton operator is given by a linear combination

$$H_\gamma = \gamma H_1 + (1 - \gamma)H_0 , \qquad \gamma \in [0, 1] . \tag{5.6}$$

How do they differ from each other? Consider the difference $H_\gamma - H_{\gamma'}$. By means of Heisenberg's commutation relations we obtain

$$H_\gamma - H_{\gamma'} = (\gamma - \gamma')\frac{\lambda\hbar^2}{2}\delta(x) . \tag{5.7}$$

Therefore the H_γ differ in the strength of a contact potential. This term beeing proportional to $\hbar^2$ is a quantum correction.

But what we want to quantize are not separate theories for different particle numbers N but a appropriate field theory with a particle interpretation after quantization. This quantum field theory should lead to a N-particle amplitude with a definite ordering prescription for the quantum analogon of $g^{\alpha\beta}(y)\xi_\alpha\xi_\beta$. For this the hydrodynamic field theory described in Section 4A is not a suitable starting point. It lacks a particle interpretation after quantization.

B) The Pseudo-Classical Schrödinger Field

We know of only one complex-valued field which after quantization possesses a particle interpretation in nonrelativistic physics: the Schrödinger field $\psi(x, t)$. With the Madelung representation

$$\psi(x, t) = \sqrt{\rho(x, t)}e^{\frac{i}{\hbar}\theta(x,t)}, \tag{5.8}$$

our hydrodynamic field theory (Section 4A)) becomes a Schrödinger field theory by adding to the Lagrangian density in (4.7) a quantum correction term $0(\hbar^2)$

$$-\frac{\hbar^2}{8\rho(x,t)}(\partial_x \rho(x,t))^2 . \tag{5.9}$$

The resulting Lagrangian L_{red} may be expressed in terms of ψ and $\psi^\dagger$ as follows

$$L_{red} = \int dx \left[\frac{i\hbar}{2}(\psi^\dagger \partial_t \psi - (\partial_t \psi^\dagger)\psi) - \mathcal{H}_{\rm red}\right] , \tag{5.10}$$

with

$$\mathcal{H}_{red}(x,t) := \frac{\hbar^2}{2}|\partial_x \psi|^2 + \frac{\lambda}{2}\pi^2(x,t) + vJ(x,t) , \tag{5.11}$$

where

$$J(x,t) := -\frac{i\hbar}{2}(\psi^\dagger \partial_x \psi - (\partial_x \psi^\dagger)\psi) , \tag{5.12}$$

and

$$\pi(x,t) := -\frac{1}{2}\int dy \epsilon(x-y) J(y,t) , \tag{5.13}$$

$\psi(x,t)$ has to be understood as a classical field – because of the appearance of $\hbar$ in (5.10) we call it 'pseudo-classical'.

By varying the action S with respect to $\psi^\dagger$ we obtain the following nonlinear, nonlocal Schrödinger equation for $\psi(x,t)$

$$i\hbar(\partial_t - e(x,t)\partial_x - \frac{1}{2}(\partial_x e(x,t)))\psi(x,t) = -\frac{\hbar^2}{2}\partial_x^2 \psi(x,t) , \tag{5.14}$$

where

$$e(x,t) := -v + \frac{\lambda}{2}\int dy|x-y|J(y,t) . \tag{5.15}$$

On the other hand the variation of S with respect to v leads to the constraint

$$\int dy J(y,t) = 0 . \tag{5.16}$$

Symplectic Hamiltonian analysis leads to the PB

$$i\hbar\{\psi(x,t), \psi^\dagger(y,t)\} = \delta(x-y) \tag{5.17}$$

with all other PB's vanishing.

C) Quantization and the N-particle Amplitude

The quantization of the Schrödinger equation (5.14) is obtained by means of the following canonical recipe:

i) Substitute

$$[\psi(x,t), \psi^\dagger(y,t)] = \delta(x-y) \tag{5.18}$$

for the PB (5.17). All other commutators vanish.

ii) Keep all operator products in Wick-ordered form (i.e. shift all annihilation operators ψ to the right)

iii) Consider the constraint (5.16) as a subsidary condition for the physical Hilbert space $\mathcal{G}$

$$\int dy J(y,t)\phi = 0 \qquad \forall \phi \in \mathcal{G}\,, \tag{5.19}$$

i.e. ϕ has vanishing total particle-momentum.

The N-particle wave function is defined as usual by

$$\chi_N(x_1, \ldots, x_N, t) := \frac{1}{\sqrt{N}} < 0|\psi(x_1,t)\ldots\psi(x_N,t)|N > \ . \tag{5.20}$$

The subsidary condition (5.19) applied to $|N>$ is then equivalent to

$$\sum_{i=1}^{N} \partial_i \chi_N = 0\,, \tag{5.21}$$

i.e. χ_N is only a function of the relative coordinates. This agrees with the classical situation described by (4.12-14).

Finally by means of (5.14), (5.18) and (5.21) we obtain the following Schrödinger equation for χ_N as a function of relative coordinates $\{\eta_i\}_{i=2}^N$

$$\begin{aligned} i\hbar\partial_t\chi_N &= \left(-\frac{\hbar^2}{2}\partial_\alpha g^{\alpha\beta}(\eta)\partial_\beta - \frac{\hbar^2\lambda}{4}\sum_\alpha \delta(\eta_\alpha) \right. \\ &\left. -\frac{\hbar^2\lambda}{4}\sum_{\alpha<\beta}\delta(\eta_\alpha - \eta_\beta) \right) \chi_N \ . \end{aligned} \tag{5.22}$$

This result corresponds to the following quantization rule for the classical N-particle Hamiltonian (4.16)

$$\mathrm{H}_{class} = \frac{1}{2} g^{\alpha\beta}(\eta)\xi_\alpha\xi_\beta \tag{5.23}$$

$$\Downarrow$$

$$\begin{aligned} \mathrm{H}_{quant} &= -\frac{\hbar^2}{8}(\partial_\alpha \partial_\beta g^{\alpha\beta}(\eta) + g^{\alpha\beta}(\eta)\partial_\alpha \partial_\beta \\ &\quad +2\partial_\alpha g^{\alpha\beta}(\eta)\partial_\beta) \,. \end{aligned} \tag{5.24}$$

The agreement of (5.24) with (5.22) is easily seen by means of the relation

$$\partial_\alpha \partial_\beta (g^{\alpha\beta}(\eta)) = 2\lambda(\sum_\alpha \delta(\eta_\alpha) + \sum_{\alpha<\beta} \delta(\eta_\alpha - \eta_\beta)) \,. \tag{5.25}$$

Therefore in the case of the two-particle problem we have to choose $\gamma = \frac{1}{2}$ for the free parameter in (5.6). The quantization rule (5.23) $\Longrightarrow$ (5.24) is nothing but a generalization of this $\gamma = 1/2$ rule to a system of N identical particles.

Now we want to express our Hamiltonian operator as much as possible in terms of entities which are invariant with respect to arbitrary coordinate transformations on our (Pseudo-) Riemannian space M. This will be done by means of the following steps:

i) Define φ_N by $\chi_N = g^{1/4}\varphi_N$. Then φ_N is normalized with respect to the invariant measure $\sqrt{g}\, d\eta^1 \ldots d\eta^{N-1}$

$$(g := \det(g_{\alpha\beta}))\,.$$

ii) Decompose M into $N!$ regions specified by the signs of the x_{ik}. Then the δ-terms in (5.22) give the corresponding boundary conditions for φ_N at the border between these regions. On each of these regions our Hamilton operator takes the form

$$H = -\frac{\hbar^2}{2}\Delta + \frac{\hbar^2}{8}R + V\,, \tag{5.26}$$

where Δ is the invariant Laplace-Beltrami operator

$$\Delta = \frac{1}{\sqrt{g}}\partial_\alpha g^{\alpha\beta}(\eta)\sqrt{g}\partial_\beta \,. \tag{5.27}$$

R is the invariant scalar curvature (contraction of the Ricci-tensor) and V is an additional noninvariant quantum correction potential defined in terms of Cristoffel symbols

$$V = -\frac{\hbar^2}{8}\Gamma^\alpha_{\beta\mu} g^{\mu\nu} \Gamma^\beta_{\nu\alpha} \,. \tag{5.28}$$

6 Two–Particle Bound States

In this section we consider the bound state solutions of the stationary Schrödinger equation (5.22) for $N = 2$

$$E\chi_2(x) = \{-\hbar^2\partial_x(1+\frac{\lambda}{2}|x|)\partial_x - \frac{\hbar^2\lambda}{4}\delta(x)\}\chi_2(x) . \tag{6.1}$$

In this case M is the line which decomposes into the negative and positive half axis. With Bose-symmetry $\chi_2(x) = \chi_2(-x)$ we obtain from (6.1) on $\mathbf{R}^1_+$ the differential equation

$$E\chi_2 = -\hbar\partial_x(1+\frac{\lambda}{2}x)\partial_x\chi_2 , \tag{6.2}$$

with the boundary condition

$$\partial_x\chi_2(0) = -\frac{\lambda}{8}\chi_2(0) . \tag{6.3}$$

Let us now apply the procedure described in Section 5.C:
M is flat and therefore $R = 0$. The Laplace-Beltrami operator is given by

$$\Delta = -\hbar^2(1+\frac{\lambda}{2}x)^{1/2}\partial_x(1+\frac{\lambda}{2}x)^{1/2}\partial_x , \tag{6.4}$$

and we obtain for V

$$V = -\frac{\hbar^2\lambda^2}{64}\frac{1}{1+\frac{\lambda}{2}x} . \tag{6.5}$$

The boundary condition for $\varphi_2(x) := (2+\lambda x)^{1/4}\chi_2$ at $x = 0$ follows from (6.3) as

$$\partial_x\varphi_2(0) = 0 . \tag{6.6}$$

Because M is flat we may again perform the transformation (4.40) to Euclidean space E_1 as in the classical case. This transforms our Schrödinger equation into its normal form

$$E\tilde{\varphi}_2(y) = (-\hbar^2\partial_y^2 + V(y))\tilde{\varphi}_2(y) , \tag{6.7}$$

with

$$V(y) = -\frac{\hbar^2}{4y^2} , \tag{6.8}$$

and the boundary condition

$$\partial_y\tilde{\varphi}_2(4/|\lambda|) = 0 . \tag{6.9}$$

The inner product has now to be taken with respect to the measure dy.

In contrast to the classical case we have a nonzero potential in (6.7) which is a quantum correction.

In order to proceed we have to treat the two cases $\lambda \gtrless 0$ separately.

A) $\lambda > 0$

The solution of (6.7) which vanishes for $y \to \infty$ $(E < 0)$ is given by

$$\tilde{\varphi}_2(y) = y^{1/2} K_0\left(\frac{\sqrt{-E}}{\hbar} y\right), \tag{6.10}$$

with the boundary condition (Eq. (6.9))

$$2sK_1(s) = K_0(s), \tag{6.11}$$

where $s := \frac{4}{\hbar\lambda}\sqrt{-E}$ and $K_{0,1}$ are modified Bessel functions of the third kind. (6.11) has one fixed point only at

$$s_0 = 0,16572,$$

i.e. there exists one bound state only. This result corresponds to the well known fact that the $1d$-Schrödinger equation with a single δ-potential has one bound state only. As the coupling strength of the δ-potential in (6.1) is proportional to $\hbar^2$, the appearance of this bound state is a quantum effect. In the classical case only unbounded solutions with $E > 0$ are possible (cp. Section 4.C).

B) $\lambda < 0$

The requirement of finiteness of $\chi_2(2/|\lambda|)$ leads to

$$\tilde{\varphi}_2(0) = 0. \tag{6.12}$$

We have to consider the Schrödinger Eq. (6.7) on the finite intervall $(0, 4/|\lambda|)$, because the transformation (4.40) holds for $x \in I := (0, 2/|\lambda|)$ only. From the physical point of view the singularity of the metric at $x = 2/|\lambda|$ means that no events inside and outside of I are connected to each other. In addition we remark that outside of I our H is not bounded from below and therefore unphysical.

The solutions of (6.7) respecting the boundary conditions are ($E > 0$)

$$\tilde{\varphi}_2(y) = y^{1/2} J_0 \left(\frac{\sqrt{E}}{\hbar} y \right) ,$$

with the spectrum given by the infinite number of fixed points of

$$2sJ_1(s) = J_0(s) , \tag{6.13}$$

with $s := \frac{4}{\hbar(\lambda)}\sqrt{E}$. We obtain

$$s_0 = 0,9407705640 ,$$

and

$$s_n = \pi(\kappa_n + \frac{1}{4}), \tag{6.14}$$

with

$$\kappa_n \overset{n\to\infty}{\longrightarrow} n . \tag{6.15}$$

Already for small values of n the κ_n lie close to the asymptotic values (6.15). We obtain

$$\begin{aligned} \kappa_1 &= 1,010306991 , \\ &\vdots \\ \kappa_9 &= 9.001369775 , \qquad \text{etc.} \end{aligned} \tag{6.16}$$

This infinite tower of bound states with $E > 0$ corresponds to the classical situation which allows for bounded motion only. At the same time this tower gives us the quantum picture for our 'strings' of finite length with masses at their end points (cp. Section 4.C).

This result also presents us with a quantum picture for our geometric bag model.

7 Conclusions and Outlook

We have shown that the application of the gauge principle to $1d$-space translations for nonrelativistic point particles or fields leads to a nontrivial interaction with interesting features in both classical and quantum dynamics. But to which interaction known in nature corresponds this interaction? We don't know of any macroscopic interaction between two particles with a potential proportional to their distance and energy as in (4.38). For a gauge coupling

$\lambda > 0$ this interaction is unphysical anyway because it leads to arbitrary large velocities if time goes on. The same holds for $\lambda < 0$ if the interparticle distance is larger then $2/|\lambda|$. On the other hand we have shown that classically we obtain for $\lambda < 0$ a geometric bag for two and three particles respectively. This result has been quantum mechanically extended for $N = 2$ in Section 6. Therefore we may speculate, that our new interaction is of physical relevance for $\lambda < 0$ in the microscopic regime in connection with the confinement problem. But such a speculative statement immediately leads to two new and important questions:

i) What happens in higher space dimensions $d = 2$ or 3 ? Will geometric bag formation persist?

   Work for $d = 2$ with a Chern-Simons-like interaction for the gauge fields is in progress. Thereby we will also include the second central charge of the Galilei group in the free Lagrangian (cp. Lukierski et al. citsti6). In this context we will also take up the problem of particles and fields moving on a circle again.

ii) What is the connection with QCD? For that we have to extend our framework by supplying our particles and fields respectively with a nonabelian color charge and by gauging the corresponding $SU(N)$-group simultaneously with the space translations.

   We will take up this very interesting question as soon as possible.

References

1. R. Mills, *Am. J. Phys.* **57**, 493 (1989),
   F. Gronwald and F.W. Hehl, in "Quantum gravity" (P.G. Bergmann, V. de Sabbata and H.J. Treder Eds.), World Scientific, Singapore, 1996.
2. R. De Pietri, L. Lusanna and M. Pausi, *Class. Quant. Grav.* **12**, 219 and 255 (1995); **13**, 1417 (1996).
3. P.C. Stichel, *Phys. Lett.* **B456**, 129 (1999), (hep-th/9902207).
4. P.C. Stichel, Gauging of 1d-space translations for nonrelativistic matter – geometric bags, preprint hep-th/9911128.
5. L.D. Faddeev and R. Jackiw, *Phys. Rev. Lett.* **60**, 1692 (1988).
6. J. Lukierski, P.C. Stichel and W. Zakrzewski, *Ann. Phys.* (N.Y.) **260**, 224 (1997).